T0271899

A Foundation Course for College Organic Chemistry

To understand and improve the underlying principles that govern how organic reactions occur, *A Foundation Course for College Organic Chemistry* follows a brick-by-brick building approach. Emphasis is given to interrelating experimental facts and findings with predictions (mechanism) and inferences (results). Discussions focus on clarifying how complex organic reactions occur, which is based on electronegativity differences, movement of electrons (through σ framework or π bonds), and addition or removal of atoms (hydrogen, halogens) or groups (hydroxy, amino).

This book begins with simple rules governing the deconstruction of reactions and applies them to explain how esterification, amide, and cyanide hydrolysis reactions proceed. The importance of stereochemistry (used in drug development, biology, and medicine), aromatic electrophilic and nucleophilic substitutions, reaction kinetics, and dynamics is explained with suitable examples.

Features:

- A systematic and structured approach is used to study all aspects of reactive intermediates (generation, structure, geometry, and reactions of carbocations, carbanions, and carbon-free radicals).
- This book incorporates scientific methods to deduce reaction mechanisms with simple and relevant explanations, and limitations.
- A proper explanation is given to understand the influence of functional groups on the stability and reactivity of intermediates, pKa, HSAB principles, structure–activity relations, and how these can be exploited in organic chemistry.
- Information is presented in an accessible way for students, teachers, researchers, and scientists.

B. S. Balaji is Chairperson of Special Center for E-learning and a Professor in the School of Biotechnology at Jawaharlal Nehru University. He is an award-winning chemistry educator, has published various research papers, has developed 3D-printed Braille models to teach chemistry concepts to BLV students, is a resource person (ICT) for Malaviya Mission Teacher training program of UGC, is a Member of Technical Expert Committee, and is in the Interface Expert Committee to assess the NON-SWAYAM Learning Platforms of HEI for UGC-DEB.

A Foundation Course for College Organic Chemistry

B. S. Balaji

CRC Press
Taylor & Francis Group
Boca Raton London New York

CRC Press is an imprint of the
Taylor & Francis Group, an **informa** business

Designed cover image: © Shutterstock

First edition published 2025
by CRC Press
2385 NW Executive Center Drive, Suite 320, Boca Raton FL 33431

and by CRC Press
4 Park Square, Milton Park, Abingdon, Oxon, OX14 4RN

CRC Press is an imprint of Taylor & Francis Group, LLC

Library of Congress Cataloging-in-Publication Data
Names: Balaji, B. S., author.
Title: A foundation course for college organic chemistry / B. S. Balaji.
Description: First edition. | Boca Raton, FL : CRC Press, 2024. |
Includes bibliographical references and index.
Identifiers: LCCN 2024010463 (print) | LCCN 2024010464 (ebook) |
ISBN 9781032631141 (hbk) | ISBN 9781032631288 (pbk) | ISBN 9781032631165 (ebk)
Subjects: LCSH: Chemistry, Organic—Textbooks.
Classification: LCC QD251.3 B353 2024 (print) | LCC QD251.3 (ebook) |
DDC 547—dc23/eng/20240509
LC record available at https://lccn.loc.gov/2024010463
LC ebook record available at https://lccn.loc.gov/2024010464

ISBN: 978-1-032-63114-1 (hbk)
ISBN: 978-1-032-63128-8 (pbk)
ISBN: 978-1-032-63116-5 (ebk)

DOI: 10.1201/9781032631165

Typeset in Times
by codeMantra

Dedication

My parents – for allowing me to breathe oxygen

My teachers – for guiding me in search of knowledge and wisdom

My wife and children – for their relentless support and caring

Contents

Acknowledgments

The course was developed with support from UGC and the Ministry of Education, Government of India. I thank Jawaharlal Nehru University for its administrative, infrastructure, and organizational support. I extend my sincere gratitude to Prof. S. Murugan, HOD, S. T. Hindu College, Nagercoil and Prof. H. Surya Prakash Rao, Pondicherry University for painstakingly going through the course material to improve its quality. I also thank Dr. Renu Upadhyay and Ms. Jyotsna Jangra of Taylor & Francis Group, LLC/ CRC Press for their support in the preparation of this manuscript.

Foreword

Prof. Balaji has been known to me for more than two decades. While I was serving as the Executive Director of Shasun Chemicals and Drugs in Chennai, I selected Prof. Balaji as a senior manager to lead a research team. His approach to identifying and synthesizing impurities for ANDA filing, troubleshooting in process scale-up, and so on exemplified his in-depth knowledge of synthetic organic chemistry.

Prof. Balaji recently created a SWAYAM massive open online course in organic chemistry that is fundamental in nature with broad applicability. Over the past 4 years, his course has received over 20,000 enrollments and was translated into eight Indian languages. It was rated the second-best course globally (Class Central). The well-organized course material has now taken the shape of a book, covering many topics in greater depth. Prof. Balaji had invested a lot of time to carry out an in-depth survey of literature in organic chemistry and focused more on teaching–learning aspects suitable for modern-day learners.

This book delves deep into the intricate world of deconstruction of organic reactions, stereoisomerism, symmetry, topicity, stereochemical course of reactions, aromatic compounds and their reactions, reaction dynamics and kinetics, reactive intermediates, enolates, enamines, enol ethers, and name reactions, all coalescing to create a comprehensive and mesmerizing organic chemistry narrative.

He has followed an inquiry-based approach where he poses several questions before explaining the concepts. For example, can we have cis-trans isomerism on sp^3 carbons? In benzene, why do we relate heat of hydrogenation to resonance stability? In Pinacol rearrangement, which is more important, the intermediate stability or the final product stability? Why does anchimeric assistance of carbocations occur faster than intermolecular or unimolecular substitutions? Carbocations and carbenes have six electrons each. Yet they react differently. Why? These are some of the questions for which one can find answers in this book.

I can confidently say that this book will serve as a key to unraveling the profound mysteries of organic chemistry, and I would strongly recommend this book to under- and postgraduate students of chemistry and to students aspiring to clear the national competitive exams like CSIR-NET, IIT-JAM, NIPER-JEE, etc.

I hope topics like photochemistry, pericyclic reactions, spectroscopy, heterocyclic chemistry, biomolecules, and so on will be covered in the subsequent edition of this book.

Author Biography

Prof. B. S. Balaji is an award-winning chemistry educator from Jawaharlal Nehru University and received his PhD from NCL, Pune. He spent eight years in Japan and the United States. He has over 24 years of teaching and research experience and published several articles. He has four patents, three design registrations, and one copyright registration. As a UGC-HRDC resource person for FDP and FIP, he had conducted more than 200 sessions and had trained more than 8000 faculty members, thereby impacting/improving the teaching skills and careers of many promising educators. He designed and developed a 3D-printed lock and key Braille model to teach fundamental chemistry concepts, valency, hybridization to blind and low-vision students.

1 Deconstruction of Organic Reactions

1.1 ELECTRON CONFIGURATION

An electron configuration is the representation of the distribution of electrons of an atom or molecule in their respective atomic or molecular orbitals. It may be used to describe the electrons in the orbitals of an atom in its ground state, excited state, or ionized state. In other words, it is an "address" that can represent where the electrons can be found in an atom. Electrons exist around the nucleus of an atom in discrete, specific orbits. These orbits are called levels. The first level is the orbit closest to the nucleus. We will focus on three principal energy levels. The maximum number of electrons in each energy level is 2, 8, and 18, respectively. There are different sublevels for each principal energy level. The principal energy level n=1 has $1s$ orbital, the principal energy level n=2 has $2s$ and $2p$ orbitals, and the principal energy level n=3 has $3s$, $3p$, and $3d$ orbitals. There are one 's' orbital, three 'p' orbitals (degenerate, p_x, p_y, and p_z), and five 'd' orbitals. The 's' orbital can accommodate a maximum of two electrons, the 'p' orbital can accommodate a maximum of six electrons, and the 'd' orbital can accommodate a maximum of ten electrons.

In the first row, we will study hydrogen, and in the second row, carbon, nitrogen, oxygen, fluorine, and boron. In the organic reactions that we will study, the second-row elements generally follow the octet rule (to attain the octet configuration, lower energy, or better stability), and for hydrogen, it can have a maximum of two electrons in the filled stable orbital.

The chemical reactions result in the formation of bonds. Valence bond theory or molecular orbital theory is used to explain how the bonds are formed. The head-to-head overlap of atomic orbitals results in the formation of bonding orbitals. Bond length is defined as the distance between the nuclei of the two atoms. The bond axis is the imaginary line that connects the two bonding nuclei.

When we travel by car or bus on a busy road, we always hear honk. But this is a little different HONC we encounter in organic chemistry. The HONC in organic can have 1, 2, 3, and 4 bonds, respectively (Figure 1.1).

The four bonds of carbon can be four single bonds (as in alkanes), two single bonds and one double bond (as in carbonyl compounds), two double bonds (as in CO_2), one single bond and one triple bond (as in cyanide. (Figure 1.2).

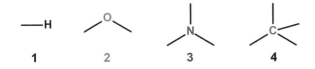

FIGURE 1.1 Bonding in HONC.

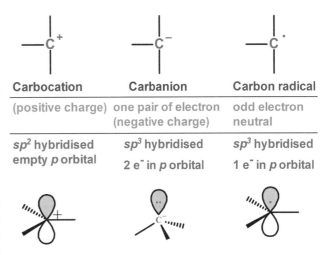

FIGURE 1.2 Bonding pattern in C.

Carbocation	Carbanion	Carbon radical
(positive charge)	one pair of electron (negative charge)	odd electron neutral
sp^2 hybridised empty p orbital	sp^3 hybridised 2 e⁻ in p orbital	sp^3 hybridised 1 e⁻ in p orbital

FIGURE 1.3 Representation of reactive species of carbon.

In reactive species, carbocation is electron deficient. Its geometry is sp^2 trigonal planar, and it has an empty "p" orbital which is perpendicular to the trigonal plane (Figure 1.3).

In the case of carbanions, they are electron-rich and have sp^3-hybridized structure. They are neither tetrahedral like normal tetravalent carbon nor planar-like carbocations but are trigonal pyramidal in shape, like ammonia. The lone pair occupy the unhybridized 'p' orbital. In the case of carbon radicals, they are odd electrons, neutral species, and have sp^3-hybridized geometry. For simplicity we can say, they are in between carbocation and carbanions, so their geometry is sometimes planar and sometimes pyramidal. The single electron occupies the unhybridized 'p' orbital.

The other two atoms that we will encounter most are oxygen and nitrogen. The neutral oxygen has two lone pairs (4 e⁻. and can form two covalent bonds (4 e⁻), totally eight electrons to complete the octet. In the case of nitrogen, it has one lone pair (2 e⁻. and can form three covalent bonds (6 e⁻), eight electrons to complete the octet. Hydronium ion is a positively charged oxygen, which we will encounter quite often in reaction mechanisms. It has three bonds. Since it has lost one of its lone pair to form the third bond, it is positively charged.

DOI: 10.1201/9781032631165-1

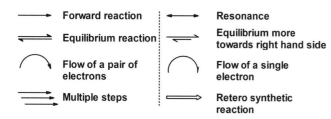

FIGURE 1.4 Representation of different types of arrows in a chemical reaction.

We will briefly learn the different types of arrows that are generally used in depicting an organic reaction. The following chart gives the details. The boxed one is the important arrow that is generally used to represent the flow of electrons and is fundamental to understanding and writing a reasonable reaction mechanism. If you familiarize yourself with this arrow pushing, then you can easily attempt to give a reasonable mechanism for an unknown reaction (Figure 1.4).

1.2 HOW DOES A CHEMICAL REACTION OCCUR?

A chemical reaction is the transformation of one compound into another. During this process, some bonds are broken, and some new bonds are formed. Since a bond means a pair of electrons, there are two ways the bond can be broken. When the bond is broken, both the electrons are retained by one entity and the other loses the electrons, or both the entities keep one electron each. The former is called a heterolytic cleavage, and the latter is called a homolytic cleavage. During the bond formation, we have two scenarios: (1. both the entities donate one electron each, and (2. both electrons are donated by one entity. The first one is homogenic bond formation, and the second one is heterogenic bond formation.

Many organic reactions follow either heterolytic cleavage or heterogenic formation and are called as polar reactions. The other one is called free radical reactions. Other than these two reactions, we also have pericyclic reactions. That is a special case where the transition state of the molecule has a cyclic geometry, and the reaction progresses in a concerted fashion.

For organic reactions to occur, we need an electron donor and an electron acceptor. The electron donors are generally called bases, nucleophiles, or reducing agents, and the electron acceptors are called acids, electrophiles, or oxidizing agents. We can say organic reactions are basically movement of electrons between molecular orbitals.

For the deconstruction of a reaction based on frontier molecular orbitals, we have four simple rules to follow for organic reactions:

1. There is an electron donor and an electron acceptor. So we need to identify the electron donor and electron acceptor orbitals.

2. The highest occupied molecular orbital will be the donor, and the lowest unoccupied molecular orbital will be the acceptor.
3. We need to look at the alignment of molecular orbitals in space. The orientation is very crucial for maximum overlap.
4. The shapes of the orbital will decide selectivity.

1.2.1 EASE OF ELECTRON DONATION

We will look at the use of electron donation from the nucleophiles which includes lone pair, π bonds and σ bonds. In the case of systems with lone pairs, we have negative ions and neutral species. The negative ions can easily donate electrons. Examples include hydroxides or cyanide. Neutral species are water and ammonia. In the case of 'p' bonds, we have alkenes, and for the σ bonds, hydride ion and borohydride are good examples. Electron donation follows the trend shown in Figure 1.5. Negative nucleophiles (hydroxide ion, cyanide ion, etc.. are stronger compared to neutral species like water, ammonia, and so on.

Why σ bonds are poor electron donors? You think σ bond is a strong bond, so it should also be the strong donor. But in reality, tightly the electrons are held by the bonding atoms and so are not freely available for donation. If you look at the π bonds, there is a greater number of electrons between the bonded atoms and are loosely held. It is fine with the π bond to donate electrons. But when you reach the lone pairs or the non-bonding electrons, they are readily available on the individual atom and can be easily donated.

1.2.2 EASE OF ELECTRON ACCEPTANCE

In the case of electrophiles, they readily accept electrons, so the use of electron acceptance is given by these figures. Empty orbitals can accommodate electrons far better than π* antibonding orbitals or σ* antibonding orbitals. Some examples of empty orbitals are carbocations which have empty "p" orbitals. For the π* antibonding, we have carbonyls, and for the σ* antibonding, we have HX (Figure 1.6).

1.2.3 POLARIZATION OF BONDS

For understanding reaction mechanism, we need to know how bonds can be polarized if there is a heteroatom. Whichever is more electronegative tends to keep the electron cloud toward itself. Some of the examples are given here with their polarized structures (carbonyl, alkyl halide, nitrile, and nitro groups. (Figure 1.7).

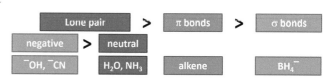

FIGURE 1.5 Ease of electron donation.

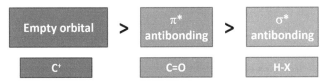

FIGURE 1.6   Ease of electron acceptance.

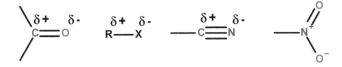

FIGURE 1.7   Polarization of bonds.

In organic chemistry, we will encounter the following elements quite often. It will be important to remember this series for future reference.

$$F > O > Cl > N > Br > I > S > C > H$$

The above series gives an overall view of trends in electronegativity. In general, electronegativity increases from left to right in a row in the periodic table. Electronegativity increases across a period because the number of charges on the central nucleus increases, which in turn attracts the bonding pair of electrons more strongly toward itself. As we go down a group, electronegativity decreases because the bonding pair of electrons is increasingly moving away from the attraction of the nucleus.

1.3   COMMON ERRORS TO AVOID

We need to pay attention to some of the errors which we need to avoid. Basically, H^+ can exist only in the gas phase. That means we cannot simply write H^+ in any reaction that occurs in solution. H^+ is always associated with something. H^+ is always written as HA (Brønsted acid), and water is written as hydronium ion (H_3O^+. in an acidic medium and hydroxide ion ($^-$OH. in a basic medium.

1.4   RULES TO REMEMBER

When we write reaction mechanisms, we need to follow these rules. Although this list is neither exhaustive nor inclusive, trying to adhere to it may help us understand the reaction mechanism in a better way.

1) Always maintain the octet rule. Mainly for carbon, nitrogen, oxygen, and hydrogen, we have a maximum of two electrons.
2) The overall reaction consists of many elementary steps. What is an elementary step? It's a single reaction step with a single transition state.
3) When we write the reaction mechanism, every step is in equilibrium, and they follow the principle of microscopic reversibility.
4) If oxygen is lost in a reaction, most probably it can be removed as water molecules. If nitrogen is lost, then it can be removed as ammonia or nitrogen gas. You already know alcohol can be dehydrated to give alkenes. This is an example of oxygen being removed as water. Similar to that in the amide hydrolysis, nitrogen is removed as ammonia. Diazonium salts release nitrogen gas.

Here we will look at the important distinctions between intermediate and transition states (Table 1.1).

Intermediates generally will have a discrete lifetime. That means, they can be detected even if they are short-lived. They may be isolable and stable or unstable. They are generally represented by the depression on the potential energy curve, and intermediates are generally for the whole reaction. If you look at the transition state, they are impossible to detect because they generally do not exist. It is practically impossible to isolate them, and it is the saddle point on the reaction coordinate. It has the local energy maximum and partial bonds formed, and the transition states are generally applicable to elementary reaction steps only.

TABLE 1.1
Comparison Between Intermediates and a Transition State in a Reaction

Intermediate	Transition State
May have a discrete lifetime	Impossible to detect
May be stable or unstable	Does not exist
Represented by the depression on potential energy curve	Is a saddle point on reaction coordinate and has local energy maximums and have partial bonds
It is for a reaction as a whole	Applicable to elementary step only
Can be isolated	Cannot be isolated

1.5 HETEROLYTIC CLEAVAGE

We will now look at the bond dissociation, mainly heterolytic fission (cleavage). If water is heterolytically cleaved, it leads to H^+ and ^-OH. In the same way, when HX is cleaved heterolytically, we get H^+ and X^-. So, in both cases, we see H^+ is generated and is an acid. The other part is the base.

> **DID YOU KNOW?**
>
> If you look at alcohol and how is it cleaved, we have two possibilities. One it can be cleaved into R^+ and ^-OH. The other one is H^+ and ^-OR. The functional group in alcohol is –OH. The reason behind why it is not cleaving into R^+ and ^-OH is that between R^+ and H^+, proton is more acidic and so it easily cleaves into H^+ and ^-OR.

Alcohol can be converted to alkyl halide by the R–O bond breaking. However, in the nucleophilic substitution, the leaving group is generally the H_2O^+ ion. It is not a ^-OH ion. In effect we can say, R-O-H can be broken between R-O bond (as in nucleophilic substitution.   and O-H bond (as in salt formation with a strong base). But ROH is not broken as R^+ and HO^-.

> **DID YOU KNOW?**
>
> How does RX cleave? It can break into R^+ and X^- or X^+ and R^-. In the case of RX, X is more electronegative than R so it cleaves into X^- and R^+.
>
> It should be stressed here that it is the acidity/electronegativity that determines how heterolytic cleavage will occur.

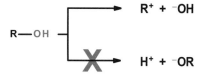

SCHEME 1.1 Within the box (dissociation of alcohol).

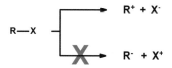

SCHEME 1.2 Dissociation of alkyl halide.

1.6 ARROW PUSHING

In organic chemistry, we use two types of arrows to describe the flow or the movement of electrons. Robert Robinson first introduced the concept of arrow pushing to depict the flow of electrons in his paper (Robinson, 1924). Robinson was trying to explain the *o/p*-directing nature of the nitroso group in aromatic electrophilic substitution (as given in the original paper).

Sir Robinson was awarded the 1947 Nobel Prize in Chemistry for his work on the synthesis of natural products, especially alkaloids.

We are moving into the very crucial arrow-pushing mechanism. There are a few important things to look at:

- Polarization of bond means that the electronegativity difference will decide how a bond will polarize
- Attack of electron-rich part, that is, lone pairs, π bonds, or even σ bonds
- Movement of protons
- Charge reversal
- Departure of leaving group

We take a simple example of esterification to find out how we can do the arrow pushing (flow of electrons). Here an acid reacts with an alcohol to give an ester with the loss of water molecule.

If you look carefully, the carboxylic acid group has two oxygen atoms, each having two lone pairs. The acid is written as HA. As it was mentioned earlier, we cannot write H^+ in the solution, and we will be writing the Brønsted acid HA. Now we have seen the arrow pushing. The lone pair of electrons from hydroxyl oxygen goes and attacks the proton.

1.6.1 IDENTIFY THE ERRORS

Can you identify the errors, if any, in the above formulation?

There are two errors. The first one is when the hydroxyl oxygen gives its lone pair for, the bond formation with a proton. But it was written in such a way that the oxygen is having three bonds and two lone pairs. That is obviously wrong. Because this oxygen now has ten electrons violating the octet rule. Three bonds mean six electrons and the two lone pairs will mean 4 e⁻, so in total 6+4 =10 electrons, which is violating the octet rule. So, this is the error number 1. Moreover, when oxygen gives one of its lone pairs to form a bond, logically it will not have additional electrons. This is similar to "you cannot have the cake and eat it."

So, what is the second error? The second error is oxygen having a plus charge. Although it is correct to write oxygen with a plus charge when it loses the electron, it is wrongly depicted as having its two lone pair electrons along with a plus charge. So that is the second error.

The carbonyl oxygen which is more electron-rich than the hydroxyl oxygen will actually attack the proton so that is the last error.

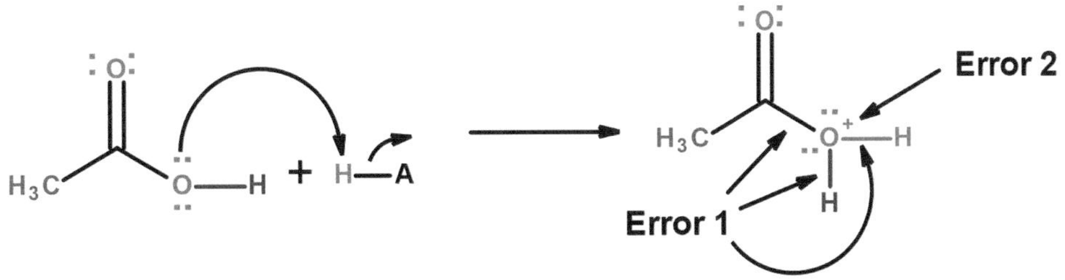

SCHEME 1.3 Errors in the flow of electrons in protonation of acid.

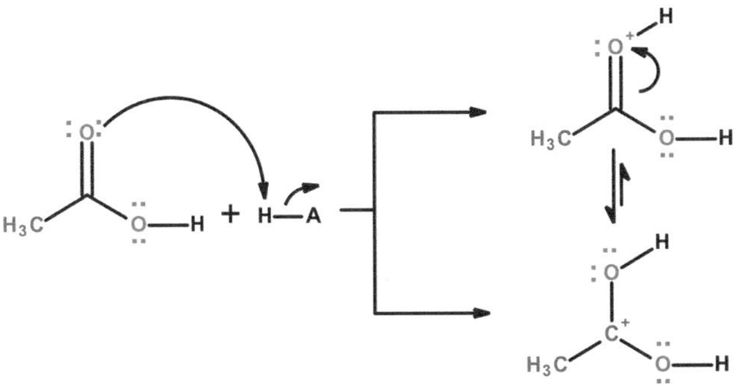

SCHEME 1.4 Flow of electrons in protonation of acid.

We know that hydroxyl oxygen is not the oxygen that attacks the proton. Now it is the carbonyl oxygen that will attack the proton if that is the case, we have two possibilities and both are given here. Oxygen being more electronegative than carbon does not want to have a positive charge so it pulls the electron toward itself creating a carbocation.

We will look at simple esterification of a carboxylic acid by an alcohol. This reaction here is catalyzed by an acid (HA). The addition of a proton to the carbonyl oxygen results in the formation of positively charged species (Int-1). The electrophilic carbon is attacked by the lone pair of electrons from the alcohol oxygen to form Int-2. There is an internal shift of the proton from the alcohol unit to one of the hydroxyl units resulting in Int-3. This intermediate loses a water molecule to give Int-4. The final loss of a proton from the intermediate four gives the ester.

1.6.2 AMIDE HYDROLYSIS MECHANISM (UNDER ACIDIC CONDITIONS)

Let us see amide hydrolysis under acidic conditions. In the esterification reaction, the initial step is the attack of the carbonyl oxygen on the hydronium ion. There are two possibilities for the attack: (1. the lone pair from carbonyl oxygen is attacking, and (2. the lone pair from the amine nitrogen is attacking.

Oxygen is more electronegative than nitrogen; moreover, oxygen has two lone pairs compared to one lone pair on nitrogen. So, oxygen is the one that can easily donate two electrons to form a bond with a proton compared to nitrogen. The following are the various steps involved:

1. Protonation of amide carbonyl (acidic medium)
2. Electronegative oxygen pulls the π electrons from the carbonyl bond
3. The nucleophilic **O** (from H_2O. attacks C^+
4. Proton transfers from H_2O^+ to NH_2
5. Loss of NH_3, O donates a lone pair to form C=O
6. Deprotonation of oxonium ion (NH_3 abstracts it)
7. Product is formed

We have seen in the esterification reaction that the oxygen pulls the π electron toward itself making the carbon δ^+. It is not a full-scale carbocation. There is only slight polarization of charges. In the next step, the solvent water molecule, being a neutral nucleophile, attacks the electropositive

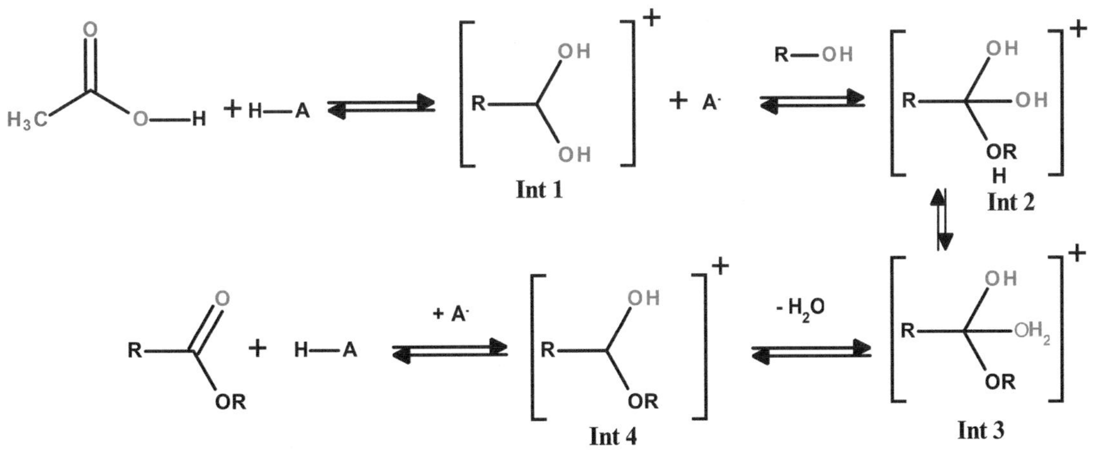

SCHEME 1.5 Esterification of carboxylic acid.

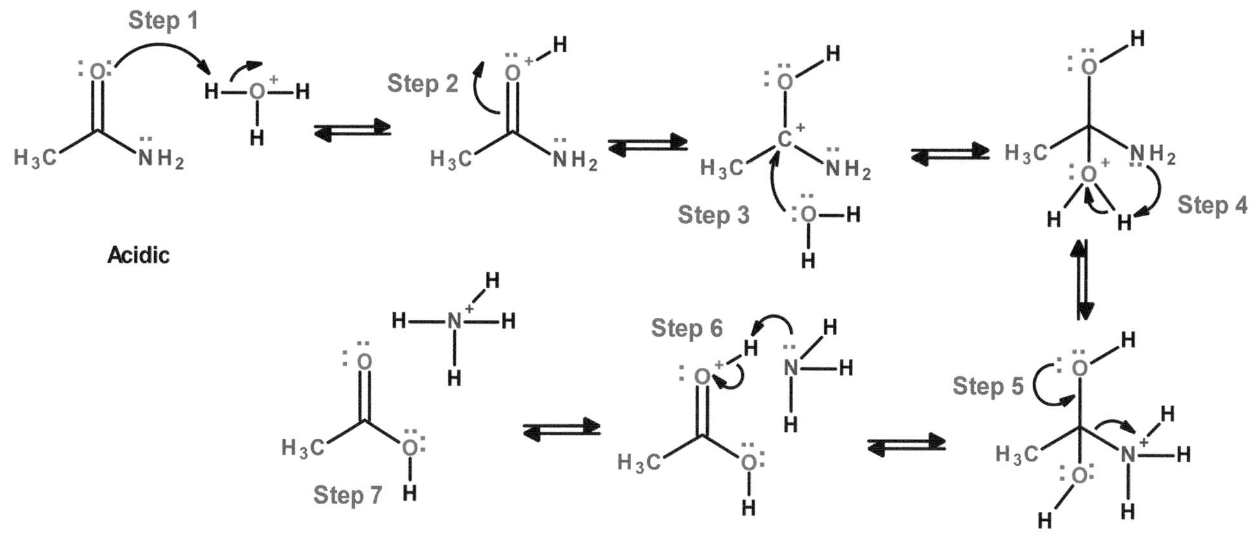

SCHEME 1.6 Amide hydrolysis mechanism under acidic conditions.

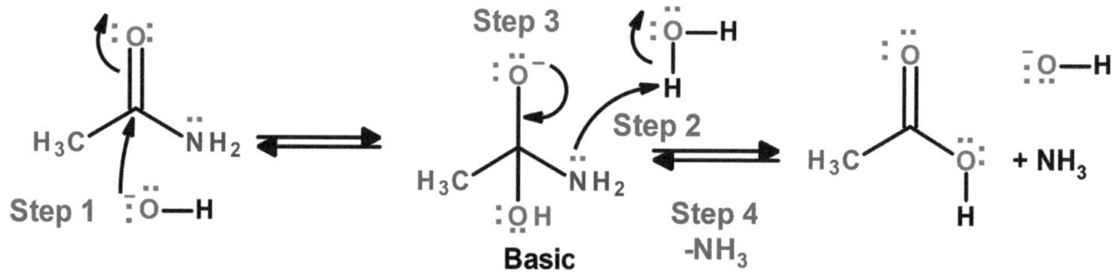

SCHEME 1.7 Amide hydrolysis mechanism under basic conditions.

carbon with its lone pair. In the next step, there is a proton transfer. Nitrogen with its lone pair attacks the hydrogen on the oxonium ion. The hydrogen is lost as a proton with oxygen retaining both the electrons from the bond. NH$_2$ is a very poor leaving group. But when a proton is added to it to make it ammonia, it leaves easily. It leaves with its bonding pair of electrons making the carbon electron deficient. Now to compensate for that, a lone pair from oxygen is given to this carbon. Now there is an oxonium ion. The ammonia molecule which was lost in the previous step (is a base. abstracts the proton from the oxonium ion. Finally, the acid is formed and the ammonia is converted into ammonium ion. As it was mentioned earlier, the proton does not exist as a proton, but it exists as an ammonium ion ($^+$NH$_4$). If you carefully look at the whole mechanism, every step is a reversible process. Every step represents one transition state. In other words, all these steps are called **elementary steps**. So, in effect, the reaction mechanism may consist of several elementary steps. Another important thing to know is that H$^+$ does not exist as proton but rather as a hydronium ion (H$_3$O$^+$. or ammonium ion ($^+$NH$_4$). It is always associated with something.

1.6.3 Amide Hydrolysis Mechanism (Under Basic Conditions)

Let us see amide hydrolysis under basic conditions. The hydroxide ion ($^-$OH. attacks the carbonyl carbon. The

arrows show how the attack takes place. In the next step, the lone pair from the nitrogen attacks the proton from the solvent water molecule. It is followed by charge reversal, and ammonia is lost to produce the corresponding carboxylic acid. But the free carboxylic acid cannot exist in a basic medium. It will exist only as a carboxylate anion. Upon acidification, we get the acid.

In the above example, we have combined three steps into one: (1. the attack of nucleophilic nitrogen on the solvent water molecule, (2. the charge reversal, and (3. the loss of ammonia. Although it is customary to write individual steps, you can also follow the mechanism by combining multiple steps in a single step. However, it is important to show all the electron flow, in the right direction to completely depict the actual course of the reaction.

1.6.4 Cyanide Hydrolysis (Acidic)

Let us understand the cyanide hydrolysis under acidic conditions. This is an example of the nitrogen lone pair attacking the hydronium ion. In the second step, a lone pair from the solvent water molecule attacks the electron-deficient carbon (C$^{\delta+}$). There is a carbon–nitrogen double bond, and a new bond between carbon and oxygen is formed. This results in the formation of imidic acid.

Among carbon and nitrogen, nitrogen is more electron deficient so it pulls the π electrons toward itself to neutralize the positive charge. (It is because an electronegative atom

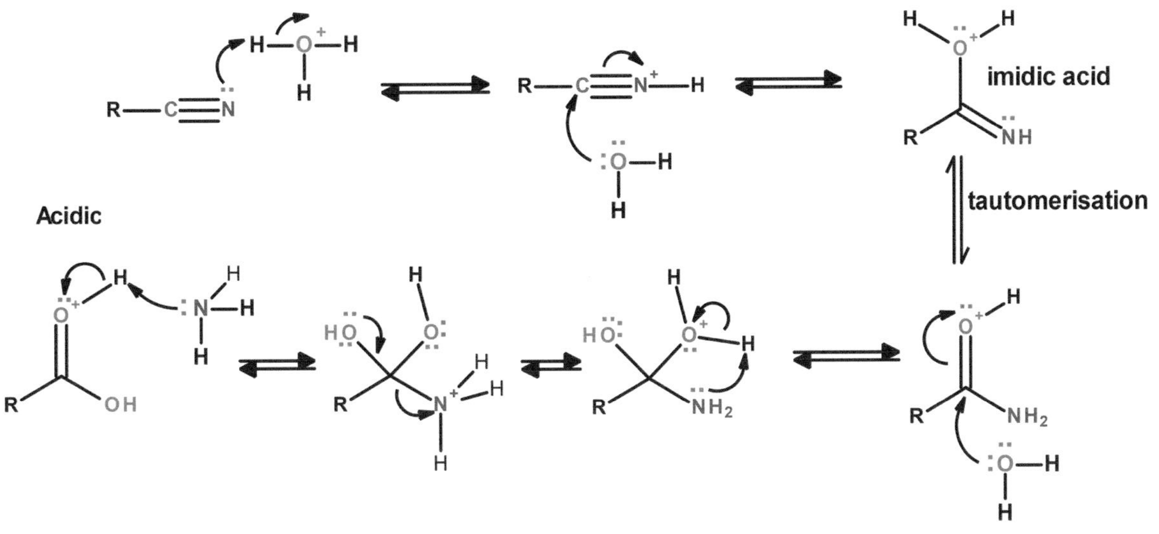

SCHEME 1.8 Cyanide hydrolysis under acidic conditions.

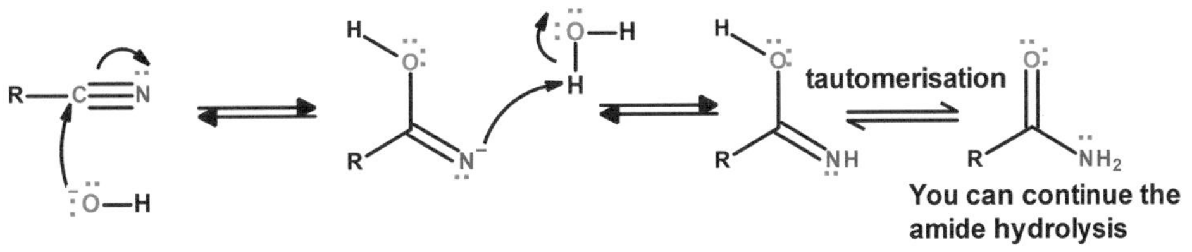

SCHEME 1.9 Cyanide hydrolysis under basic conditions.

with a positive charge is quite unlikely and is unstable. But carbon can accommodate a positive charge.)

Carbon is ok with electron deficiency. It readily takes the electrons given by the oxygen of solvent water molecule. Now, this intermediate undergoes tautomerization, there is a double bond shift from imidic acid, and a proton moves to give the amide. Now the amide oxygen pulls the electron toward itself, and the solvent water molecule attacks the electron-deficient carbon. Next, the amide nitrogen with its lone pair attacks the proton on the hydronium ion. Lone pair donation from one of the hydroxyl oxygens produces a C=O (double bond. with concomitant loss of ammonia. Finally, a proton is abstracted by the ammonia from the oxonium ion to give ammonium ion and carboxylic acid.

1.6.5 CYANIDE HYDROLYSIS (BASIC)

In cyanide hydrolysis under basic conditions, the hydroxide ion attacks the electron-deficient carbon. The nucleophilic nitrogen now attacks the proton of solvent water molecule. In the next step, tautomerization produces amide. From here onward, we can follow the same amide hydrolysis mechanism.

With this, we conclude the arrow-pushing mechanism. We need to remember the rules as to how to move the electrons during a chemical reaction. You can practice this multiple times to familiarize yourself with electron flow.

1.7 QUESTIONS

True or False

1. Hydroxide ion has three lone pairs of electrons.
2. Hydronium ion has a negative charge.
3. Ammonia (NH_3. is a better nucleophile than amine (-NH_2).
4. In any reaction mechanism, the overall charge must be the same for every elementary step.
5. The π bonds can be a better nucleophile than lone pairs.
6. Carbanion is planar in shape.
7. Carbocation is sp^2 hybridized.
8. Carbon radicals can have either planar or pyramidal structure.
9. The highest occupied molecular orbital can act like both an electron donor and an acceptor.
10. Orbital orientation is important for selectivity.

Short and Long Answer Questions

11. What are the rules related to the deconstruction of reactions?
12. Write briefly about the ease of electron donation and acceptance.
13. What are the important distinctions between intermediate and transition states.

14. Write elaborately about the amide hydrolysis mechanism under acidic conditions.
15. Write elaborately about the amide hydrolysis mechanism under basic conditions.
16. What are the important distinctions between cyanide hydrolysis mechanisms under acidic and basic conditions?

BIBLIOGRAPHY

Bruice, P. Y. *Organic Chemistry*, 4th Ed. Prentice Hall, 2003, 121–124.

Clayden, J., & Greeves, N. *Organic Chemistry*, 2nd Ed. Oxford University Press, 2012, 113–133.

Robinson, R. (1924). CCXC.—An accessible derivative of chromonol. *Journal of the Indian Chemical Society*, 43, 1297

Robert C. Neuman, Jr. (University of California, Riverside), Chapter 1: Organic Molecules and Chemical Bonding in Organic Chemistry (https://people.chem.ucsb.edu/neuman/robert/orgchem-byneuman.book/01%20OrganicMolecules/01FullChapt.pdf. (accessed 15-Apr-2024)

2 Geometric Isomerism

Here we study about spatial arrangements of atoms/groups in a molecule. We will start with a very brief historical beginning of isomerism. In 1826, *Justus von Liebig* prepared *fulminic acid* having an elemental composition HCNO. In the following year (1827), *Friedrich Wöhler* prepared *cyanic acid* having an elemental composition HOCN. It was found that the elemental composition of cyanic acid was identical to fulminic acid. But its properties were quite different.

The silver fulminate studied by Liebig was an explosive, whereas the silver cyanide studied by Wöhler was not. Both were surprised to see that compounds having similar compositions had different properties. It was Berzelius who named these types of compounds as isomers. In the following year, Wöhler prepared urea by heating ammonium cyanate. That is the beginning of organic chemistry.

> I can no longer, so to speak, hold my chemical water and must tell you that I can make urea without needing a kidney, whether of man or dog; the ammonium salt of cyanic acid is urea.
>
> *(Wöhler to Berzelius)*

Do you know that there is a distant cousin to isomers called allotrope? Allotropes are different types of compounds of the same element but have different chemical compositions and different arrangements. Allotropy is the structural modification of an element, whereas isomers are modification of molecules. Molecular oxygen (O_2), ozone (O_3), and octaoxygen (O_8), also known as ε-oxygen or red oxygen, are few allotropes of oxygen. Graphite, diamond, carbon nanotubes, and buckminsterfullerene (buckyballs) are few allotropes of carbon.

The flowchart in Figure 2.1 shows how different types of isomers are interrelated.

Isomers are compounds with the same molecular formula. If the connectivity is the same, they are stereoisomers, and if the connectivity is not the same, they are constitutional isomers. One type of stereoisomers where interconversion is possible through single-bond rotation are conformational isomers. If interconversion is not possible by a single-bond rotation, then they are configurational isomers. Within the configurational isomers, if there is an isomerism about the double bond, they are geometric isomers, but if not, they are optical isomers. Optical isomers having non-superimposable mirror images are enantiomers; otherwise, they are diastereomers.

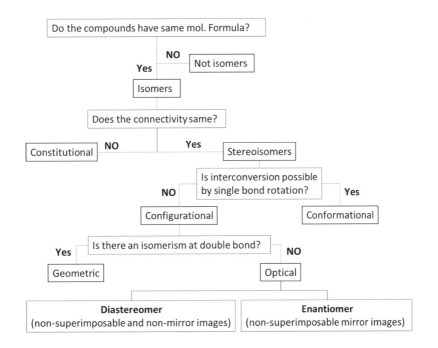

FIGURE 2.1 Interrelationship between different types of isomers.

DOI: 10.1201/9781032631165-2

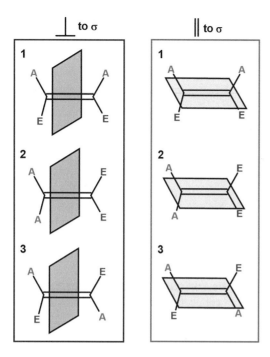

FIGURE 2.2 Planes bisecting the molecule.

2.1 CRITERION FOR GEOMETRICAL ISOMERISM

In geometric isomers, each carbon of the alkene must be bonded to two different groups. If any of the carbon has both the groups on the double bond, then geometric isomerism is not possible, because the basic definition of geometric isomerism states that the isomers need to be different in the position of atoms or groups relative to a reference plane. Let us see how we can identify the reference plane.

In Figure 2.2, we have two planes bisecting the molecule. One is perpendicular to σ bond axis, and the other one is parallel to σ bond axis. For our reference plane, we need to consider the plane that is parallel to the σ bond axis. Among compounds 1, 2, and 3, compound 2 has similar groups on both sides of this reference plane. Hence, it cannot form geometrical isomers. For geometrical isomerism to exist, the general rule is that there should be no mirror image on the reference plane (parallel to the σ bond axis). Even in one of the carbons if there is a mirror image, in other words, if both the groups are the same on any one of the carbon atoms across the double bond, then that compound cannot have geometrical isomers.

2.2 TYPES OF ISOMERISM

Let us look at the overview of different isomerism. The main classes of isomers are structural (constitutional) and stereo (spatial). We shall study stereoisomerism. Within stereoisomerism, we have different types and we will discuss more about them later.

Geometrical isomers have different chemical and physical properties, and they may be optically active or inactive

TABLE 2.1

Comparison between Geometrical and Optical Isomers

Geometric Isomerism	Optical Isomerism[a]
Different chemical properties	Same chemical properties
Different physical properties	Same physical properties
Optically inactive	Optically active

[a] For enantiomers.

depending on the molecular geometry. Optical isomers (enantiomers) will have the same physical and chemical properties, but they will have opposite optical rotation (Table 2.1).

Geometric isomerism arises because of the presence of a double bond. Let us look at sp^2 hybridization. We take the example of ethylene. In this, each carbon has 3 σ bonds and one electron in the unhybridized 'p' orbital. The lateral or sidewise overlap of 'p' orbital electron gives the π bond.

2.2.1 CIS-TRANS ISOMERISM

In Figure 2.3, two isomers are written. Unlike the C-C single bond, the C-C double bond cannot be rotated easily. There is a barrier to rotation. That leads to two different sets of arrangements for the same formula leading to geometric isomers. When both groups are on the same side, it is called *cis* isomer, and when they are on the opposite side, it is called *trans* arrangement.

Let us see a few examples. There are six compounds given in Figure 2.4 (**A**, **B**, **C**, **D**, **E**, and **F**). In the first one (**A**), the hydrogens are on opposite sides, so it is a *trans* arrangement. In the next one (**B**), both chloro groups are on the same side, so it is a *cis* arrangement. In the third compound (**C**), we have a problem. When we say the same side, is it across the double bond or on the same carbon? For geometric isomers, we need to consider groups across the double bond and not on a single carbon. So, this compound is not a geometric isomer. Because exchanging the position

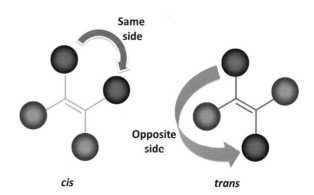

FIGURE 2.3 Pictorial depiction of atoms across double bond in *cis-trans* isomerism.

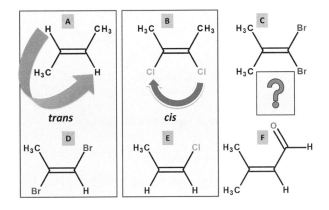

FIGURE 2.4 Different example of *cis-trans* isomerism.

of the methyl group or bromine does not lead to different arrangements. So, this compound is not a geometric isomer.

When we go to the fourth compound (**D**), the bromine atoms are on the opposite side, so it is the *trans* isomer. The next compound (**E**) has both the hydrogens on the same side so it is a *cis* compound. The next one (**F**) is not a geometric isomer. As we mentioned earlier, replacing or exchanging the methyl group does not produce a new compound.

2.2.1.1 Ambiguity in *cis-trans* Isomerism

Let us go to a new set of compounds (Figure 2.5).

In the first compound, both the amino groups are on the opposite side so is it the *trans* isomer? In the next one both the hydroxyl groups are on the same side. So, is it a *cis* isomer? If I name the first isomer as *trans* and the second one as *cis*, it is wrong. These compounds are not like the compounds we have seen earlier. In the previous sets, we had the smallest atom hydrogen. So, our assignments were correct. But when the atoms are different, we cannot apply the same arguments. Should we go with the size of the atoms, electronegativity, degree of unsaturation, and so on? To solve this problem, CIP rule or Cahn Ingold Prelog rule was introduced.

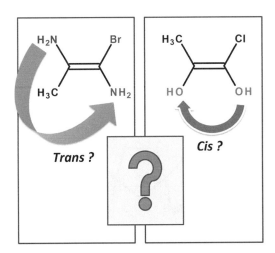

FIGURE 2.5 Ambiguity in *cis-trans* isomerism.

2.2.2 CIP Rule

A simplified version of the CIP rule is given here. The important points to consider are as follows:

1. The atom which has the higher atomic number is given the higher priority.
2. For isotopes, the one with the higher relative atomic mass has the higher priority.
3. When there is a tie, look for the first point of difference to assign priority.
4. Each multiple bonds are counted as a single bond connected to multiple atoms while assigning priority.

Analyze the two groups at each end of the double bond. Rank the two groups on each end of the double bond, using the CIP priority rules. If higher priority groups are on the same side across the double bond, then the configuration is Z (zusammen) If higher priority groups are on the opposite side across the double bond, then the configuration is E (entgegen = opposite).

A pictorial representation of the same is given in Figure 2.6 for your easy understanding.

Using this CIP rule, now we will try to identify the E/Z notation for the compounds we have taken for our discussion. The previous assignment of *trans* is wrong and is to be called the Z isomer. In the same way, the second compound is not a *cis* isomer but is the E isomer (Figure 2.7). When you need to assign E/Z notation, you need to follow the CIP rule and it will help you identify the correct notation.

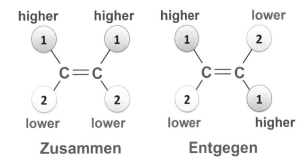

FIGURE 2.6 A pictorial representation of E/Z configuration.

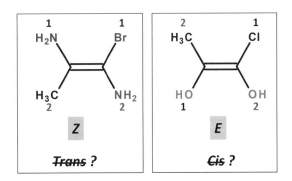

FIGURE 2.7 Assigning E/Z configuration.

So far, we have seen *cis/trans* isomerism around double bonds. Do we really need *sp²* carbons for *cis-trans* isomerism? Can we have *cis-trans* isomerism on *sp³* carbon? To answer this, we need to know the meaning of *cis* and *trans*. *Cis* and *trans* means whether the two atoms or groups that are under consideration are on the same side or not, respectively. In effect, the geometry of the bond is not the deciding factor. That means we can also have *cis-trans* isomerism on a *sp³*-hybridized carbon. Since it is a comparative method, it is not confined to double bonds only. It applies to both cyclic and acyclic systems alike.

We will study more about the *cis-trans* isomerism in cyclic systems in conformational isomerism of cyclic systems.

2.2.3 *Cis-trans* Isomerism in Heteroatom Double Bonds

Let us start with double bonds with one or more hetero atoms. The first one is C=N. Some of the examples include imines and oximes. Although we cannot use *cis* and *trans*, we can use Y as our reference point and say whether W or X is *cis* or *trans* with respect to it. We will see more about oximes when we study Beckmann rearrangement under carbocation. In the case of diazene, we have *E* or *Z* isomer. It is also called diimide. When we talk about the aromatic diazene, the simplest one is azobenzene. *trans*-azobenzene is planar, and the *cis*-azobenzene is nonplanar. The *trans* isomer is more stable. Due to the extension of conjugation, this compound is colored (Figure 2.8).

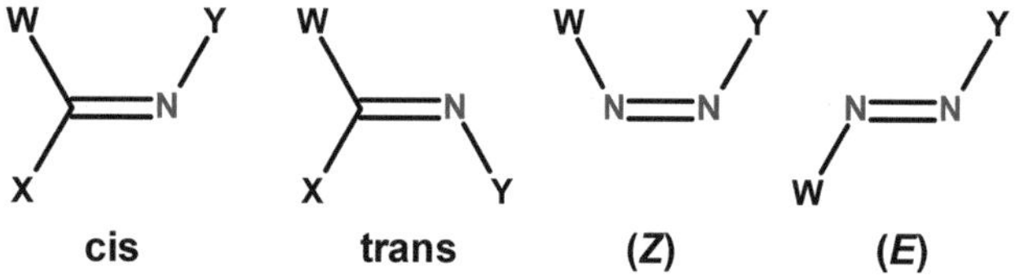

FIGURE 2.8 Assigning configuration to double bonds with one or more heteroatoms.

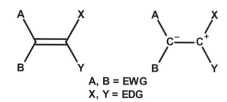

A, B = EWG
X, Y = EDG

FIGURE 2.9 Bonding in captodative ethylenes.

2.2.4 Captodative Ethylenes

We know that rotation about formal C=C is having high energy barrier, and hence, they do not occur readily, and the orientation of the molecule is fixed. But there are certain types of compounds where there is a possibility of rotation about C=C. When two electron-withdrawing groups are present in one carbon and two electron-donating groups on the other carbon across a double bond, they are called as push-pull or captodative ethylenes. The flow of electrons from the electron-donating substituents to the electron-withdrawing substituents through the double bond as shown in Figure 2.9 is possible. Because of the reduction in double-bond character and increase in single-bond character, there is a possibility of rotation between the atoms attached to the formal double bond.

2.2.5 *Cis-trans* Isomerism in Cyclic Systems

In the presence of a ring, which prevents free rotation between the atoms of the ring, we can expect *cis* and *trans* isomers whenever there are two different groups present across the two carbons on a ring. The two carbons may be adjacent or may not be adjacent (Figure 2.10).

2.2.6 *Cis-* and *trans*-Decalin

Chemical studies of steroids are very important to our present understanding of the configurations and conformations of six-membered rings. In the next chapter, we will study more about conformational analysis. If the cyclohexane

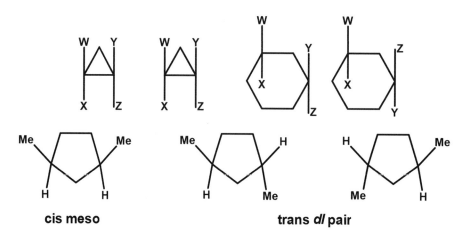

cis meso   trans *dl* pair

FIGURE 2.10 *Cis-trans* isomerism in cyclic systems.

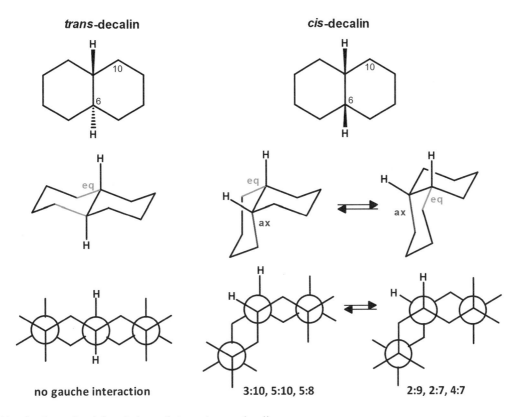

FIGURE 2.11 Conformational descriptions of *cis*- and *trans*-decalin.

rings are drawn in their chair conformations, the hydrogens are in the axial position at the ring junction. In other words, both substituents are in the equatorial position for the *trans* isomer. The fusion of rings A and B creates a rigid, roughly planar, structure.

Conformational descriptions of *cis*-decalin are complicated. Here we have two energetically equivalent fusions of chair cyclohexanes, which are in rapid equilibrium. Unlike the *trans* case, here each of these fusions has one axial and one equatorial bond, and the overall structure is bent at the ring junction. The *cis* isomer will have one substituent in the equatorial position and one substituent in the axial position. *Trans*-fused cyclohexane rings, therefore, are more stable than *cis*-fused cyclohexane rings (Figure 2.11).

2.2.7 MOLECULAR ELECTRICAL POTENTIAL SURFACES OF *CIS* AND *TRANS* ISOMERS

Here we will see some special properties of *cis* and *trans* isomers. The dipole moment of *trans* isomers is always zero, and the *cis* isomers will have some dipole moment. Let us consider two examples. One is 2-butene (*cis* and *trans*), and the other is 1,2-dichloroethylene (*cis* and *trans*).

Molecular electrical potential surfaces illustrate the charge distributions of molecules three-dimensionally. For our easy understanding, blue color indicates electron

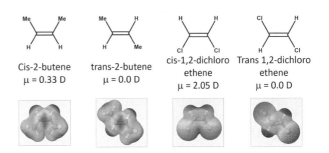

Cis-2-butene $\mu = 0.33$ D

trans-2-butene $\mu = 0.0$ D

cis-1,2-dichloro ethene $\mu = 2.05$ D

Trans 1,2-dichloro ethene $\mu = 0.0$ D

FIGURE 2.12 Molecular electrical potential surfaces of *cis* and *trans* isomers.

deficiency and red color indicates high electron density. In both the *cis* isomers, we have distinct blue color contour map indicating the polarization of the molecule. In the case of *trans* isomers, the electron distribution is still non-uniform in the case of dichloroethylene. But both sides cancel out each other. So, the net dipole moment is zero (Figure 2.12).

2.3 APPLICATION OF GEOMETRIC ISOMERISM

In the medical field, many drugs that are *cis* or *trans* are used. For example, the very famous anti-cancer drug *cis*-platin

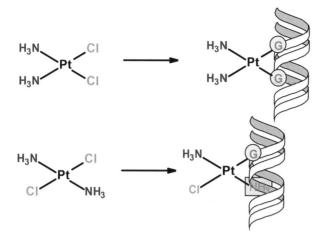

FIGURE 2.13   Structure of *cis*-platin.

got its name *cis* based on the placement of two chlorine atoms in the *cis* orientation. When this drug is administered to cancer patients, *cis*-platin forms crosslinks with DNA in a stepwise manner. In the first step, one chlorine is replaced by a deoxyguanosine residue, and then in the next step, the second chlorine in *cis*-platin is replaced. If the two chlorines are replaced by guanines from the same DNA strand, then it is called intrastrand crosslinks, and if the guanines are from different strands, then it is called as interstrand crosslinks. These crosslinked products have many effects like DNA unwinding, DNA bending, replication inhibition, and transcription inhibition to name a few. These effects further cause DNA strand breaks. The *trans*-isomer cannot form the required bond, and hence, it was found to be chemotherapeutically inactive (Figure 2.13).

2.4   QUESTIONS

True or False

1. When connectivity differs in isomers, it is called constitutional isomerism.
2. Isomers arising out of double bonds are called geometrical isomers.
3. For geometrical isomerism to occur, the reference plane is perpendicular to the sigma bond axis.
4. Geometrical isomers and optical isomers will have the same physical properties.
5. According to the CIP rule, between hydrogen and deuterium, the priority is given to deuterium.

6. According to the CIP rule, multiple bonds are considered as a single unit while assigning priority.
7. Both *sp²* and *sp³* carbons can show geometrical isomerism.
8. Due to partial double-bond character, amides show atropisomerism.
9. Generally *cis* isomers have high melting points.
10. When saturated and *trans* fat compounds are present in cell membranes, they allow easy transportation of nutrients.

Short and Long Answers Questions

11. What are the criteria for geometric isomerism?
12. What are the different types of isomerism?
13. What are the Cahn Ingold Prelog rules that are related to assigning priority?
14. Write briefly about *cis-trans* isomerism in *sp³*-hybridized carbon.
15. Write briefly about *cis-trans* isomerism in hetero-atom double bonds.
16. Write briefly about *cis-trans* isomerism in decalins.

BIBLIOGRAPHY

Carey, F. A. *Organic Chemistry*, 5th Ed., McGraw-Hill, 2004, 192–197.

Charles Ophardt et al., 27.6: Steroids, in Organic Chemistry (OpenStax) (https://chem.libretexts.org/Bookshelves/Organic_Chemistry/Map%3A_Organic_Chemistry_(McMurry)/Chapter_27%3A_Biomolecules_-_Lipids/27.06_Steroids) (accessed, 15-Apr-2024)

Craig Wheelock, Stereochemistry: biological significance of isomerism (https://metabolomics.se/Courses/CEW_Isomer%20lecture_Part%20I.pdf) (accessed, 15-Apr-2024) https://www.chemguide.co.uk/basicorg/questions/a-geomisomerism.pdf (accessed, 15-Apr-2024)

Jim Clark 2000 (last modified February 2020), STEREOISOMERISM - GEOMETRIC ISOMERISM (https://www.chemguide.co.uk/basicorg/isomerism/geometric.html) (accessed, 15-Apr-2024)

Smith, M. B., & March, J. *March's Advanced Organic Chemistry Reactions, Mechanisms and Structure*, 5th Ed. John Wiley & Sons, Inc., 2001, 156–163.

Stereochemistry (https://www.siue.edu/~tpatric/ster.pdf) (accessed, 15-Apr-2024)

William Reusch, Professor Emeritus (Michigan State University), Steroids in Virtual Textbook of Organic Chemistry (https://chem.libretexts.org/Bookshelves/Organic_Chemistry/Supplemental_Modules_(Organic_Chemistry)/Lipids/Properties_and_Classification_of_Lipids/Steroids) (accessed, 15-Apr-2024)

3 Conformational Isomerism
Open Chain and Cyclic Systems

What is conformational analysis? Conformation can be defined as the different spatial arrangements that a molecule can adopt due to rotation about σ bonds. Conformational analysis is the study of the relationship between different conformations and the energy levels of a molecule, or it is the study of how some properties (particularly the free energy and reactivity) of a molecule are related to its shape.

Before we start with conformational analysis, we will briefly look at the nature of electrons, how does Schrödinger wave equation predicts the shapes of orbitals, and next the relationship between hybridization and shapes of orbitals. We will then study the representation of molecules.

3.1 TYPES OF MOLECULAR REPRESENTATION

3.1.1 NATURE OF ELECTRON

Electron has both particle and wave nature (duality principle). When an electron interacts with something, it behaves like a particle. When it is allowed to move freely, it behaves like a wave.

Energy of a particle (electron) is given by the following equation (Figure 3.1)

V(x) – Hamiltonian
ψ – Wave function
E – Total energy
M – mass of the particle
ℏ – Planck's constant

The solution to this equation gives the shapes of orbitals. How Schrödinger equation gives shapes of orbitals? (Find the particle)?

Quantum mechanics states that "electron can only exist in well-defined energy levels".
Energy depends on the position of the particle.
Solution to the Schrödinger equation gives the probability of finding the particle. (position)

$$-\frac{\hbar^2}{2m}\frac{d^2\psi}{dx^2} + V(x)\psi = E\,\psi$$

FIGURE 3.1 Schrödinger equation.

When we extend the same to an atom, we can find the probable position of electron. Thus, orbital is the shape where there is maximum probability of finding the electron. (shape)

3.1.2 SHAPES OF ORBITALS

s orbitals are spherical in shape (no orientation)
p orbitals are dumb-bell in shape (axial orientation) (p_x, p_y, p_z)
d orbitals are having lobes different of shapes (d_{xy}, d_{yz}, d_{zx}, d_{x2-y2}, d_{z2})

When orbitals combine, we get hybridized orbitals. The VSEPR theory treats the mixing of orbitals to give equivalent orbitals (hybridization), all having the same energy. The MO theory treats the mixing of orbitals to form bonding and antibonding orbitals. Half of the resulting orbitals are lowered in energy, and half of them are raised in energy.

3.1.3 REPRESENTATION OF MOLECULES

For a better understanding of conformational analysis, we need to know how to draw different conformations and represent them pictorially, because depicting three-dimensional structures in two-dimensional plain paper is a difficult task. Organic chemists use wedge bonds and broken lines or hash bonds. By repeatedly practicing the 3D representation of 3D molecules in 2D, one can get to know the technique.

Let us start with wedge bonds. It is to represent the groups or atoms pointing toward the observer. It may also be used to represent the groups which are above the plane.

When we move to hash bonds, it is to represent the groups or atoms pointing away from the observer. It may also be used to represent the groups that are below the plane (Figure 3.3).

It is sometimes customary to represent only the heteroatoms in representing the three-dimensional structure. That is, we can conveniently drop the hydrogen atoms in drawing the three-dimensional structure. The main reason is, it gives a simple structure not cluttered with too many

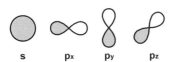

s   px   py   pz

FIGURE 3.2 Shapes of s and p orbitals.

wedge bonds   **hash bonds**

FIGURE 3.3 Wedge and hash bonds.

DOI: 10.1201/9781032631165-3

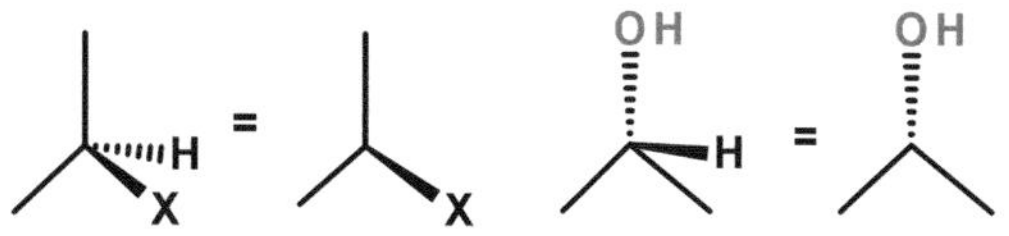

FIGURE 3.4 Heteroatoms in the three-dimensional structure.

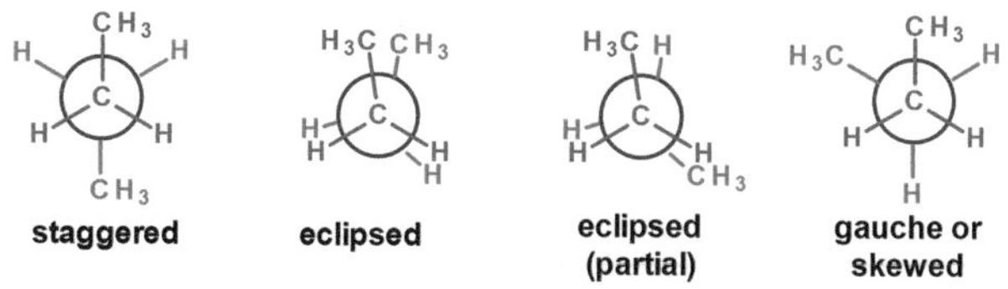

FIGURE 3.6 Newman projection of n-butane structure.

bonds, because it is obvious that of the four bonds of the tetrahedral carbon, two are represented on the plane, while the other two are out of plane. One is pointing toward the observer, while the other one is away from the observer. So, it is understandable that if the heteroatom is toward the observer, then the hydrogen has to be away from the observer (Figure 3.4).

3.1.4 SAWHORSE REPRESENTATION

This is a perspective representation of the molecules. We need to imagine that we are looking at the molecule by keeping the first carbon toward us, and the rest of the molecule is behind that particular carbon. So, to represent the bond between the two carbons, we use a slanting line. Each carbon has three atoms/groups attached to it. For the first carbon, we represent them occupying the vertices of an equilateral triangle. The atoms/groups on the second carbon are also represented in the same way. Various conformational isomers like eclipsed, gauche, anti, or staggered can all be easily represented by Sawhorse representation (Figure 3.5).

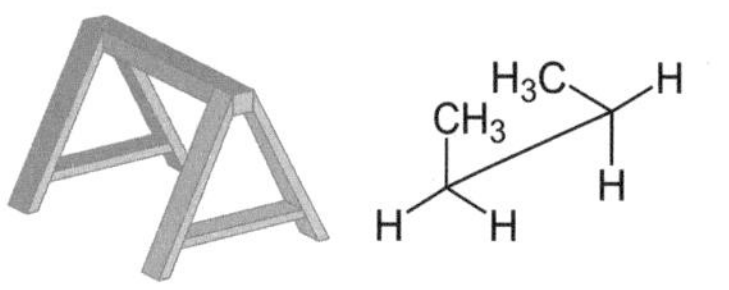

3.1.5 NEWMAN PROJECTION

Newman projection is another way of representing the molecule. Here we are looking at the molecule directly through the front carbon keeping it in such a way that it completely blocks the view of the second carbon. The front carbon is represented as a dot with three atoms/groups attached to it at 120° angles or the vertices of an equilateral triangle. The second carbon is represented by a circle, and the three groups or atoms are represented similar to the first carbon. That is, they are placed at the vertices of an equilateral triangle (Figure 3.6).

3.1.6 FISCHER PROJECTION

Although this representation is not being practiced regularly, we will briefly look at how to represent the molecule using Fischer projection. Mainly sugars are represented by this representation. In this representation, the entire carbon framework is written as a straight line. The substituents on the carbon atom are pointing toward the observer. The top and bottom carbons have the groups that are pointing away from the observer. A simple representation is given below. In fact, D and L notations are generally used in the case of sugars. Nowadays, Fischer projections are not generally used. We have moved to R and S nomenclature (Figure 3.7).

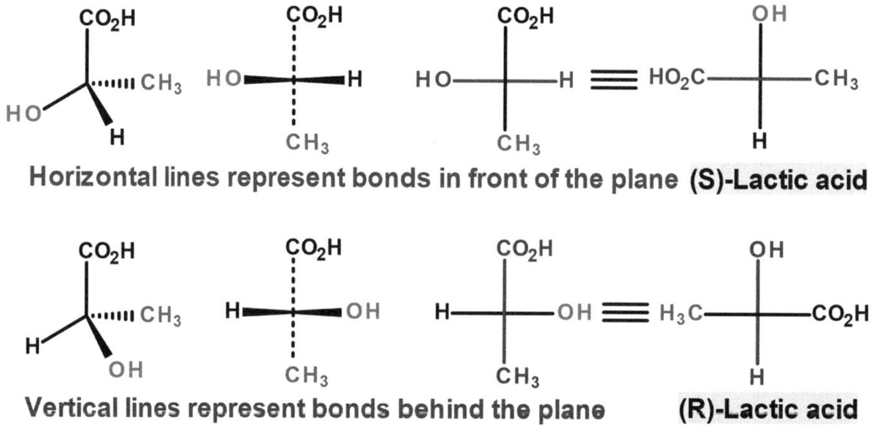

FIGURE 3.5 Sawhorse representation of n-butane structure.

FIGURE 3.7 Fischer projection of R and S-lactic acids.

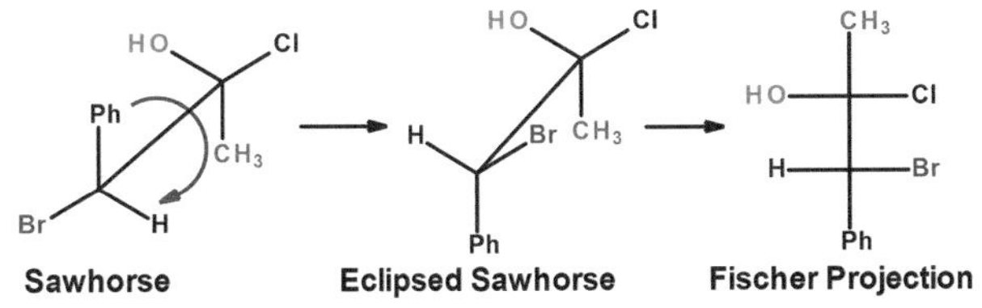

FIGURE 3.8 Interconversion of Sawhorse to Fischer projection.

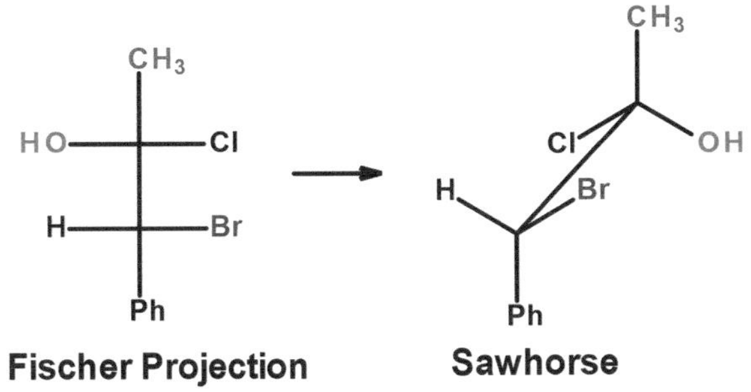

FIGURE 3.9 Interconversion of Fischer projection to Sawhorse.

3.1.7 INTERCONVERSION OF SAWHORSE AND FISCHER PROJECTION

In Figures 3.8 and 3.9, we can see how Sawhorse and Fisher projection can be interconverted.

3.1.8 CONVERSION OF SAWHORSE TO FISCHER PROJECTION

1. Front C-atom (Sawhorse)=lower C-atom (Fischer)
2. Convert Sawhorse to eclipsed Sawhorse
3. Flatten

3.1.9 CONVERSION OF FISHER TO SAWHORSE PROJECTION

1. Lower C-atom (Fischer)=Front C-atom (Sawhorse)
2. Groups that are anti or trans in Fischer are drawn on the same side of the Sawhorse

3.2 CONFORMATIONAL ISOMERISM IN ACYCLIC SYSTEMS

Derek H. R. Barton and Odd Hassel got the Nobel Prize in Chemistry (1969) for their contributions to the development of the concept of conformation and its application in chemistry.

Conformational isomerism is a form of stereoisomerism in which the isomers can be interconverted by rotations about single bonds (σ bonds). These isomers are often rapidly interconverting at room temperature. Due to the repulsive interactions between the electron clouds of C–H bonds, not all orientations are freely possible. We will soon see the arrangements and strains associated with them leading

to high-energy conformations. This repulsive interaction is termed a torsional strain.

3.2.1 ROTATIONAL BARRIER

As you know, the carbon–carbon single bond can rotate at a faster rate at room temperature, yet there are some barriers to free rotation. In other words, the carbon–carbon single bond is not completely free to rotate, that is there is no frictionless rotation. A small barrier to rotation exists even for the smallest molecule like ethane.

You may be wondering hydrogen is very small, so the carbon–carbon single-bond rotation should be very easy in ethane. There is extremely rapid internal rotation about the C–C bond. Ethane passes from a staggered through an eclipsed to a staggered conformation at a rate of about 10^{11} times per second at room temperature. Although this rotation is extremely fast, it does not qualify to be called a free rotation. Similar to the non-existence of a perpetual machine (second law of thermodynamics), we also do not have a frictionless free rotation about carbon–carbon σ bond. There is a subtle difference between frictionless free rotation and faster rotation. There is a 3 kcal/mol barrier to free rotation in ethane. This is not an absolute value, but a relative one that is obtained from the heat of combustion. This value is a relative one. That is, the energy difference between the staggered and eclipsed conformations of ethane is 3 kcal/mol.

Now you will be wondering (1) when the rotation is so fast, how do we even calculate the energy of a particular conformation? (2) Can we actually isolate any of the conformation or freeze a particular conformation, to find out the heat of combustion? (3) How accurate will be our estimation? An answer to these questions is provided by the molecular mechanics calculations. The beauty of molecular mechanics calculation is that we can virtually freeze the molecule in any of the conformations and apply different force fields to calculate its total energy, in this case, the steric energy. How can we do it? A force field is a set of parameters for the bond lengths, bond angles, torsional parameters, electrostatic properties, van der Waals interactions, etc.

$$E_{-steric} = E_{-stretch} + E_{-bend} + E_{-torsional} + E_{-vanderWaals}$$
$$+ E_{-electrostatic} + E_{-solvent}$$

Molecular mechanics is well suited for calculation of the relative energies (steric energies) of different conformers or stereoisomers. It is not generally suitable for calculating the relative energies of tautomers, double-bond isomers, or other constitutional isomers.

Figure 3.10 gives a rough idea about the barrier to rotation among different bond types.

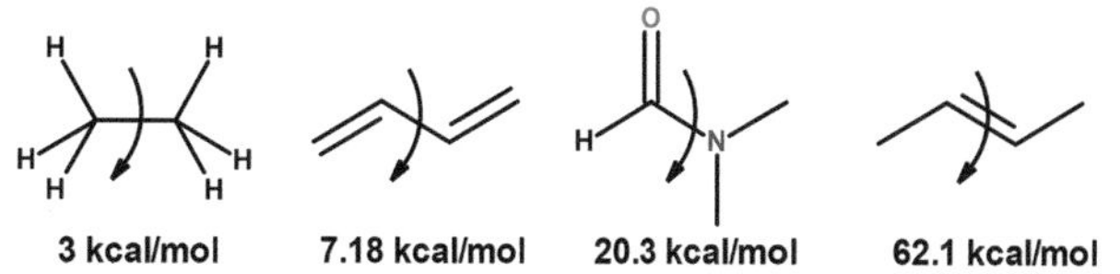

FIGURE 3.10 Barrier to rotation in few molecules.

The C–C single bond in ethane has the lowest rotational barrier of 3 kcal/mol. It gradually increases from a single bond to a partial double bond (amide) to a double bond (62.1 kcal/mol).

3.2.2 CONFORMATIONAL ISOMERS OF ETHANE

The C–C bond in ethane has a rotational barrier of 2.875 kcal/mol. This is due to two effects. (1) There is a steric repulsion in the eclipsed conformation (destabilization effect). (2) There is an enhanced stabilization of the staggered conformation due to hyperconjugation.

How do we know that the eclipsed conformation is higher in energy and not the staggered arrangement? We know, in the case of ethane, the inter-atomic distance between the vicinal hydrogen atoms of the two neighboring methyl groups in eclipsed conformation (torsional angle 0°) is significantly shorter (2.26 Å) than in the gauche conformation (torsional angle 60°) (2.50 Å). The shorter distance in eclipsed conformation puts the electronic cloud of the hydrogen atoms in close proximity. This, in turn, increases the repulsion and hence higher energy for the eclipsed conformation. In fact, various researchers who systematically studied the internal rotation barrier of many molecules concluded that the origin of the rotation barrier in ethane is due to the fact that the eclipsed conformer is possessing a large steric repulsion than the staggered conformer. They even used different approximation methods like the "kinetic energy pressure" in atoms and molecules or the Pauli Exclusion Principle are used to explain the observed phenomenon. Despite various approaches, the most popular explanation is the steric repulsion and it remains the important reason for the rotational energy barrier.

The stabilizing effect is hyperconjugation. This interaction is between σ-bonds and an unfilled or partially filled π- or p-orbital. This stabilizing interaction results from the interaction of the electrons in a σ-bond (usually C–H or C–C) with an adjacent empty or partially filled p-orbital or a π-orbital to give an extended molecular orbital.

3.2.3 ROTATIONAL BARRIER FOR SIMPLE ALKANES

Figure 3.11 represents the staggered and eclipsed conformations of simple alkanes.

In the case of acyclic systems, the relative energy of any conformation is the sum of its torsional strain and steric strain. The rotational barriers for ethane, propane, and n-butane are 3.0, 3.4, and 3.8 kcal/mol, respectively. Torsional strain is

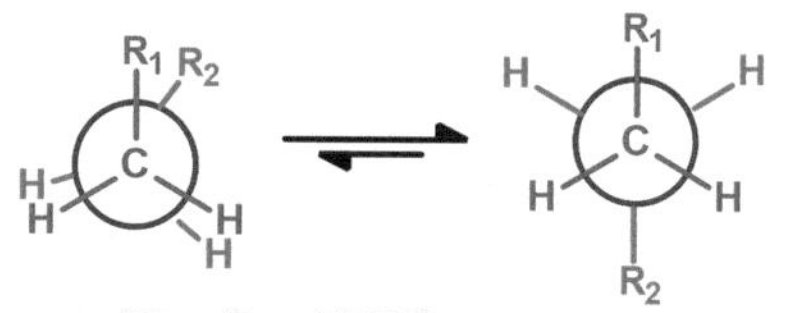

R₁ = R₂ =H, Ethane
R₁ = CH₃, R₂ =H, Propane
R₁ = R₂ =CH₃, n-Butane

FIGURE 3.11 Staggered and eclipsed conformations of simple alkanes.

responsible for the higher energy of eclipsed conformations compared to staggered conformations of a molecule. On the other hand, steric strain is responsible for the difference in energy among a group of eclipsed conformations like fully eclipsed and partially eclipsed conformations, or among a group of staggered conformations like anti or gauche conformations for a molecule.

3.2.4 NOTATIONS USED IN DIHEDRAL ANGLE CHANGES

We will now learn about the Klyne–Prelog system for describing conformations about a single bond based on torsional angles in stereochemistry. This protocol is a general unambiguous method to describe a particular conformation, where the torsional or dihedral angles are not found to occur in 60° increments. In this, the substituents are placed in front and back carbons. The placement of the substituent on the front atom is in regions of space called anti/syn and clinal/periplanar relative to a reference group on the rear atom. A plus (+) or minus (–) sign is placed at the front to indicate the sign of the dihedral angle. Anti or syn indicates the substituents are on opposite sides or on the same side, respectively. Clinal substituents are found within 30° of either side of a dihedral angle of 60° (from 30° to 90°), 120° (90°–150°), 240° (210°–270°), or 300° (270°–330°). Periplanar substituents are found within 30° of either 0° (330°–30°) or 180° (150°–210°) (Figure 3.12).

When R=methyl, we can apply this notation to butane to see the various conformations. When the dihedral angle, τ=60°, the conformer is gauche (staggered or skewed), but according to Klyne–Prelog proposal, it is either ±syn-clinal (±sc) or ±anti-clinal (±ac). When the torsional (dihedral) angle between the two R groups is between +30° and +90°, the conformer is +syn-clinal (+sc), and if it is between –30° and –90°, the conformer is –syn-clinal (–sc). When the torsional (dihedral) angle between the two R groups is between +90° and +150°, the conformer is +anti-clinal (+ac), and if it is between –90° and –150°, the conformer is –anti-clinal (–ac). When the torsional angle between two R groups is between –30° and +30°, the conformer is ±syn-periplanar (±sp), and if it is between –150° and +150°, the conformer is ±anti-periplanar (±ap). When the torsional (dihedral) angle between two R groups is 0°, the conformer is eclipsed and when the torsional (dihedral) angle between

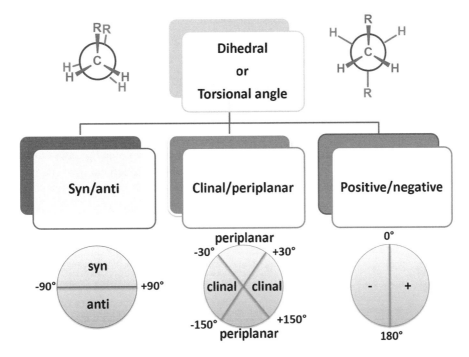

FIGURE 3.12   Notations used in dihedral angle changes.

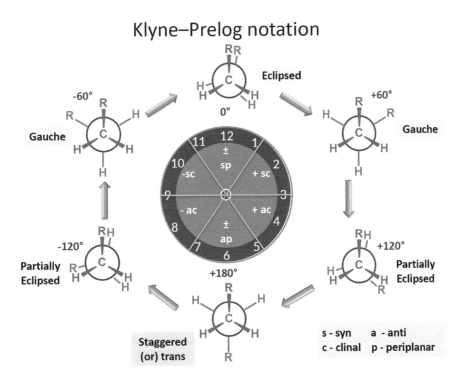

FIGURE 3.13   Klyne–Prelog notations.

two R groups is 180°, the conformer is staggered or trans or anti or anti-periplanar (Figure 3.13).

There is an easy way to understand this complicated figure. Replace the central dial with a picture that you are very familiar with. In a clock, every minute is 6° (the clock is a circle, so the total angle is 360°. There are 60 minutes in an hour. So, 1 minute is 360°/60=6°. From this, we can easily understand 5 minutes is 30°, 10 minutes is 60°, and so on. Using this, we can represent the Klyne–Prelog notation as given in Table 3.1.

TABLE 3.1
Klyne–Prelog Representation Using Clock

Time	Angle	Notation
11:00–1:00	−30° to +30°	syn-periplanar
1:00–3:00	+30° to +90°	+syn-clinal
3:00–5:00	+90° to −150°	+anti-clinal
5:00–7:00	+150° to −150°	anti-periplanar
7:00–9:00	−150° to −90°	− anti-clinal
9:00–11:00	+30° to +90°	−syn-clinal

3.2.5 POTENTIAL ENERGY DIAGRAM FOR VARIOUS CONFORMATIONS OF ETHANE

The energy required to rotate a C–C bond in ethane is called the torsional energy. The instability of conformation is called torsional strain. In Figure 3.14, we can see the difference between anti and eclipsed conformations of ethane. The difference is 3 kcal/mol.

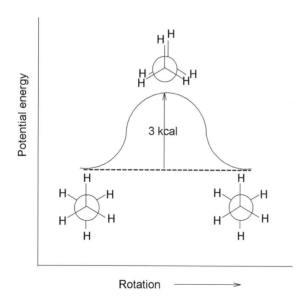

FIGURE 3.14 Potential energy diagram for various conformations of ethane.

3.2.6 CONFORMATIONAL ISOMERS OF BUTANE

Ethane has two major conformations. But butane has two additional conformational isomers. They are gauche and partially eclipsed conformations. The potential energy diagram in Figure 3.15 depicts the various conformations and their relative energies for n-butane.

In the following figure, we can see the different stabilization energies for different conformers. Assuming eclipsed conformation as zero energy, the other conformations are stable by the indicated energy. The negative value means that the molecule is more stable than the conformation with zero energy.

3.2.7 ROTATIONAL ENERGY BARRIERS FOR OTHER SIMPLE SYSTEMS

The eclipsed conformation of ethane has three pairs of C–H bonds, and its rotational barrier is close to 3 kcal/mol. You may be tempted to assume that each C–H bond has 1 kcal/mol torsional strain. Can we check this out? The ideal situation would be, if we can replace one hydrogen at a time and find out what is the total strain of the molecule, then it may be possible for us to predict an approximate value for the C–H bond rotational barrier.

While doing this, we need to keep a few things in mind: (1) the replacement of a group/atom should result in a compound whose geometry is very similar to the ethane geometry (sp^3 hybrid). (2) The C–H bond distance in the new compounds should not deviate much from the value of C–H bonds in ethane. (3) The carbon-heteroatom-hydrogen (C-X-H) bond angle should not deviate much. Figure 3.16 is an answer to this. Here we are replacing one of the methyl groups with amine (–NH$_2$ with a lone pair), hydroxyl (–OH with two lone pair), and fluorine (three lone pair). By doing so, we are trying to address all the three concerns we just mentioned.

As per the expectation, rotational barrier for methylamine is 1.98 kcal/mol, and for methanol, it is 1.07, and for methyl fluoride, it is 0 kcal/mol. So this follows the trend of replacement of one C–H bond is approximately equal to a rotational barrier of 1 kcal/mol. Replacement of hydrogen (change in size) with electron actually relieves some eclipsed interactions. So there is a reduction in the rotational energy barrier.

When we move to another set of simple alkyl halides, their rotational barrier is more or less the same. This is contrary to our expectations, because the halogens are of different sizes. Iodine is bigger than bromine which in turn is bigger than chlorine. Fluorine is the smallest one. The biggest size atom should lead to the biggest steric strain, and the molecule has to experience a higher strain (Figure 3.17).

But on the contrary, the rotational barriers of halogen are in the range of 3.3–3.8 kcal/mol only. In the case of

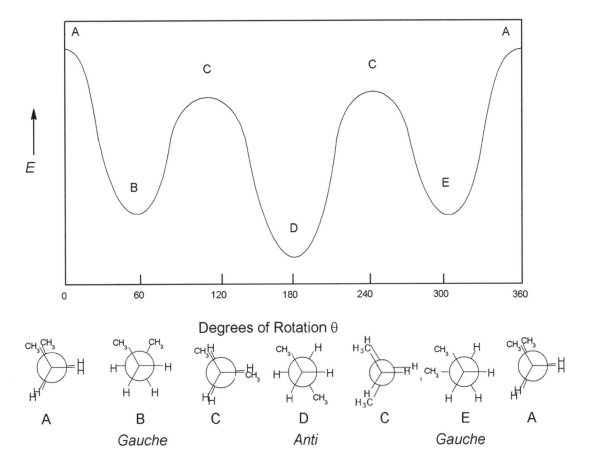

FIGURE 3.15 Potential energy diagram for various conformations of n-butane.

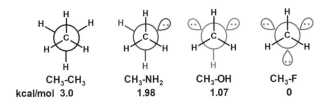

FIGURE 3.16 Rotational energy barriers for other simple systems.

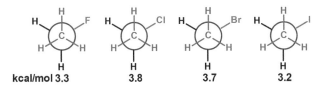

FIGURE 3.17 Rotational energy barriers for simple alkyl halides.

halogens, there is an increase in size as well as bond length. An increase in bond length reduces the effect of the size. Hence, the rotational energy barrier does not change much.

3.2.8 CONFORMATIONAL ANALYSIS OF ACYCLIC SYSTEMS: FLUORINE GAUCHE EFFECT

When we have disubstituted halo alkanes, we can have various conformations like eclipsed, partially eclipsed, gauche, and anti-conformations. For dichloro and dibromo derivatives, the anti-conformation is the most stable one (Figure 3.18).

Surprisingly for the difluoro derivative, the gauche conformation is the most stable one. The difluoro compound in the gauche conformation can have hyperconjugation.

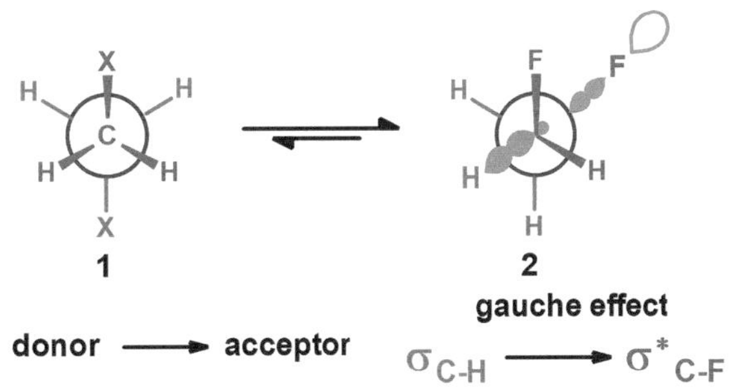

FIGURE 3.18 Fluorine Gauche effect.

In the gauche conformation, there are two C–H sigma (σ) bonding orbitals having interactions with sigma star (σ*) antibonding orbitals of C–F bond. Gauche conformation has the maximum number of interactions between the adjacent electron pairs and/or polar bonds (hyperconjugation).

3.2.9 Conformational Analysis of Acyclic Systems: Allylic Strain Effect

Allylic strain arises due to the interaction between a substituent on one end of an olefin and an allylic substituent on the other end. Allylic strain plays a vital role in controlling the stereochemical outcome of reactions (Figure 3.19).

The following derivatives can have allylic strain, and it is responsible for the unique reactivity of these substrates (Figure 3.20).

As shown in Figure 3.21, the amide hydrolysis is influenced by the nature of substituents that are present in the allylic position.

3.2.10 Application of Conformational Analysis

In this section, we will study briefly about the application of conformational analysis.

Polyketides are a class of secondary metabolites produced by certain living organisms to help them survive from their predators or to live on hosts. Many fungi produce polyketides that are toxic. Structurally, polyketides are highly active biologically complex organic compounds. Some examples of polyketides are given below.

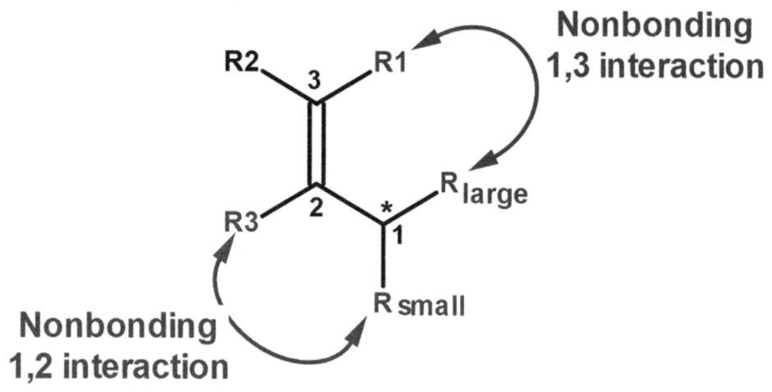

FIGURE 3.19 Allylic strain effect.

These compounds are structurally unique. They have multiple methyl side chain units. These methyl groups actually put a restriction of rotation about a C–C sigma (σ) bond, thereby giving a particular shape. These molecules are sometimes called as conformationally flexible molecules with a defined shape (Figure 3.22).

Nature is an excellent manipulator. To synthesize the right conformation, nature uses destabilizing effects of the undesired conformations. It avoids 1,3-allylic strain and syn-pentane interaction.

The methyl group in the side chain forces the molecule to avoid the 1,3-allylic strain and the syn-pentane interaction (Figure 3.23).

3.3 CONFORMATIONAL ANALYSIS OF CYCLIC SYSTEMS

Let us study the various aspects of conformational analysis. We will start with the rotation of bonds. According to the IUPAC gold book, the rotational barrier about a bond is defined as the potential energy barrier between two adjacent minima of the molecular entity as a function of the torsional angle.

3.3.1 Factors Affecting Conformation

Three major strains are present in cyclic systems: (1) angle strain, (2) torsional strain, and (3) steric strain. In contrast to these three destabilizing effects, dipole–dipole interactions sometimes stabilize a particular conformation.

3.3.2 Angle Strain or Baeyer Strain

Baeyer is a notable German scientist. He got the Nobel Prize in 1905 for his work on dyes. He studied cyclic systems carefully and made a lot of observations pertaining to the stability of cyclic systems. Hence, this strain is named after him.

Let us see what is meant by the Baeyer strain. It is the strain that arises due to the deviation in C–C bond angle from its ideal value. Here we show four cyclic systems, namely, three-, four-, five-, and six-membered ring systems.

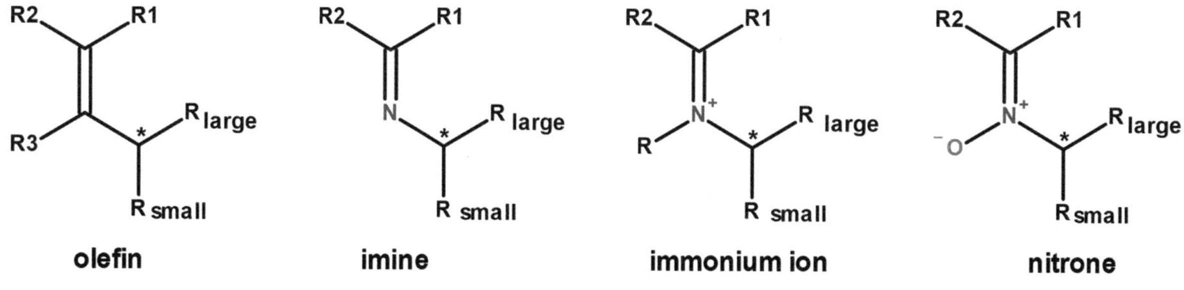

FIGURE 3.20 Some examples of allylic strain effect.

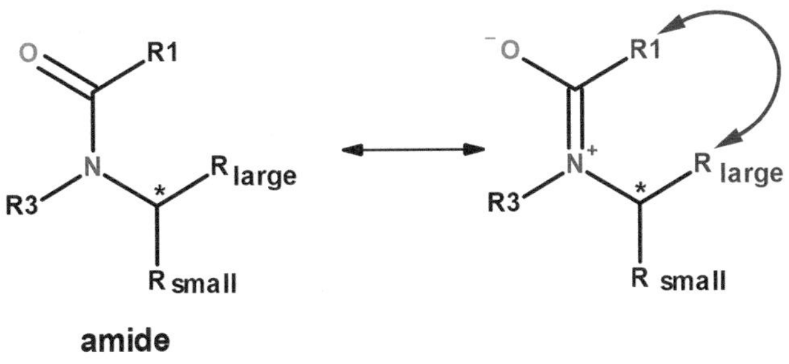
O
R1
N
R3
R large
R small
amide
O
R1
N+
R3
R large
R small

FIGURE 3.21 Influence of allylic strain on amide hydrolysis.

The three-membered system is an equilateral triangle, the four-membered one is a square, the five-membered is a pentagon, and the six-membered is a hexagon. As we know, the angle in an equilateral triangle is 60°. The cyclopropane consists of three CH_2 groups connected with a bond angle of 60°. But we know the ideal sp^3-hybridized carbon has a bond angle of 109°5′. So in cyclopropane, there is a huge deviation of 49°5′ from the ideal value; this leads to enormous destabilization. Figure 3.24 depicts the deviation in bond angle from its ideal value for the cyclic planar systems.

3.3.3 Limitations of Baeyer Theory

According to Baeyer, cyclopentane has the bond angle close to the ideal value, and compounds of higher cyclic systems will have higher bond angles which can lead to increased angle strain. However, his approximation was applicable to small ring systems only. He assumed the ring system to be planar. But in reality, cyclopropane is the only planar cyclic (ring) system, and his prediction is true only for the cyclopropane molecule. All the other ring systems are nonplanar. Although Baeyer was not correct in predicting angle strain, his attempts were in the right direction. With the advent of new spectroscopic techniques, his theory was proved years later to be incorrect regarding the planarity of ring systems.

When we look at the bond angle changes, more precisely it keeps increasing when we move from three-membered rings to ten-membered. For three-membered ring, it is 60°, it goes all the way up to 135° for eight-membered ring, and for ten-membered ring, it is 144°. For the 15-membered ring, the bond angle is 156°.

3.3.4 Angle Strain for Various Ring Systems

As we can see, the bond angle is close to the ideal value for cyclopentane (the difference between the bond angle and 109°5′ is 1°5′), and it keeps on increasing with ring size. For a cyclic system, ring strain comprises several strains: angle strain, torsional strain, and steric strain. In fact, the strain in six or higher cyclic systems is more due entirely to torsional and steric strain and not at all to angle strain, as these rings can bend and pucker to make the bond angles close to or equal to 109°5′. Bond angle is one of the factors, and it is not the only factor responsible for ring strain (Figure 3.25).

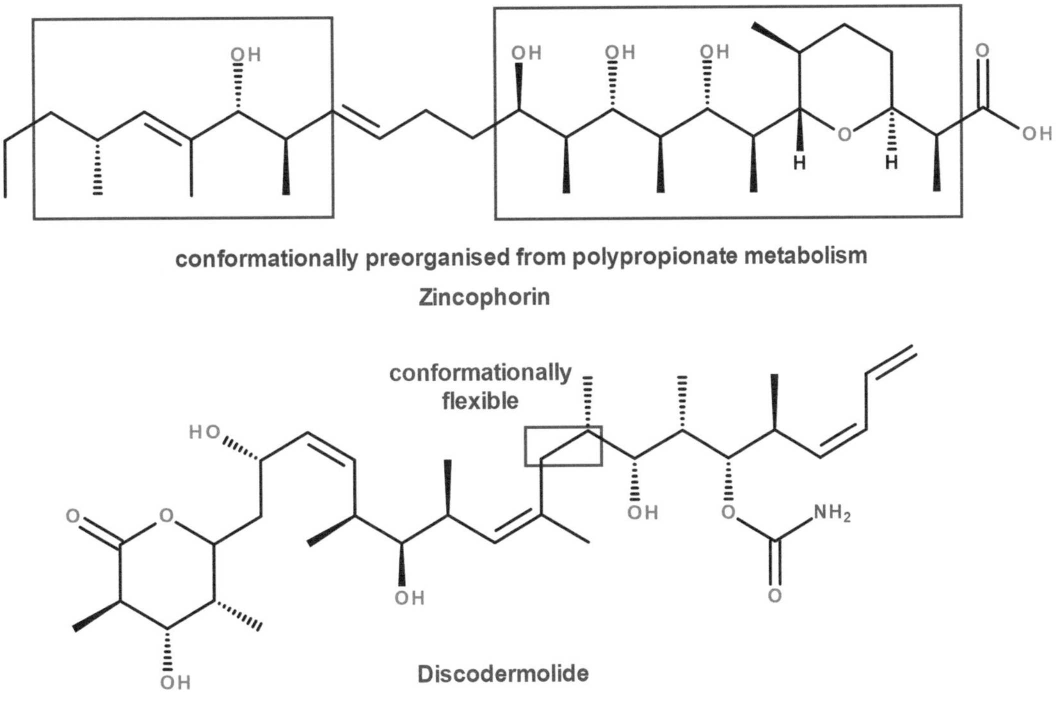
OH
OH
OH
OH
O
O
OH
H
H
conformationally preorganised from polypropionate metabolism
Zincophorin
conformationally flexible
HO
O
O
OH
OH
OH
O
NH2
O
Discodermolide

FIGURE 3.22 Some examples of polyketides.

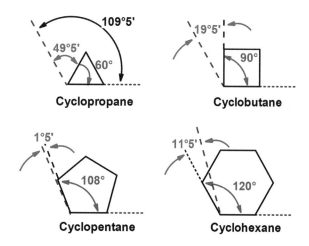

FIGURE 3.23   1,3-Allylic strain and syn-pentane interaction.

3.3.5   HEAT OF COMBUSTION IN CYCLOALKANES

It is challenging to calculate the angle strain alone. Because ring strain is a combination of various strains, the total strain of the system can be calculated by burning them with oxygen, measuring the amount of heat that is released, and comparing this value to the amount of energy released during the formation of the cyclic system having the same number of carbon atoms. The heat of combustion keeps on increasing for all the cycloalkane. However, the heat of combustion per CH_2 remains close to 158 kcal/mol for cycloalkanes from C_5 to C_{10}. Only for the small ring systems it is higher (164–166 kcal/mol).

Cyclohexane has the minimum heat of combustion (157.4 kcal/mol) value per CH_2 which is the same as the heat of combustion per CH_2 for open-chain systems 157.4 (kcal/mol).

3.3.6   TOTAL STRAIN IN CYCLOALKANES

The total destabilizing energy for 3, 4, 5, 6, 7, 8, 9, and 10 membered rings are 27.5, 26.3, 6.2, 0.1, 6.2, 9.7, 12.6, and 12.4 kcal/mol, respectively. In the case of small ring systems, both bond angle distortion and torsional strains are very high. Due to these two strains, the small ring systems are highly destabilized. In the case of a normal size ring that is from 5 to 7, both the torsional and angle strains are at a minimum. In fact, the six-membered ring system, namely, cyclohexane, is the most stable one. It has a destabilization energy value that is close to 0 kcal/mol.

FIGURE 3.24   Angle strain.

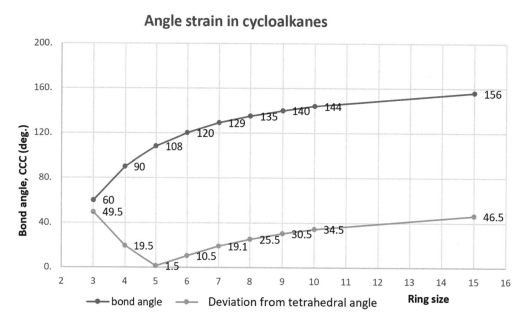

FIGURE 3.25   Angle strain in cycloalkanes.

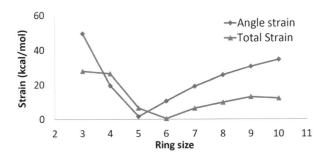

FIGURE 3.26   Angle and total strains in cycloalkanes.

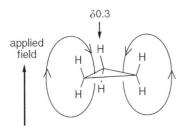

FIGURE 3.28   Magnetically anisotropicity of cyclopropane.

When we move up to intermediate-size rings that is from 8 to 12, all the systems will have either torsional strain or Prelog strain. In these systems, the molecule has the flexibility to orient in such a way that one of the strains is either reduced completely or minimized. During this process, only the other strain operates. When we move to larger ring systems beyond 13-membered rings, they are strain free.

3.3.7   Angle Strain in Cycloalkanes

The total destabilizing energy for cyclopropane is very high which is close to 28 kcal/mol. It reaches zero at cyclohexane and then again increases up to 9-membered ring, and then it reaches zero at 14-membered ring. This difference is due to different strains, namely, angle strain and torsional strain operating on different ring systems. We have already seen that small rings have a considerable amount of angle strain. Two additional strains also operate in cyclic compounds. Similar to acyclic systems, cyclic systems also have torsional strain and steric strain. If the dihedral angles are not perfectly staggered, that leads to torsional strain, and if non-bonding groups are forced to come closer to each other in space, then it leads to steric strain.

In graph in (Figure 3.26), we compare the stability of various ring structures. Three and four are very unstable, five and six are comparatively more stable, and seven and eight have either angle strain or torsional strain. Higher members are more or less strain free.

3.4   CYCLOPROPANE

Cyclopropane is the smallest ring system. Earlier we had seen the bond angle deviation (49°5′) which results in high strain. The C–C and C–H bond lengths in cyclopropane are shorter than in ethane, but higher than in ethylene (Figure 3.27).

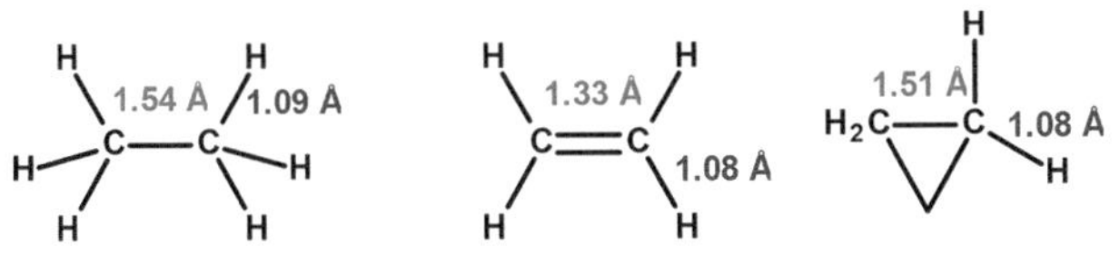

3.4.1   Bonding in Cyclopropane

The electron density distribution of cyclopropane is unique. Electron density distribution is found outside the internuclear axis, at the center, and at the ring. This suggests that the normal mode of bonding (either VB theory or MO theory) cannot explain the unique characteristic of cyclopropane, because, for normal systems, the electron density is generally high at the internuclear axis region. One can think of bent bonds to explain the higher electron density outside the internuclear axis, but that cannot explain the electron density found in the center of the ring.

Some of the characteristic features of the cyclopropane include a nearly planar (similar to sp^2 and not sp^3) structure and highly distorted bond angles from tetrahedral (49° difference); hydrogens are eclipsed and trans arrangement is preferred for disubstitution.

3.4.2   NMR of Cyclopropane

The CH_2 protons are magnetically anisotropic. Compared to the open-chain CH_2 group, there is a 1 ppm upfield shift of the protons in cyclopropane. The anisotropy is due to the presence of overlap from three sets of orbitals having a total of six electrons in them (Figure 3.28).

3.4.3   Reactivity of Cyclopropane

The C-C σ bonding in cyclopropane is somewhere between σ bonding (head-on overlap) and π bonding (sideways overlap). Due to the shorter C–C bond length in cyclopropane than the regular single-bond length, cyclopropane is much more reactive than ethane but not as reactive as ethylene.

3.4.3.1   Pitzer Strain (Torsional Strain)

In cyclic molecules, Torsional strain (resistance to bond twisting) is called as Pitzer strain. Torsional strain occurs when atoms separated by three bonds are placed in an eclipsed conformation instead of the more stable staggered conformation.

3.5   CYCLOBUTANE

When we look at cyclobutane, the actual bond angle is 90° and the ideal one is 109°5′. So there is a difference of 19°5′ degrees. It is not highly strained like cyclopropane, but still, it is unstable.

FIGURE 3.27   C–C and C–H bond length in cyclopropane.

Some of the characteristics of cyclobutane are given below:

- Its structure is twisted from planar
- It has slightly distorted bond angles from tetrahedral
- Its substituents are eclipsed
- Trans arrangement preferred for 1,2-disubstitution
- Cis arrangement preferred for 1,3-disubstitution

Cyclobutane has a dubious nature. The energy difference between the planar and puckered conformations of many substituted cyclobutanes is of the order of <1 kcal/mol. Due to this very small energy difference between the different conformations, it is often difficult to predict which conformation will be adopted by a particular molecule. We will take two examples: (1) trans-1,3-cyclobutane dicarboxylic acid and (2) trans-1,3-dibromocyclobutane (Table 3.2).

Trans-1,3-dibromoderivative is also puckered since it shows a dipole moment of 1.10D which corresponds to a puckering angle of 143°. As can be seen from the table, it is challenging to predict the geometry of substituted cyclobutane (Figure 3.29).

TABLE 3.2
Geometry of Various Substituted Cyclobutanes

Compound	State	Geometry
trans-1,3-cyclobutane dicarboxylic acid	Solid	Planar
trans-1,3-cyclobutane dicarboxylic acid	Solution	Puckered
trans-1,3-cyclobutane dicarboxylic acid	Dianion in solution	Planar
cis-1,3-cyclobutane dicarboxylic acid	Solid	Puckered
trans-1,3-dibromocyclobutane		Puckered

3.6 CYCLOPENTANE

Cyclopentane is the next cycloalkane. Here the bond angles are very close to the ideal value of sp^3-hybridized carbon. This shows that cyclopentane is not having very high-angle strain. Some of the characteristics of cyclopentane are given below (Figure 3.30):

- Four atoms are planar, and the fifth one is not in the plane
- Slightly distorted bond angles from tetrahedral
- Substituents are eclipsed
- Trans arrangement preferred for 1,2-disubstitution
- Cis arrangement preferred for 1,3-disubstitution

There is some torsional strain in planar cyclopentane. To gain stabilization, the planar cyclopentane ring can distort itself in a couple of ways. It can twist, or it can buckle. That results in two conformations having similar energies; one of them is having C_2 axis of symmetry (half-chair), and the other one is having a C_s plane of symmetry (an envelope). Unfortunately, both conformations do not have full staggering of the C–H bonds, and also, they do not have the C–C–C angles close to the ideal 109.47° angle. Figure 3.30 depicts the various conformations of cyclopentane.

The planar cyclopentane has a C–C–C bond angle of 108° and is having a minimum angular strain. But there are many C–H eclipsing interactions, due to which it is having high energy. The overall strain energy of 6.2 kcal/mol is mainly due to the torsional or the eclipsing strain.

To reduce this, cyclopentane has to move from its planar structure. This results in one of the carbons moving away from the planar structure. It can either go above the plane or below the plane. This results in an envelope conformation having mirror plane symmetry.

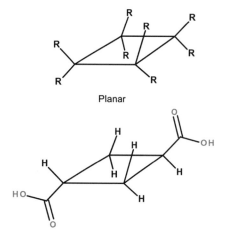

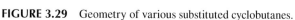

FIGURE 3.29 Geometry of various substituted cyclobutanes.

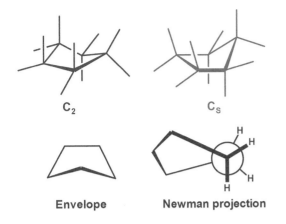

FIGURE 3.30 Various conformations of cyclopentane.

The other one is the twisting of C–C bond which leads to a half-chair conformation. Due to the half-chair form, it is having two-fold rotation axis of symmetry.

Since the energy difference between the envelope conformation and half-chair conformation is around 2 kcal/mol, this process is also called pseudo rotation.

3.7 CYCLOHEXANE

It is the most special case of cycloalkanes. It is the only small ring cycloalkane that is having the least strain energy.

You will be really surprised to learn that the diamond structure actually has chair conformation. The six atoms covered by the purple box show the chair conformation.

According to Baeyer, the planar cyclohexane has a bond angle of 120°. There is a difference of 11° from the ideal value. However, it was observed that cyclohexane is the most stable cycloalkane. Later it was found that cyclohexane does not exist in a planar form as proposed by Baeyer; rather, it exists in chair conformation. In this conformation, three carbons lie in one plane, and the other three carbons lie in another plane. Figure 3.31 gives the reason for the stability of cyclohexane. Cyclohexane chair conformation is similar to the staggered conformation of ethane, and this is responsible for its greater stability. On the other hand, the half-chair conformation of cyclohexane is the most unstable form.

3.7.1 CONFORMATIONAL ANALYSIS OF CYCLOHEXANE

Boat conformations are also possible with cyclohexane systems, but they are of high energy than the chair conformations. This is due to the fact that in this conformation, all the hydrogens are eclipsed. Moreover, the flagpole hydrogens also destabilize this structure as shown below.

Since the molecule is in random motion, the eclipsed conformation of boat cyclohexane moves slightly to reduce the eclipsing orientation. That leads to twist boat conformation. As marked in Figure 3.33, the hydrogens labeled a and

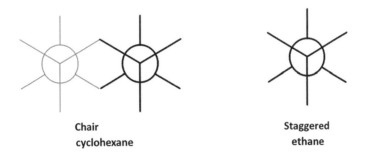

FIGURE 3.31 Comparison of cyclohexane chair conformation with staggered conformation of ethane.

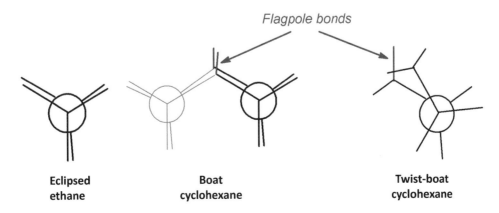

FIGURE 3.32 Comparison of cyclohexane boat and twist-boat conformation with eclipsed conformation of ethane.

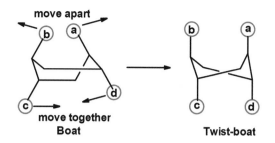

FIGURE 3.33 Movement of flagpole bonds in cyclohexane boat conformation to cyclohexane twist-boat conformation.

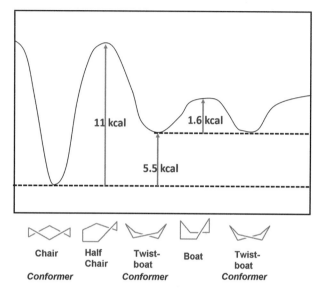

FIGURE 3.34 Potential energy diagram of various conformations of cyclohexane.

b move apart and c and d move closer. This leads to twist boat conformation which is better stabilized by 1.6 kcal/mol (Figure 3.33).

The potential energy diagram in Figure 3.34 depicts the different stabilization energies of various conformations of cyclohexane.

3.7.2 Axial and Equatorial Bonds

The molecular formula of cyclohexane is C_6H_{12}. Out of the 12 hydrogens in cyclohexane, 6 are called axial hydrogens and 6 are called equatorial hydrogens. Figure 3.35 on the

left represents the positions of axial hydrogens. Three of them are pointing upwards and three of them are pointing downwards. Figure 3.35 on the right represents the positions of equatorial hydrogens. Three of them are pointing slightly upward, and three of them are pointing slightly downward.

3.7.3 1,3-Diaxial Interaction

If we look at the simple methyl-substituted cyclohexane, there are two orientations the methyl group can have. In one, it can occupy the equatorial position, and in the other, it can occupy the axial position. When it occupies the axial position, the hydrogens on the methyl group are in close proximity to the axial hydrogen (3rd position). So, this type of interactions is called 1,3-diaxial interactions. Figure 3.36 gives a pictorial representation of 1,3-diaxial interactions.

3.8 A-VALUE OF VARIOUS SUBSTITUENTS

The A value for various substituents is given in Table 3.3. The fluorine substituent has the minimum A value (0.15 kcal/mol), and the *tert*-butyl group has the maximum A value (>4.5 kcal/mol). When we carefully look at Table 3.3, we can find that the highly electronegative atoms have smaller A value compared to alkyl substituents. F, O, and N substituents all have lower A value than even the smallest alkyl group namely methyl. In other words, we can say that A value mainly depends not only on the size but also on the C-X bond length as well.

3.9 TRANSANNULAR OR PRELOG STRAIN (STERIC STRAIN) OR VAN DER WAALS INTERACTIONS

It arises due to the unfavorable interactions of ring substituents on non-adjacent carbons.

These interactions, called transannular interactions, arise from a lack of space for the substituents in the interior of the ring. It is an attempt by the substituent to avoid angle and torsional strain. In medium-sized cycloalkanes (8–11 carbon), this strain is a major source of the overall strain.

For the bicyclic rings, trans decalin is a more stable form than the cis decalin.

FIGURE 3.35 Axial and equatorial bonds of cyclohexane chair conformation.

TABLE 3.3
The A Value for Various Substituents

Electronegativity vs Steric Interactions

Group	A-Value (kcal/mol)	Group	A-Value (kcal/mol)	Group	A-Value (kcal/mol)	Group	A-Value (kcal/mol)
F	0.15	O-NO$_2$	0.59	NO$_2$	1.05	Et	1.75
Br	0.38	O-Me	0.6	NH$_2$	1.2	CF$_3$	2.1
C≡C–H	0.41	O-Ac	0.6	CH=CH$_2$	1.35	i-Pr	2.15
Cl	0.43	C(=O)H	0.8	CO$_2$H	1.35	Ph	3
O-Ts	0.5	O-H	0.87	Me	1.7	t-Bu	4.5

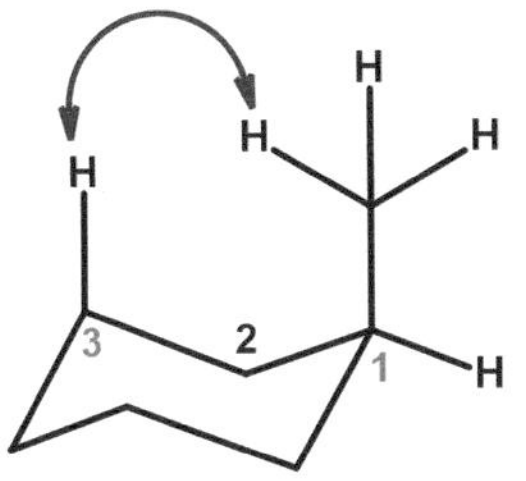

FIGURE 3.36   1,3-Diaxial interactions in cyclohexane.

3.10   COMPARISON OF VARIOUS STRAINS

Table 3.4 gives a comparison of the three different strains that operate in cyclic systems.

3.11   APPLICATION OF CONFORMATIONAL ANALYSIS

Conformational analysis can be used to explain the product formation in the following reactions (cyclic systems)

TABLE 3.4
Comparison of Angle, Torsional, and Steric Strain

Angle Strain	Torsional Strain	Steric Strain
Arise due to bond angles deviating from the ideal values	Repulsion due to electrostatic forces between electrons in adjacent MO	Repulsion when two bulky groups become too close (which are not directly bonded to each other)
	Exists between atoms separated by three bonds	Exists between atoms separated by four or more bonds
	Hyperconjugation stabilizes anti or staggered conformation	Difficult to reduce this strain
	Deviation from planar structure tends to reduce this strain	Applicable only for bulky substituents

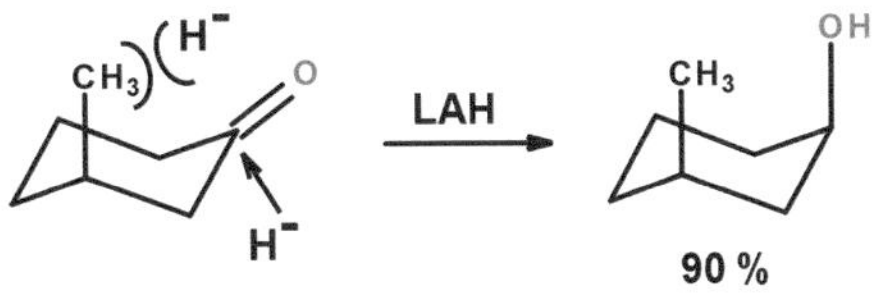

FIGURE 3.37   LAH reduction of 3-methyl cyclohexanone.

like acetylation, elimination, oxidation, amide hydrolysis, reduction, and anomeric effect of sugars.

As an example, we consider the LAH reduction of 3-methyl cyclohexanone produces the axial alcohol, although the product will suffer from 1,3-diaxial interaction. Before the product is formed, the hydride needs to attack the electrophilic carbon. The approach of the hydride is hindered by the axial methyl group (the top-side approach is hindered). On the other hand, the sideways approach of the hydride is still possible leading to the axial alcohol. Figure 3.37 depicts the approach diagrammatically.

3.12   QUESTIONS

True or False

1. Electron has both particle and wave nature.
2. According to Quantum Mechanics, "electron can only exist in any energy levels."
3. The σ orbital has a directional orientation.
4. The sp^2-hybridized orbital has restricted rotation.
5. Hash bonds represent groups or atoms pointing away from the observer.
6. In Sawhorse representation, the slanting line represents the front and back carbon atoms.
7. In the Fischer projection, groups or atoms on the vertical line represent bonds that are toward the observer.
8. Total energy of a molecule can be calculated using molecular mechanics calculation.
9. Di halo derivatives including fluro derivatives prefer staggered conformation.
10. In a disubstituted alkane, when both the substituents have 180° dihedral angle, it is the most stable conformation.

Short and Long Answers Questions

11. What are the factors that affect conformation?
12. Write briefly about the different representations of molecules.
13. Write briefly about the conformation of ethane.
14. Write briefly about the different conformations of n-butane.
15. Write briefly about the fluorine Gauche effect and allylic strain effect.
16. Write briefly about the angle strain.
17. Write briefly about A value with respect to shapes of substituents.
18. Write briefly about the conformational analysis of cyclohexane.

BIBLIOGRAPHY

Calculation of Relative Energies of Conformers and Stereoisomers and Equilibrium Ratios of Products using Molecular Mechanics (https://www.chem.uci.edu/~jsnowick/groupweb/Maestro/molecular_mechanics.pdf) (accessed, 15-Apr-2024)

Carey, F. A. *Organic Chemistry*, 5th Ed., McGraw-Hill, 2004, 105–124.

Chapter 6 Stereochemistry (https://www.askthenerd.com/NOW/CH6/CH6_5-8.pdf) (accessed, 15-Apr-2024)

Chapter 4 Cyclohexane (https://www.t.soka.ac.jp/chem/iwanami/stereo/stereoCh4.pdf) (accessed, 15-Apr-2024)

Hoffmann, R. W. (2000). Conformation design of open-chain compounds. *Angewandte Chemie International Edition*, 39, 2054–2070.

Smith, M. B., & March, J. *March's Advanced Organic Chemistry Reactions, Mechanisms and Structure*, 5th Ed. John Wiley & Sons, Inc., 2001, 167–178.

William Reusch, Acyclic Conformational Analysis, 05/05/2013 (https://www2.chemistry.msu.edu/faculty/reusch/virttxtjml/rotconf1.htm) (accessed, 15-Apr-2024)

4 Basics of Symmetry Elements and Point Groups

If you want to be an expert in organic chemistry, there are only two major concepts you need to master. Both are related to arts and not to science. We have already learned one of the arts, which is the art of arrow-pushing mechanism (reaction mechanism), and the second one is the visualization of molecules in 3D (stereochemistry). Although both are a little difficult to understand in the beginning, once you get the grip of it, you will be really surprised to see that organic chemistry is not at all a difficult subject to learn.

4.1 SYMMETRY ELEMENTS

4.1.1 STEREOCHEMISTRY AND SYMMETRY

One way to explain stereochemistry is by invoking symmetry. Why are we fascinated by symmetry? The question invariably inspires creative minds. We, humans look at beauty through symmetry. When we talk about organic chemistry, molecular symmetry is studied by crystallographers. Some of the myriad examples of compounds where symmetry is essential are molecular assemblies, DNA origami, electronic chips, organic light-emitting diodes, thin-film transistors, and so on.

4.1.2 LINEAR SYMMETRY, POINT SYMMETRY, AND RADIAL SYMMETRY

If an imaginary line passes through the object in such a way that it divides the objects into two identical parts, then the object is said to possess linear symmetry. Instead, if the line passes through a central point in such a way that when every part has a matching part at the same distance from the central point but in the opposite direction, then the molecule is said to be having point symmetry.

Radial symmetry is rotational symmetry around a fixed point (the center). Cyclic and dihedral are the two types of radial symmetry.

Sunflowers have radial symmetry (pinecones, pineapples, and artichokes). The number of seed spirals in the sunflower follows the Fibonacci series. Nature follows the golden ratio, or Phi (Φ1·6180339887), for the angle between each seed, leaf, petal, or branch.

In the case of Egyptian pyramids, the ratio of the length of a face of the Great Pyramid (from center of the bottom of a face to the apex of the pyramid) to the distance from the same point to the exact center of the pyramid's base square is about 1.6. Leonardo da Vinci used divine proportion in his paintings (Mona Lisa, Last Supper, and Vitruvian Man).

Snowflakes exhibit radial symmetry. In other words, symmetry is omnipresent.

4.2 TYPES OF SYMMETRY

The major components of symmetry are operations and elements. Symmetry operation is an action performed on an object. After the operation is performed, there is no change to the initial orientation.

The other part is symmetry elements. Here we talk about the object itself. Now the object is transformed using the symmetry operations. We can conclude that by performing a symmetry operation on the molecule, we can identify the symmetry elements that are present (Figure 4.1).

There are two conventions used for symmetry.

1. The Schoenflies notation (used by spectroscopists).
2. Hermann–Mauguin or international notation (used by crystallographers)

Table 4.1 depicts the various symbols used in these two conventions.

4.2.1 SCHOENFLIES NOTATION

Let us look at some additional terms that we may encounter when we study symmetry. The term point groups include reflection, rotation, and inversion (and hence leave one point unchanged), and space groups, include the above operations and additionally translational symmetry elements.

Schoenflies notation is a method to describe point group symmetry. Those who study molecular symmetries

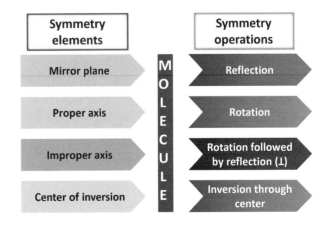

FIGURE 4.1 Symmetry elements and symmetry operation.

DOI: 10.1201/9781032631165-4

TABLE 4.1

Various Symbols Used in Schoenflies and Hermann–Mauguin Notation

	Symmetry Element	Notation	
		Hermann–Mauguin	Schoenflies
Point symmetry	Identity	1	C
	Rotation axes	n	C_n
	Mirror plane	m	σ
	Center of inversion	I	i
	Improper axis		S_n

(chemists and spectroscopists) use this notation. The large letter refers to the family: C for rotational axes; S for rotation-reflection axes; and D, T, O, and I for dihedral, tetrahedral, octahedral, and icosahedral symmetries. The subscripts h, v, or d are used to indicate the presence of 'horizontal' (perpendicular to the primary axis) and 'vertical' (containing the primary axis) mirror planes or dihedral/diagonal mirror planes.

4.2.1.1 Identity (E)

This operation consists of no change in structure. Every object has an identity. It is represented as "E" (Einheit-German-unity) (Figure 4.2)

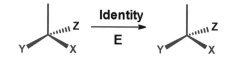

FIGURE 4.2 Identity operation.

4.2.1.2 Proper Rotation (C_n)

This operation consists of rotation about an axis through 360°/n. The principal axis is the axis of highest n. For n = 2, C_2 (180°); n = 3, C_3 (120°); n = 4, C_4 (90°); and n = 6, C_6 (60°).

4.2.1.3 Reflection (σ)

This operation consists of reflection about a plane (similar to placing a mirror). We have three types: vertical (σ_v) (passing through C_n), horizontal (σ_h) (perpendicular to C_n), and dihedral (σ_d) (passing between two C_2). This is just a special case of σ_v.

Some of the examples are given in Figure 4.3.

4.2.1.4 Inversion (i)

Inversion through the center of mass. For any atom in the molecule, an identical atom exists diametrically opposite from the center at an equal distance (Figure 4.4).

4.2.1.5 Improper Rotation (S_n)

An improper rotation consists of two operations: (1) a proper rotation (360°/n) around the principal (S_n) axis and (2) reflection across a plane perpendicular to the principal (S_n) axis (Figure 4.5).

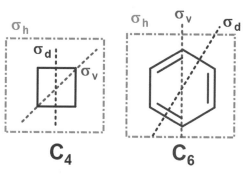

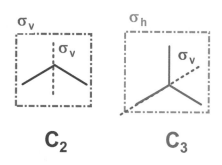

FIGURE 4.3 Some of the examples of reflection operations.

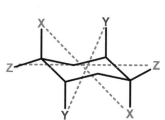

FIGURE 4.4 Example of reflection operations.

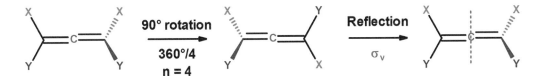

FIGURE 4.5   Example of improper rotation.

4.3   INTERRELATIONSHIP BETWEEN THE SYMMETRY ELEMENTS AND POINT GROUP

Figure 4.6 depicts the various interrelationships between the symmetry elements and the point group.

You can choose any molecular structure. Is a molecule linear? If yes, does it contain an inversion center, then it belongs to point group $D_{\infty h}$. If it does not contain an inversion center, then it belongs to the $C_{\infty v}$ point group. If the molecule is linear, then it will belong to either $D_{\infty h}$ or $C_{\infty v}$. If it is not linear, then we need to see does it contains C_n axis, if so the next one we check is, does it have any perpendicular rotation axis. Finally, we need to find out does it have a horizontal reflection plane, if the answer is yes, then the molecule belongs to D_{nh}. The molecule does not contain σ_h but contains σ_d, and then it belongs to the point group D_{nd}. If it neither contains σ_d nor σ_h, then the molecule has the point group D_n.

Let us now come back to the case where the molecule does not have a perpendicular axis. If it has σ_h plane, then the molecule has C_{nh} symmetry. If it contains σ_v instead of σ_h, then it belongs to C_{nv}. Finally, it neither has σ_h nor σ_v plane, but contains 2n fold S_n axis, and then it belongs to S_{2n}. If it does not contain S_n axis, it belongs to C_n.

Finally, we go back to C_n rotation axis. If the molecule is neither linear nor contains C_n axis, but has σ plane, then it belongs to C_s. If it does not have a reflection plane but

contains an inversion center, it belongs to C_i. When the molecule does not have any of the above-mentioned symmetry elements, it belongs to C_1 group.

4.4   POINT GROUP CLASSIFICATION

Let us now look at how these point groups can be classified into similar categories. The first one is the linear point group. Here we have a point group belonging to C_∞ axis. Compounds with C_n axis are always chiral. The next one is C point group. All compounds with C_n principal axis but no C_2 perpendicular axis fall under this category.

In point group D, we have nC_2 axes perpendicular to a principal axis (C_n): D_n, D_{nh}, D_{nd} (no D_{nv}), and finally the point group S. In this category, the point groups will have a principal axis (C_n) and S_{2n} axis but no perpendicular C_2 axes and no mirror planes S_{2n}.

Figure 4.7 categorizes various point groups. They are linear, C, D, and S point groups.

4.4.1   How to Identify Point Groups

First, look at the molecule and identify whether it is symmetric or un-symmetric. If it is un-symmetric, it probably belongs to one of the special groups (low symmetry: C_1, C_s, C_i, or linear $C_{\infty v}$, $D_{\infty h}$). If it is highly symmetric, then it may belong to high symmetry point groups like (T_d, O_h).

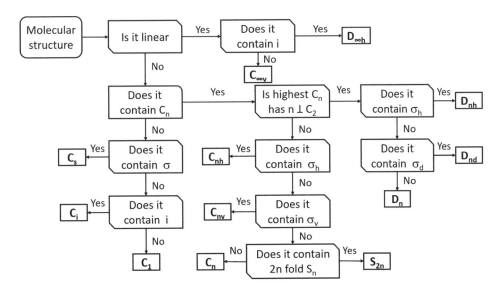

FIGURE 4.6   Interrelationship between the symmetry elements and point group.

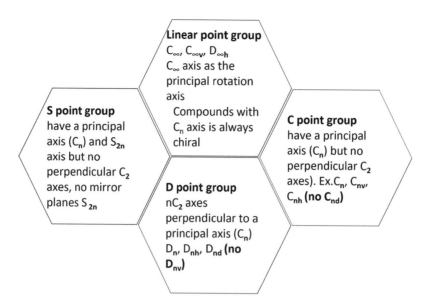

FIGURE 4.7   Details of various point groups.

Step 1:

We will start with identifying the rotational axis. For all molecules, find the rotation axis with the highest n. This will be the highest order C_n axis of the molecule.

Step 2:

We need to find whether there is any C_2-axis perpendicular to the C_n axis? If such an axis exists, we have to confirm whether there is 'n' such C_2 axes. If it does, then the molecule belongs to the point group D. If there is no C_2 axis perpendicular to the C_n axis, it will belong to the point groups either C or S.

Perpendicular C_2 axis = the point group D

No Perpendicular C_2 axis = the point group C or S

Step 3:

Next, we need to look for reflection (mirror) planes. We need to specifically look for horizontal planes. If they exist, then we need to find out whether this mirror plane (σ_h) is perpendicular to the C_n axis or not? If there is mirror plane (σ_h), then the molecule belongs to either the point group C_{nh} or D_{nh}.

Step 4:

Does it have any mirror plane (σ_d, σ_v)? If so, it is C_{nv} or D_{nd}

If E, C_n and S_{2n} are the symmetry elements present, then the molecule belongs to S_{2n}

If E and σ are the symmetry elements present, then the molecule belongs to C_s

If E and I are the symmetry elements present, then the molecule belongs to C_i

If the only symmetry element present is E $\Rightarrow$ C_1 (essentially, no symmetry)

To quickly recap the different symmetry elements present in different point groups, we can refer to Table 4.2.

Remembering the above table is not at all difficult if we can identify a few patterns.

1. If there is a C_2 axis perpendicular to the principal axis, then they all belong to the point group D.
2. Whether the mirror plane is on the principal axis or perpendicular to the principal axis.
3. Point group C does not have a diagonal plane.

4.4.2   RELATIONSHIP BETWEEN VSEPR GEOMETRY AND POINT GROUPS

According to VSEPR, we have various geometry for molecules. The following table depicts the relationship between the geometry of a molecule and the corresponding point group.

For a quick revision of point groups, Table 4.3 can be used.

*All the point groups have identity (E) as one of their symmetry elements.

4.5   POINT GROUP FOR ETHANE

4.5.1   STAGGERED ETHANE

Ethane is a very good example for us to see the various point groups. Let us see the point group for staggered ethane. The highest principal axis is C_3. There is a C_2 axis along the C_3 axis. It also has an inversion center. Then it has a diagonal mirror plane. With this information, if we look at the points

TABLE 4.2
Details Different Symmetry Elements

Point Group	C_2 Axis $\perp$ C_n Axis	Mirror Plane $\perp$ to C_n	Mirror Plane Containing C_n	Vertical Plane (v)	Mirror Plane (h)	Mirror Plane (d)
C_n	No	No	No	No	No	No
C_{nv}	No	No	Yes	Yes	No	No
C_{nh}	No	Yes	No	No	Yes	No
D_n	Yes	No	No	No	No	No
D_{nh}	Yes	Yes	No	No	Yes	Yes
D_{nd}	Yes	No	Yes	Yes	No	Yes
S_n	No	No	No	No	No	No

All have rotation axis (C_n).

TABLE 4.3
Geometry of Molecules and Point Groups

Geometry of Molecule	Point Group
Linear	$D_{\alpha h}$
Bent or V-shape	
Sawhorse or see-saw	C_{2v}
T-shape	
Trigonal pyramidal	C_{3v}
Square pyramidal	C_{4v}
Trigonal planar	D_{3h}
Square planar	D_{4h}
Pentagonal bipyramidal	D_{5h}
Trigonal bipyramidal	
Tetrahedral	T_d
Octahedral	O_h

group table we see that this molecular arrangement belongs to the point group D_{nd}. Since the highest principal axis is C_3, this is D_{3d} point group.

Explanation: Let us look at ethane staggered conformation and we will work out, what are the symmetry elements present in this molecule and it belongs to which point group.

First let us find out what is the principal axis. First, look at the staggered conformation of ethane and find out the point group to which it belongs. Ethane may be represented in Newman's projection. The first carbon eclipses the second carbon. Now consider the bond axis passes through both the carbon atoms. The protons attached to the first carbon occupy the vertices of an equilateral triangle with 120°. When we do a 360° rotation, we get identical arrangements three times. This implies that we have a C_3 axis.

In addition to C_3 axis, there are three C_2 axes containing the C_3 axis, that is, 180° rotation produces the same orientation of the molecule. In Figure 4.8, the molecule is rotated in such a way that the hydrogen H_6 occupies the position occupied by H_1 during the 180° rotation. We have three such C_2 axes for the molecule. H_7 rotated to H_4 and H_8 rotated to H_3 are the other instances. For the principal axis, we will only take the axis with the highest n. Since C_3 is the highest rotational axis, we will take this as the principal axis.

Keep in mind, do not double-count. For example, consider the 180° rotation of H_6, and it occupies the position occupied by H_1 which is one C_2 axis. Now 180° rotation of the H_1 and H_6 should not be considered. This will amount to double counting. That means instead of three C_2 axes, we will have six C_2 axes. So, this is not correct.

Let us move on to see what other symmetry elements are present. We will now look at whether this staggered ethane has an inversion center. If you look at H_6 and H_1, they are present on the opposite side and at equidistance from the center. For the whole molecule to have an inversion center as a symmetry element, all the atoms should have the same atoms in the opposite direction at the same distance from the central point. H_4 and H_7, and H_3 and H_8 are the other two sets of hydrogens that belong to the inversion center. We can conclude that this molecule has a center of inversion.

Finally, we will look for the presence of reflection symmetry. There are three diagonal planes which pass through the principal axis. The first reflection plane is passing through H_6-C_b-C_a-H_1, the second one is passing through H_7-C_b-C_a-H_4, and the last one is passing through H_8-C_b-C_a-H_3. These reflection planes cut the molecule into two identical halves. The net result we have is three σ_d planes.

All the above information can be put in the form of a table. From the table, we can deduce the point group of staggered ethane. Since the principal axis is C_3, we say this molecule has the point group D_{3d}. This has an S_6 axis also (Table 4.4).

Summary: The point group of staggered ethane is D_{3d}.

4.5.2 ECLIPSED ETHANE

Let us see the point group of eclipsed ethane. The highest principal axis is C_3. There are three C_2 axes perpendicular to C_3 axis. It has a horizontal mirror plane. It also has a diagonal mirror plane. With this information, if we look at the points group table (Table 4.5), we see that this molecular arrangement belongs to the point group D_{nh}. Since the highest principal axis is C_3, this is the point group D_{3h}.

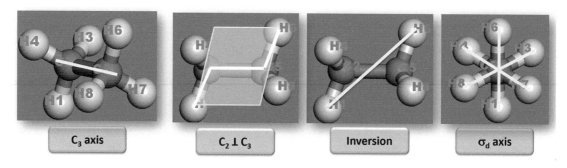

FIGURE 4.8 Various symmetry elements that are present in staggered ethane.

TABLE 4.4
Point Group and Symmetry Elements in Staggered Ethane

Point Group	C_2 Axis $\perp C_n$ Axis	Mirror Plane $\perp$ to C_n	Mirror Plane Containing C_n	Vertical Plane (v)	Mirror Plane (h)	Mirror Plane (d)
D_{nd}	Yes	No	Yes	Yes	No	Yes

TABLE 4.5
Point Group and Symmetry Elements in Eclipsed Ethane

Point Group	C_2 Axis $\perp C_n$ Axis	Mirror Plane $\perp$ to C_n	Mirror Plane Containing C_n	Vertical Plane (v)	Mirror Plane (h)	Mirror Plane (d)
D_{nh}	Yes	Yes	No	No	Yes	Yes

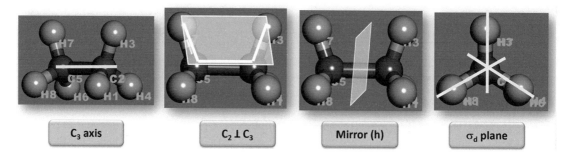

FIGURE 4.9 Various symmetry elements that are present in eclipsed ethane.

Explanation: Let us look at ethane eclipsed conformation, and we will work out what are the symmetry elements present in this molecule and it belongs to which point group.

First let us find out what is the principal axis. First, look at the eclipsed conformation of ethane and find out the point group to which it belongs. Ethane may be represented in Newman's projection. The 1st carbon eclipses the 2nd carbon. Now consider the bond axis passes through both the carbon atoms. The protons attached to the first carbon occupy the vertices of an equilateral triangle with 120°. When we do a 360° rotation, we get identical arrangements three times. This implies that we have a C_3 axis.

In addition to C_3 axis, there are three C_2 axes perpendicular to the C_3 axis, that is, 180° rotation produces the same orientation of the molecule. In Figure 4.9, the molecule is rotated in such a way that the hydrogen H_3 occupies the position occupied by H_7 during the 180° rotation. We have

three such C_2 axes for the molecule. H_4 rotated to H_6 and H_1 rotated to H_8 are the other two instances.

Let us move on to see what other symmetry elements are present. We will now look at whether this eclipsed ethane has any inversion center. If you look at H_3 and H_7, they are present on the same side in equidistance from the center. There is no hydrogen atom in the opposite direction. So, this molecule does not have an inversion center.

Next, we will look for the presence of reflection symmetry. There is one horizontal reflection plane perpendicular to the C_3 axis. It cuts the molecule into two equal halves at the C-C bond. This is the σ_h plane.

Let us look for the other reflection planes. We have three vertical reflection planes containing (or passing through) the C_3 axis. Two hydrogens and two carbon atoms are present in each of the vertical planes. The first vertical reflection plane passes through H_3-C_b-C_a-H_7, the second one passes through

H_6-C_b-C_a-H_4 and the last one passes through H_8-C_b-C_a-H_1. These reflection planes cut the molecule into two identical halves. The net result is we have three σ_v planes.

Information that we have seen can be conveniently written in Table 4.5. From this, we can deduce the point group of eclipsed ethane. Since the principal axis is C_3, we say this molecule has the point group D_{3h}. This has S_3 axis also.

Summary: The point group of eclipsed ethane is D_{3h}.

Let us compare three compounds obtained by the replacement of hydrogens with the OH group. Now we look to see how the symmetry changes and what is the resultant point group. We already saw D_{3d} and D_{3h} point groups, respectively, in staggered and eclipsed ethane.

4.6 POINT GROUP FOR ETHANOL

4.6.1 STAGGERED ETHANOL

We will now look at the staggered conformation of ethanol. Because of the presence of one OH group, there is no higher order principal axis. In other words, there is no symmetry axis higher than C_1 in this molecule.

Next, we will look at reflection planes. It has only a plane that cuts the OH group vertically. The arrangements are HO–C_a–C_b–H. Apart from this vertical plane, there is no horizontal plane or diagonal plane. There is no inversion center. So, the staggered conformation of ethanol belongs to the point group C_S.

4.6.2 ECLIPSED ETHANOL

In eclipsed ethanol because of the presence of one OH group, there is no higher order principal axis. In other words, there is no symmetry axis higher than C_1 in this molecule.

Next, we will look at the reflection planes. It has only a plane that cuts the OH group vertically. The arrangements are HO–C_a–C_b–H. Apart from this vertical plane, there is no horizontal plane or diagonal plane. There is no inversion center. So, the eclipsed conformation of ethanol belongs to the point group C_S.

4.7 POINT GROUP OF 1,2-ETHANEDIOL

4.7.1 STAGGERED 1,2-ETHANEDIOL

Here the principal axis is C_2. The principal axis is perpendicular to the C-C single bond. In Figure 4.10, the molecule is rotated in such a way that the back OH occupies the position occupied by front OH during the 180° rotation. We have three such C_2 axes for the molecule. The other two instances are related to C_2 axes with respect to hydrogen atoms.

Similar to the staggered conformation of ethane, staggered ethanediol also has center of inversion. Except for the C_n axis, all the other symmetry elements are similar to ethane. So, the staggered 1,2-ethanediol has the point group C_{2v}.

4.7.2 ECLIPSED 1,2-ETHANEDIOL

Now, let us look at the eclipsed compound. Here the principal axis is C_2. The principal axis is perpendicular to the C-C single bond. In Figure 4.10, the molecule is rotated in such a way that the back OH occupies the position occupied by the front OH during the 180° rotation. We have three such C_2 axes for the molecule. The other two instances are related to C_2 axes with respect to the hydrogen atoms.

Ethane, Ethanol, ethylene diol

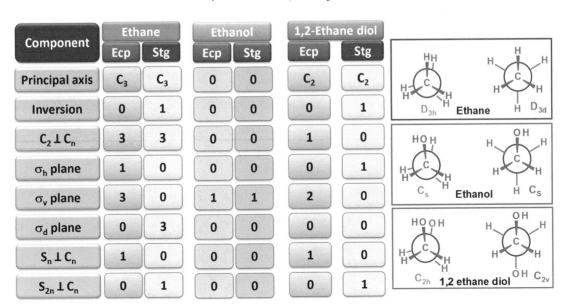

Component	Ethane		Ethanol		1,2-Ethane diol		
	Ecp	Stg	Ecp	Stg	Ecp	Stg	
Principal axis	C_3	C_3	0	0	C_2	C_2	
Inversion	0	1	0	0	0	1	
$C_2 \perp C_n$	3	3	0	0	1	0	
σ_h plane	1	0	0	0	0	1	
σ_v plane	3	0	1	1	2	0	
σ_d plane	0	3	0	0	0	0	
$S_n \perp C_n$	1	0	0	0	1	0	
$S_{2n} \perp C_n$	0	1	0	0	0	1	

FIGURE 4.10 Various symmetry elements that are present in ethane, ethanol, and 1,2-ethanediol.

Except for the C_n axis, all the other symmetry elements are similar to ethane. So, the eclipsed 1,2-ethanediol has the point group C_{2h}.

We can compare the three compounds ethane, ethanol, and 1,2-ethanediol. The following Figure 4.10 compares the various symmetry elements present in the three compounds.

4.8 TYPES OF CHIRALITY

Like symmetry, we also have three major types of chirality. One is the point or center of chirality, axial chirality, and planar chirality. Compounds belonging to the last two types of chirality may not have any asymmetric carbon or chiral carbon.

4.8.1 AXIAL CHIRALITY

There are two types of axial chirality, namely atropisomerism and isomerism, in allenes.

4.8.1.1 Atropisomerism

This is caused by a high barrier to free rotation around a single bond. The inhibition or high barrier to free rotation is mainly observed in the biaryl compounds having four ortho-substituents or some sterically hindered carboxamides.

Here we see some examples of axial chirality. Ortho-substituted biphenyls are a good example that represents axial chirality. It is given either R_a or S_a similar to RS nomenclature. The subscript a represents axial chirality.

Axial chirality will be stable at room temperature when the rotation barrier is above 30 kcal/mol.

4.8.1.2 Isomerism in Allenes

Another example is the disubstituted allene system. Both the images are non-superimposable. The chirality is seen in some spiro bicyclic compounds also (Figure 4.11).

4.8.2 PLANAR CHIRALITY

Planar chirality appears when a planar molecule loses its symmetry plane by bridging with a short tether or forming π-complex. The chirality is seen in some cyclophanes, metallocenes, and trans-cycloalkenes.

Here we see some examples of planar chirality. *Trans*-cyclooctene is the simplest example that represents planar chirality. It is given either R_p or S_p similar to RS nomenclature. The subscript p represents planar chirality. Or it is called P (plus) or M (minus). Another example is the spiro system. Both the images are non-superimposable.

The following sets of spiro compounds also belong to planar chirality (Figure 4.12).

4.9 SYMMETRY TERMINOLOGY

Dissymmetry is a term used when the molecule does not have reflective symmetry elements. Asymmetry is used when the molecule has no symmetry elements. Both dissymmetric and asymmetric molecules are chiral. What is a chiral object? Chiral object is not superimposable on its mirror image. If the molecule has reflective symmetry elements, then it is said to be achiral. Finally, chiral objects may have rotational symmetry.

- Dissymmetry: The molecule does not have reflective symmetry elements. All dissymmetric objects are chiral.
- Asymmetry: The molecule does not have any symmetry elements. All asymmetric objects are chiral.
- A chiral object is not superimposable on its mirror image.
- The existence of a reflective symmetry element (a point or plane of symmetry) is sufficient to assure that the object is achiral.
- Chiral objects, therefore, do not have any reflective symmetry elements but may have rotational symmetry axes.

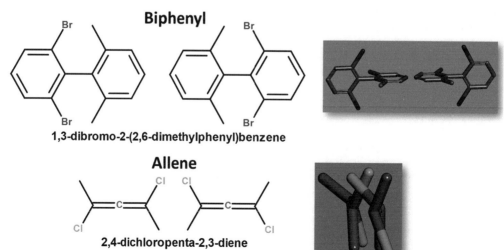

Biphenyl

1,3-dibromo-2-(2,6-dimethylphenyl)benzene

Allene

2,4-dichloropenta-2,3-diene

FIGURE 4.11 Types of axial chirality, atropisomerism, and isomerism in allenes.

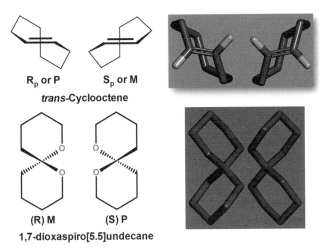

R_p or P   S_p or M

***trans*-Cyclooctene**

(R) M   (S) P

1,7-dioxaspiro[5.5]undecane

FIGURE 4.12 Types of planar chirality.

4.10 QUESTIONS

1. Identify the point group for the following compounds
   a. HCl
   b. SOFCl
   c. F_2
2. Identify and write all the symmetry elements present in the following compounds
   a. PPh_3
   b. BH_3
   c. $B(OH)_3$
   d. CH_3Cl
   e. Biphenyl
   f. $[PtCl_4]_{2-}$
3. Identify the point group for the following compounds
   Allene
   Cyclohexane chair
   PF_5
   Cyclopropane
   Cyclobutane
   Triphenylphosphine
   Hydrazine

4. Write briefly about linear symmetry, point symmetry, and radial symmetry.
5. What are the different symmetry elements and symmetry operations?
6. What is Schoenflies notation?
7. What is the interrelationship between symmetry elements and point groups?

BIBLIOGRAPHY

Kalsi, P. S. *Stereochemistry: Conformation and Mechanism*, 9th Ed. New Age International (P) Ltd., 2017, 99–108.

Nasipuri, D. *Stereochemistry of Organic Compounds, Principles and Applications*, 3rd Ed. New Age International Publishers, 2018, 15–25, 70–73.

Rauk, A. *Orbital Interaction Theory of Organic Chemistry*, 2nd Ed. John Wiley & Sons, Inc., 2001, 1–12.

Summary of Symmetry Operations, Symmetry Elements, and Point Groups (https://troglerlab.ucsd.edu/GroupTheory224/Chap1B.pdf) (accessed, 15-Apr-2024)

Tim Wallace, University of Manchester, Stereochemistry, April 2021, (https://www.stereoelectronics.org/webSC/SC_home.html) (accessed, 15-Apr-2024)

5 Stereochemical Conventions and Topicity

In this chapter, we will study the various stereochemical conventions used in organic chemistry. We will start with some basic terminologies like erythro, threo, D and L, and R and S stereo notations.

Given below is one of the top-selling anti-cancer drug taxol. Many drug molecules like this exist as optically active compounds having a specific configuration at the stereogenic center (Figure 5.1).

What is drug discovery and development?

Drug discovery and development is a process of finding new therapeutic agents that target a key enzyme, protein–protein interaction, receptor–ligand interaction, or protein–nucleic acid interaction.

Why is drug discovery and development important?

By using new drugs, we can treat many diseases and thereby trying to mitigate the severity of the diseases.

5.1 HUMANS AND CHIRALITY

5.1.1 CHIRALITY AND DRUGS

Drugs interact differently with the body depending on the handedness. This difference comes about due to several reasons:

1. Different rates of metabolism. The different isomers can be broken down at different speeds.
2. Different effects on the body. Sometimes one isomer may have no adverse effect on the patients.

Why is it important?

Although enantiomers have similar physical and chemical properties, they may produce different biological effects. Way back in the 1960s, a mixture of the (R) and (S) enantiomers of thalidomide (Contergan) (structure A and B) was given as a medicine for preventing vomiting in pregnant women in European countries. Unfortunately, many children who were born to those who took this medication suffered from birth defects (phocomelia, i.e., malformation of the limbs). Later it was found that the useful R isomer having anti-inflammatory activity was converted into toxic S isomer which is teratogenic inside the body. The thalidomide tragedy forced government regulators to redefine the importance of chiral drugs for human consumption.

Now every pharmaceutical company working on drug development must synthesize the right enantiomer. This increases the responsibility of the organic chemists working on drug development (Figure 5.2).

5.1.2 CHIRALITY AND SMELL

Not only drugs even our taste and smell are affected by the chirality of the molecules. We know chiral molecules interact differently with enzymes depending on their chirality, and they can also interact differently with receptors in our sense organs, such as the smell receptors in our noses. The two enantiomers of carvone (coriander and spearmint) have different smells (Figure 5.3).

In fact, the pioneering work by Buck L, Axel R. on discoveries of *"Odorant Receptors and the Organization of the Olfactory System"* was awarded the Nobel Prize in Physiology or Medicine in the year 2004 (Buck & Axel, 1991).

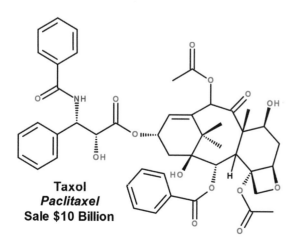

Taxol
Paclitaxel
Sale $10 Billion

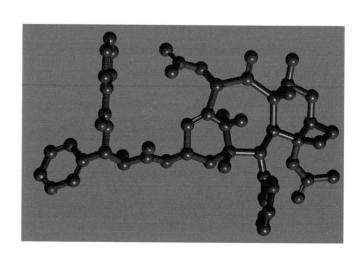

FIGURE 5.1 Stereochemical structure of taxol.

40

DOI: 10.1201/9781032631165-5

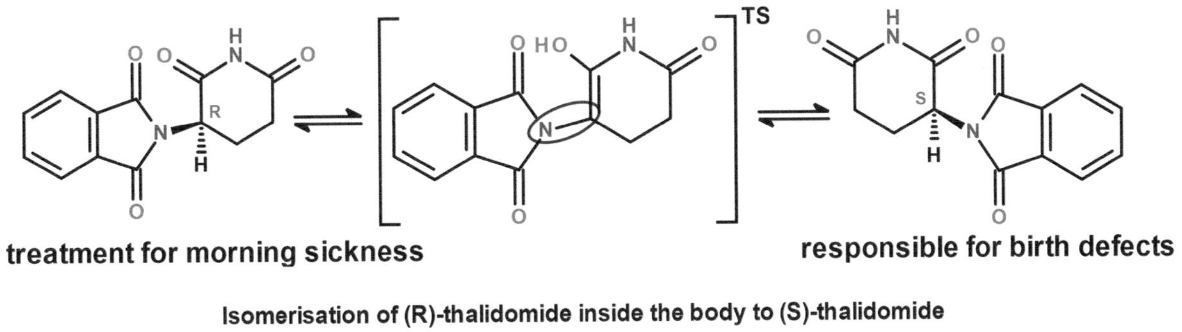

treatment for morning sickness   **responsible for birth defects**

Isomerisation of (R)-thalidomide inside the body to (S)-thalidomide

FIGURE 5.2   Racemization of thalidomide.

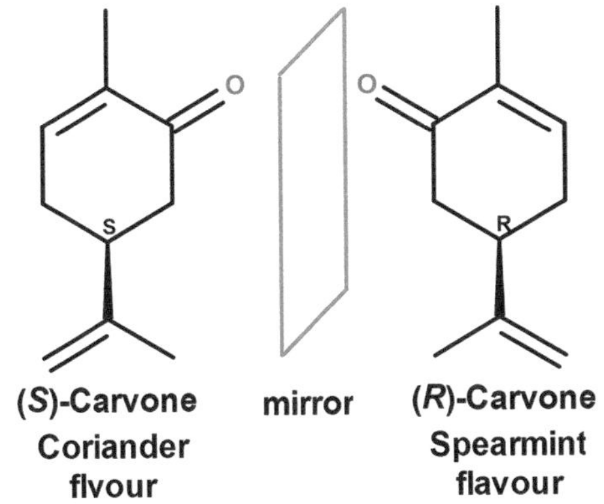

(S)-Carvone   mirror   (R)-Carvone
Coriander   Spearmint
flvour   flavour

FIGURE 5.3   R and S isomers of Carvone.

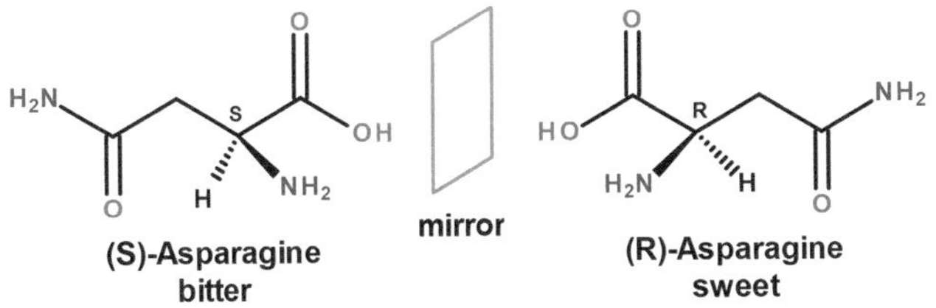

(S)-Asparagine   mirror   (R)-Asparagine
bitter   sweet

FIGURE 5.4   R and S isomers of Asparagine.

5.1.3   CHIRALITY AND TASTE

Chiral structure of proteins plays a vital role in the perception of taste as well as the perception of aroma. Depending on their configuration, all the amino acids may be neutral, sweet, or bitter, except methionine. In fact, four of them, that is, asparagine, tryptophan, tyrosine, and isoleucine, are characterized by bitter taste in their L form and sweet in the D form. Proteins also play a significant role in the process of taste perception. It is believed that the perception of bitter and sweet tastes is dependent on a receptor connected with a G-protein, gustducin (McLaughlin & Margolskee, 1994) (Figure 5.4).

5.2   STEREOCHEMICAL CONVENTIONS

What are the conventions used in stereochemistry?

They are Syn-anti, Cis-trans, D-L, d-l, R-S, P-M (for helical compounds). Table 5.1 depicts the notations used, their meaning, and their relationship to conventions.

Where is it used?

Stereoisomers are molecules whose atomic connectivity is the same but whose three-dimensional arrangement of atoms in space is different.

Among the different chiral compounds, one kind of compound has chirality due to the restricted rotation (limitation) around the C–C single bond. This is axis chirality or axial chirality, and the compounds are atropisomers.

When we say a compound is optically active, what do we mean by that? The compound can rotate the plane-polarized light. This may be due to the fact that one of the enantiomers is in excess. Other terms include racemic mixture, where we have a 50:50 mixture of both the enantiomers and their rotation gets canceled (optically inactive). If the molecule is optically inactive, it may be due to any one of the following reasons:

1. It may be an achiral molecule
2. A racemic mixture
3. The Refractive Index difference between enantiomers is very small. For example, Alkanes
4. Optical inversion (+ to −) occurs at that particular wavelength
5. A meso compound

The relationship between the configurations of stereocenters within a molecule is known as *relative stereochemistry*. Some of the terms used here are *syn/anti, threo/erythron*, and *cis/trans (for cyclic structures)*.

5.2.1   ERYTHRO AND THREO

This notation was commonly used to represent the stereochemistry of molecules with two stereogenic centers. This name was derived based on erythrose and threose (carbohydrates). Erythro means the same side, and threo means the opposite side. Let us take the erythro isomer of 3-chlorobutan-2-ol. In the first figure, both the hydrogens are on the left-hand side or on the right-hand side. The top representation is the Fischer projection, and the bottom one is the sawhorse representation. When we want to convert the Fischer

TABLE 5.1

Conventions Used in Stereochemistry

Notation	Representation	Related to
R and S	Clockwise/anti-clockwise	CIP rule
D and L	Right and left sides	Fischer projection/sugars
D or (+) and L or (−)	sign	Optical rotation
P and M	Right (plus) and left (minus) handed	Helicity
α and β	Below and above the plane	Orientation
Erythro/threo	Similar groups same and opposite sides	Fischer projection/sugars
Exo and endo	Outside and within or inside	Cyclic system
Cis and trans	Same and opposite sides	Geometrical
Syn and anti	Same and opposite sides	Geometrical
Z and E	Same and opposite sides	Geometrical
Re and Si	Clockwise/anti-clockwise	Face/side of carbonyls

projection to sawhorse, retain the top and bottom groups as it is. In the bottom carbon, arrange the groups as they appear in the Fischer projection, and interchange the groups in the 2nd carbon and put it in the Fischer. If we move on to the threo isomer of 3-chlorobutan-2-ol (as shown on the right in Figure 5.5), the hydrogens are on the opposite side.

Stereo refers to two compounds be it *cis-trans*, *syn-anti*, *E-Z,* D and L, R and S, or even enantiomers. More number of isomers are possible, if there is more than stereogenic center in the molecule. In general, if there are 'n' stereogenic centers in a molecule, then 2^n stereoisomers are theoretically possible.

I know you are confused about how to remember configuration and conformation. I will tell you a simple way. Say, you are the captain of a cricket team. You are given two important jobs:

1. To identify the team players (figure out how many spinners, pacers, right- and left-hand batsmen)
2. Decide about fielding setup (formation of attacking field).

Once you **figure out** who are all the players, you can change the team only by **replacing** a player.

But when the team is **formed,** you can keep on changing the position of the players by **rotation.** You can move a slip fielder to third man or to long on or to long off etc.

So, **configuration** can only be changed when there is **replacement** (breaking and forming/remaking of a bond), but **conformation** can be changed by simple **rotation**.

5.2.2 DL Notation or Fischer Notation

The reference is glyceraldehyde an easy way to remember this is, L for left. The L configuration has the hydroxyl

group on the left-hand side for the glyceraldehyde. For the Fischer projection, the main carbon is written as a straight line all vertical lines represent the bonds that are behind or below the plane.

The substituents that are drawn on the horizontal line are considered to be pointing toward the observer or above the plane. The convention is that the most oxidized carbon is written at the top. The last but one carbon is taken for the assignment of the configuration. If the hydroxyl is on the left it is called the L isomer and if it is on the right, it is called as D isomer.

5.2.3 Alpha-Beta Arrangements

Alpha means below, and beta means above. In this example (Figure 5.6), on the left-hand side, the hydroxyl group in the anomeric carbon is pointing downwards. It is an alpha configuration. In the other case (right-hand side) the hydroxyl is pointing upwards and it is beta. An easy way to remember this is *a equal to b.* A in Greek is *B* in English. A in Greek is alpha So b in English is below. In other words, alpha means below the plane.

5.2.4 Cis and Trans or Syn and Anti

Here we have *cis* and *trans* or *syn* and *anti*-arrangements. *Cis* or *syn* means the same side, whereas *trans* or *anti* means the opposite side. When the hydroxyls are pointing toward the same side (either toward the observer or away from the observer), it is the *cis* arrangement. When the hydroxyls are pointing in opposite directions (one pointing toward the observer and the other pointing away from the observer), it is *trans* arrangement.

In the first example of hexan-2,3,4,5-tetrol (Figure 5.7), we have all *cis* arrangement and in the last one, we have all *trans* arrangement. *Cis-trans* arrangements are not necessarily restricted to adjacent atoms, we can also use this convention for any two atoms that are not adjacent to each other. In the following example, we can use the *cis-trans* relationship for 2,3- or 2,4- or 2,5- or 3,5- substituents also.

3-chlorobutan-2-ol

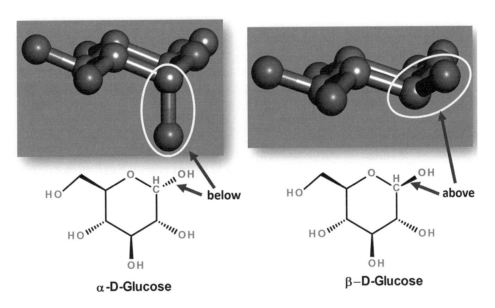

FIGURE 5.5 Erythro and threo isomers of 3-chlorobutan-2-ol.

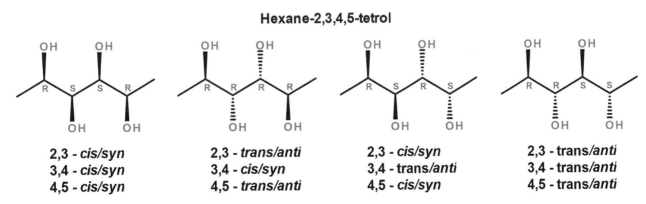

FIGURE 5.6 Alpha and beta notation of D-glucose.

Hexane-2,3,4,5-tetrol

2,3 - cis/syn
3,4 - cis/syn
4,5 - cis/syn

2,3 - trans/anti
3,4 - cis/syn
4,5 - trans/anti

2,3 - cis/syn
3,4 - trans/anti
4,5 - cis/syn

2,3 - trans/anti
3,4 - trans/anti
4,5 - trans/anti

FIGURE 5.7 cis and trans isomers of hexan-2,3,4,5-tetrol.

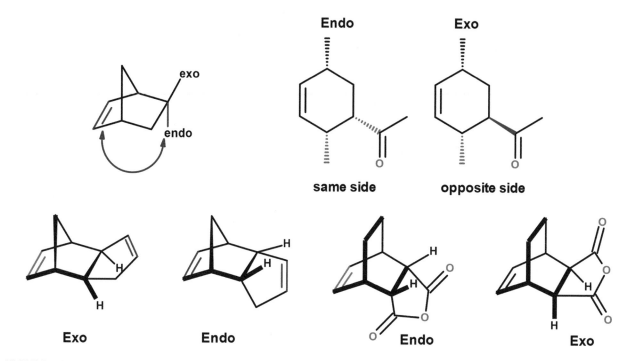

FIGURE 5.8 Exo and endo isomers.

5.2.5 EXO AND ENDO

Exo means outside and endo means inside or within. This is generally used in cyclic systems with double bonds. If a particular group is closer to the double bond, that is called endo, if it is far from the double bond, then they are called exo.

Exo/endo and **syn/anti** are used in many ways. Exo-endo is predominantly used in defining the stereochemical course of reactions such as Diels–Alder cycloadditions. In this case, endo/exo refers to the orientation of a substituent group in the transition state of the cycloaddition, leading to a product of defined geometry (Figure 5.8).

5.3 R-S-NOMENCLATURE

This is based on the CIP rule. This is also called the sequence rules.

5.3.1 CAHN–INGOLD–PRELOG RULES

Let us see how to assign the priorities to the groups/atoms that are directly attached to the center. To assign R/S configuration, we need to first identify the priority of the substituents around the stereogenic center/chiral center. After assigning the priority, we need to rotate the molecule in such a way that the lowest priority group is kept far away from the observer. Now we will be left with three groups surrounding the chiral center. Now if the groups from highest priority to the lowest priority are in the clockwise direction, then the configuration is assigned as R (Rectus) and if it is in the anti-clockwise direction, it is called S (sinister) (Figure 5.9).

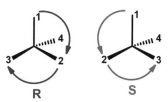

FIGURE 5.9 R and S notation.

1. Assign priority based on atomic number/mass. A higher atomic number atom takes precedence over a lower atomic number atom.

   Let us assume four atoms/elements attached to the central carbon are hydrogen, nitrogen, oxygen, and chlorine with atomic numbers 1, 7, 8, and 17, respectively. Here we have hydrogen with atomic number one as the smallest one, and chlorine with atomic number 17 as the highest one. Accordingly, we can give chlorine the highest priority (number one) and hydrogen the lowest priority (number four).

   If isotopes are present, then the priority is decided based on the mass number. Isotopes with higher mass number takes precedence over lower mass number. For example, between hydrogen, deuterium, and tritium (atomic mass 1, 2, and 3), the order of priority is 3>2>1.

   By application of the above rules, some groups in descending order of precedence are COOH, COPh, COMe, CHO, CH(OH), o-tolyl, m-tolyl, p-tolyl, phenyl, C≡CH, tert-butyl, cyclohexyl, vinyl, isopropyl, benzyl, neopentyl, allyl, n-pentyl, ethyl, methyl, deuterium, and hydrogen.

2. If the first point of attachment with the chiral center is the same, then we need to look at the next atom or group till there is a difference in priority.

If there is a tie of atoms directly attached to the stereocenter, we have to look at the immediate vicinity of the atom. That is, the next available atom is considered for comparison. For example, if −CH3 and −CH2CH3 are attached to the chiral carbon, we can work out the priority as follows. The first point of attachment is carbon. So we cannot assign priority. We need to move to the next point of attachment. In methyl group (−CH3), all the second attachments are hydrogen (atomic number 1). In the −CH2CH3 group, there are two hydrogens, but the third atom is carbon. Its atomic number is 6. So when we compare between -CH$_3$ and -CH$_2$CH$_3$, the ethyl group gets higher priority.

When we have −CH2OH and −CH2Cl, the first point of attachment is carbon. The next two atoms are hydrogens. Till this place, we have a tie. We need to move to the third atom. Here we have the difference. One is oxygen (atomic number 8), and the other is chlorine (atomic number 17). So chlorine gets the priority.

3. When there are multiple bonds, each multiple bonds are treated as 'n' single bonds.

Double and triple bonds: each multiple bond is considered as n single bonds. The vinyl group (C=C) or alkene portion is treated as two C–C bonds and gets higher priority over the alkane (C–C) part.

As given below, we can work out how aldehyde and ketone which has a carbonyl group (C=O) and cyanide which has a C triple bond N are assigned priorities. If an atom C is double-bonded to an atom O as in the carbonyl group, the C is treated as being singly bonded to two atoms of O as given below. The circled oxygen is added for assigning the order of priority. Similarly, a triple bond is handled the same way. The C is shown as attached to three N of which two nitrogens are added for assigning priority. The circled atoms are the phantom atoms. They are introduced for the purpose of assigning priorities to the multiply bonded groups (Figure 5.10).

4. After assigning the priority, the lowest priority group is kept away from the observer.

5. For the remaining three groups/atoms, if the order of priority is clockwise, it is R, and if it is anti-clockwise, then it is S.

Fischer's notation of D and L can be converted to the absolute configuration R and S. In Fischer notation, we focus only on the last but one carbon, but R and S configurations can be assigned to each and every stereogenic center in the molecule. It may be deduced by ranking the substituents according to the sequence rules, placing the **lowest ranking**

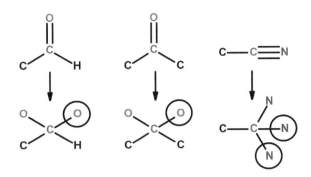

FIGURE 5.10   Phantom atoms for multiple bonds.

substituent on a vertical axis and determining whether the priority of the remaining substituents descends clockwise (R–) or anti-clockwise (S–).

5.4   HELICAL NOTATIONS P-M

Here we are looking at helical chirality, the descriptors used are P and M: P for plus and M for minus. Imagine you are climbing a spiral ladder (stairs), if the turn you take to climb up is to the right-hand side, it is P, and if you go toward the left, it is M. In this figure in the top left figure, we are climbing the ladder by walking toward the right. But in the right one, it is the left-side movement leading to the top, so it is M.

Confusion comes when the molecule is projected facing away from you as given at the bottom. To find out whether it is P or M what you need to do is to simply rotate the molecule in such a way that you are facing it. Always remember to keep the lower end of the helical chain at the bottom. We need to rotate the molecule toward you to find the helical chirality. For identifying the helical chirality P or M, you need to move up and not down (Figure 5.11).

5.5   RE-SI FACES

For visualizing three-dimensional objects in 2D, perspective drawing is an easy and convenient way to depict them.

Although stereochemical representation of drawing starting materials, reagents, ligands, or products can be done using wedge and dashed line representation, the challenge lies in depicting the direction of attack at the reacting center. If the reaction produces racemic products as in

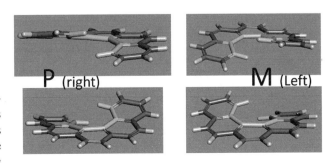

FIGURE 5.11   Helical notation.

the S_N1 mechanism or inversion as in S_N2 mechanism, we may not have difficulty in representing the reaction. But the major issue we face is depicting the direction of attack if we move from planar (sp^2) to tetrahedral (sp^3) arrangement during the reaction. Unless a universal convention is followed, we will end up with wrong descriptions based on how the starting material is stereochemically drawn. Let us take the simple example of butanone. It can be represented by both the following structures (**A** and **B** in Figure 5.12). If a nucleophile (H⁻) attacks this planar compound to give a tetrahedral product, we can have two products. The issue here will be how to identify the face/direction of the attack.

    i. Is it from the front when the methyl group is on the left
    ii. When the methyl group is on the right
    iii. Is it from the back when the methyl group is on the left
    iv. When the methyl group is on the right.

To avoid all these confusions, Re-Si convention is used.

When the carbonyl carbon is planar, we assume the front or back attack means they are perpendicular to the plane (90°). However, the attack follows the Bürgi–Dunitz angle (the Nu-C-O bond angle). The nucleophilic attack on the unsaturated trigonal center of the electrophilic carbonyl takes place at an angle of 107° in such a way that the attack leads to tetrahedral geometry for the product.

When a prochiral ketone is converted to a chiral compound in a single step, we can say the nucleophilic attack took place from either Re (pronounced as Ray) or Si (pronounced as Sigh) face. It is not enough to say, a planar sp^2 carbon is converted to tetrahedral sp^3 carbon. It is also important to say how the attack took place. Let us take the example of carbonyl reduction. Here we will get two products, so it is important to define from which side the attack or approach of the nucleophile took place. There is one more complexity. We can also draw the ketone in two different ways, the high-priority R1 group on the right side or left side. So to remove the ambiguity, we assign Re or Si face. Similar to R/S ranking, the Re face has the priority

groups in the clockwise direction and Si face has the groups in the anti-clockwise direction.

WORD OF CAUTION

Here we are focusing only on the priority of the three groups and we are not considering the 4th group. So, the reaction from Re face may produce R or S isomer when we include the 4th group according to the CIP rule.

5.5.1 Re-Si Face: Hydride Ion Attack

Let us take the butanone for our example to study the hydride ion attack. For easy understanding, we will keep the ethyl group on the right-hand side and the methyl group on the left-hand side. According to the CIP rule, we can say this is similar to R configuration. The highest priority oxygen is at the top, second priority ethyl is on the right and the least priority methyl is on the left. The arrangement is clockwise for 1 to 2 to 3. If the nucleophile attacks from the front, that is going to be Re face attack. We are going to get a product in which the oxygen goes behind and after workup leading to the hydroxyl derivative (OH away from the observer). The obtained butan-2-ol is S- isomer (refer to Figure 5.12, bottom isomer).

Let us look at the other case. If the nucleophile attacks from the back, that is going to be Si face attack. We are going to get a product in which the oxygen comes to the front side and after workup leads to the hydroxyl derivative (OH toward the observer). The obtained 2-butanol is R- isomer (refer to Figure 5.12, top isomer).

If you carefully look at both the above two examples of Re and Si face attack, we can understand that Re face attack need not lead to R isomer and Si face attack need not lead to S isomer. The final configuration of the product depends on the priority of the groups. The Re face attack leads to 2-butanol with S configuration. The Si face attack leads to 2-butanol with R configuration (Figure 5.13).

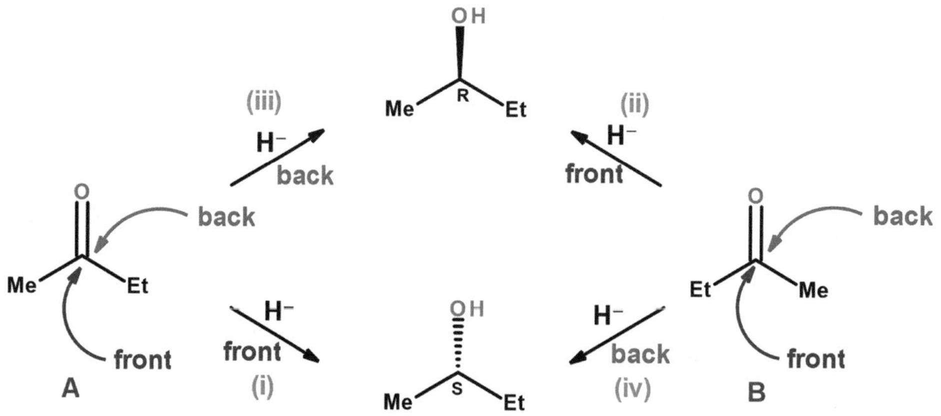

FIGURE 5.12　Front- and back-side attacks on a carbonyl carbon by nucleophile.

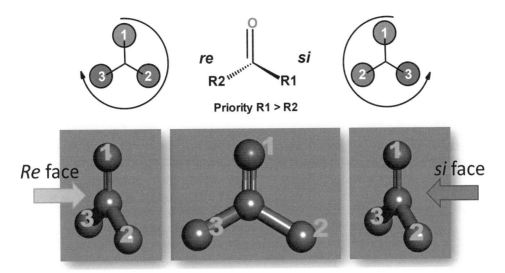

FIGURE 5.13 Re face and Si face.

5.5.2 RE-SI FACE: BROMIDE ION ATTACK

Again we consider butanone itself as an example, to study the attack of bromide ion on the carbonyl carbon. This will be easy for our comparison with hydride ion attack. Similar to the earlier example, we will keep the ethyl group on the right-hand side and the methyl group on the left-hand side. According to the CIP rule, we can say this is similar to R configuration. If the nucleophile, the bromide ion attacks from the front, that is going to be Re face attack. We are going to get a product in which the oxygen goes behind (OH is away from the observer) and bromide comes in the front. The obtained 2-bromobutanol is R isomer.

Let us look at the other case. If the nucleophile attacks from the back, that is going to be Si face attack. We are going to get a product in which the oxygen comes to the front side (OH toward the observer) and the bromide will be away from the observer. The obtained 2-bromobutanol is S isomer.

Let us compare the different products obtained from butanone. In the case of hydride ion nucleophile, the Re face attack gave S isomer and Si face attack gave R isomer. In the case of bromide ion nucleophile, the Re face attack gave R isomer and Si face attack gave S isomer. From these results, we can conclude that there is no relationship between the side of attack (Re vs Si) and the product configuration (R vs S) (Figure 5.14).

5.6 TOPICITY

It talks about the stereochemical relationship between the substituents and the structure to which they are attached. We have four types: homotopic, heterotopic, enantiotopic, and diastereotopic. They can be differentiated or identified on the basis of their reactivity or using NMR.

Ethanol is considered to be achiral (both the protons in $-CH_2-$ are chemically equivalent) for normal chemical transformations. Hence, we do not see any stereo-controlled oxidation of ethanol to acetaldehyde. On the other hand, in the famous alcohol fermentation reaction, where an enzyme, namely, alcohol dehydrogenase (brewer's yeast), is employed for the oxidation of ethanol in the presence of NAD^+ (nicotinamide adenine dinucleotide), there is a stereospecific reaction.

5.6.1 THE SUBSTITUTION TEST FOR EQUIVALENCE OF HYDROGENS

What are equivalent hydrogens? They have the same chemical environment. A molecule with 1 set of equivalent hydrogens gives one ^1H NMR signal. For example, cyclopentane gives a singlet because all the 10 protons in cyclopentane are equivalent.

For a pair of protons to be tested for topicity, replace one of the hydrogens with a heavier isotope (may be deuterium) and then do the same for the other proton. Now compare the spectra of these two structures formed.

FIGURE 5.14 Re face and Si face attacks of bromide ion.

- If the structures are identical then the hydrogens are homotopic. Homotopic atoms or groups have identical chemical shifts under all conditions.
- If the structures are enantiomeric then the hydrogens are enantiotopic. Enantiotopic atoms or groups have identical chemical shifts in achiral environments. They have different chemical shifts in chiral environments.
- If the structures are diastereomeric, then the hydrogens are diastereotopic. Diastereotopic hydrogens have different chemical shifts under all conditions.

5.6.2 HOMOTOPIC FACE

The compound will have identical substituents around the central carbon. Generally, CH_2 protons are homotopic protons. In the following example, the Grignard reaction between methyl magnesium iodide and formaldehyde gives the ethyl alcohol. The Grignard reagent can attack the aldehyde carbon from either side. Irrespective of which side the Grignard reagent attacks, we will end up with the same ethyl alcohol. Here, the two protons of the formaldehyde are said to be homotopic.

5.6.3 ENANTIOTOPIC FACE

It is the relationship between two similar atoms or groups of a molecule in which, if one or the other atoms or groups were replaced, would generate enantiomers.

Let us take the simple example of ethanol. There are two hydrogens on the second carbon. Let us label them as H_R and H_S. Now replacement of one of the hydrogens with deuterium leads to enantiomeric products. Replacement of H_R leads to R isomer and H_S leads to S isomer. We can say the two hydrogens of ethanol are enantiotopic.

Let us look at another example, the enantiotopic faces of pentan-3-ol.

Let us answer some of the questions you may have.

Q 1) what is the configuration of pentan-3-ol?

The central carbon in pentan-3-ol is not a stereogenic carbon because we cannot assign priority to either side.

We have four hydrogens on the neighboring carbon with respect to the hydroxyl. I am going to replace one proton each on the neighboring carbon with a –OH group and find out whether they are enantiotopic or not. Let us first replace one of the hydrogens that is pointing toward you (where the hydroxyl is pointing up). Let us do the same on the other side as well, that is replace one of the hydrogens that is pointing toward you. If you look at these two molecules as given below, you can see that these two are mirror images of each other. If you try to superimpose one on the other, you will find out that they are not superimposable. These two compounds are mirror images and are non-superimposable. That is, they are enantiomers (Figure 5.15).

Q 2) Why the hydroxyl carbon have R configuration in one and S in another?

The configuration of a particular carbon is assigned based on the priority of the substituent attached to it. In the left-side image, the second priority CHOH is on the right side, and in the other compound, it is on the left-hand side.

Q 3) There are two hydrogens on each neighboring carbon. So can I replace the other hydrogen and still get an enantiomer?

Before we answer this question, let us find out what happens if we replace the other hydrogens. We have four hydrogens on the neighboring carbon with respect to the hydroxyl. This time let us first replace one of the hydrogens that is pointing away from the observer. Let us do the same on the other side as well, that is replace one of the hydrogens that is pointing away from the observer. If you look at these two molecules as given below, you can see that

Homotopic

SCHEME 5.1 Grignard reaction between methyl magnesium iodide and formaldehyde.

Enantiotopic

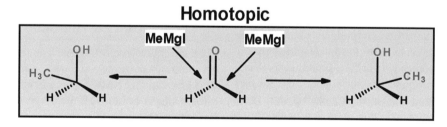

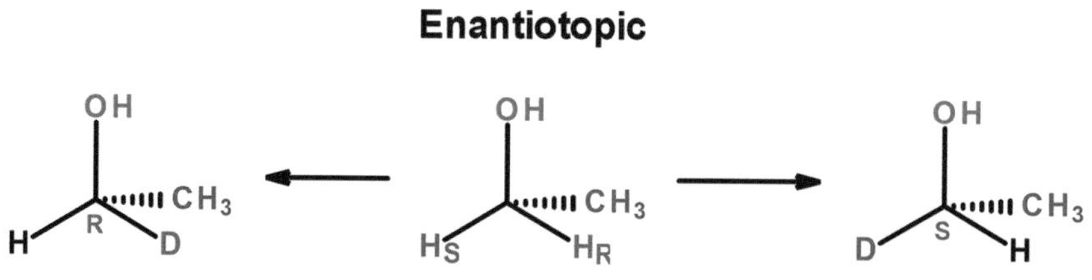

SCHEME 5.2 Replacement of hydrogens with deuterium in ethanol.

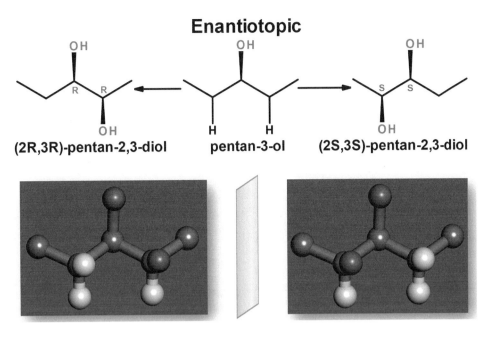

FIGURE 5.15 Enantiotopic faces of pentan-3-ol (H replacement toward the observer).

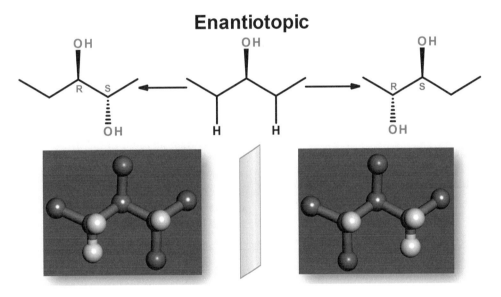

FIGURE 5.16 Enantiotopic faces of pentan-3-ol. (H replacement away from the observer)

these two are now mirror images of each other. If you try to superimpose one on the other, you will find out that they are not superimposable. These two compounds are mirror images and are non-superimposable, that is, they are enantiomers (Figure 5.16).

From these two examples, what did you find out? If I replace hydrogens from the same side (pointing toward the observer or pointing away from the observer on both sides of the hydroxyl), we are getting enantiomers, in other words, these are called enantiotopic faces. In this particular example of 3-pentanol, we have four hydrogens on the carbon adjacent to the hydroxyl and we can have two sets of enantiomers.

The following Scheme 5.4 depicts the enantiotopic faces. Replacement of H_2 and H_4 or H_1 and H_3 leads to enantiomers.

So H_1 and H_3 are one set of enantiotopic faces, and H_2 and H_4 are another set of enantiotopic faces.

5.6.4 DIASTEREOTOPIC FACE

It is the relationship between two atoms or groups in a molecule which, if one or the other were replaced, would generate diastereomers.

Let us take the example of (2S)-propane-1,2-diol. There is a stereogenic carbon labeled as S isomer. There are two hydrogens labeled as H_R and H_S on the primary hydroxyl carbon. Now replacement of one of the hydrogens with chlorine leads to diastereomeric products. Replacement of H_R leads to R isomer, and H_S leads to S isomer. We can say the two hydrogens of (2S)-propane-1,2-diol are diastereotopic.

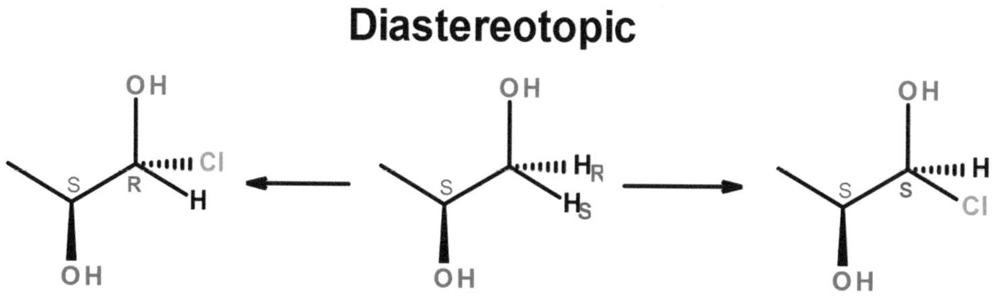

Diastereotopic

SCHEME 5.3   Two hydrogens of (2S)-propane-1,2-diol.

5.6.5   TOPICITY AND CHIRAL REAGENTS

Homotopic faces cannot be differentiated by chiral and achiral conditions. Reactions involving homotopic atoms do not lead to chiral products.

Enantiotopic faces can be differentiated under chiral conditions. But achiral conditions can not differentiate them when achiral aldehydes and ketones undergo the addition of chiral allylic reagents, they give chiral allylic alcohols.

Diastereotopic faces can be differentiated by chiral and achiral conditions. Reactions involving diastereotopic atoms lead to chiral products. Examples include Enolate additions.

5.6.6   DIASTEREOMER REPLACEMENT

We have four hydrogens on the neighboring carbon with respect to the hydroxyl. If one hydrogen is replaced on each of the neighboring carbon, do we get diastereomer? Let us first replace one of the hydrogens that is pointing toward you (where the hydroxyl is pointing up). Let us replace one hydrogen on the other side as well, that is replace one of the hydrogens that is pointing away from you. If you look at these two molecules in Figure 5.17, these two do not have mirror image relationship to each other. If we try to super-impose one on the other, we will find out that they are not superimposable. These two compounds do not have object–mirror image relationship and are also non-superimposable. They are diastereomers.

Let us try to replace the other two hydrogens on the neighboring carbon with respect to the hydroxyl. Let us first replace one of the hydrogens that is pointing away from you (whereas the hydroxyl is pointing up). Let us replace one hydrogen on the other side as well, that is replace one of the hydrogens that is pointing toward you. If we look at these two molecules in Figure 5.18, we can see that these two do not have an object–mirror image relationship to each other. If you try to superimpose one on the other, you will find out that they are not superimposable. They are diastereomers.

From these two examples, what did you find out? If I replace hydrogens from the different directions (pointing toward the observer and pointing away from the observer) on both sides of the hydroxyl, we are getting diastereomers, in other words, these are called diastereotopic faces. In this particular example of 3-pentanol, we have four hydrogens on the carbon adjacent to the hydroxyl and can have two sets of diastereomers.

The following figure depicts the diastereotopic faces. Replacement of H_1 and H_2 or H_3 and H_4 leads to diastereo-mers. So H_1 and H_2 are one set of diastereotopic faces, and H_3 and H_4 are another set of diastereotopic faces.

Diastereotopic

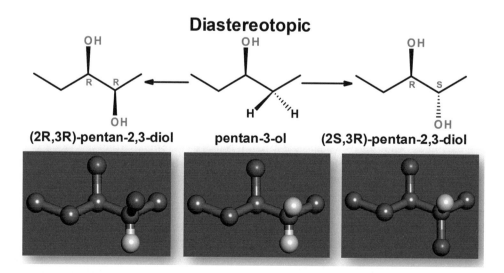

(2R,3R)-pentan-2,3-diol   pentan-3-ol   (2S,3R)-pentan-2,3-diol

FIGURE 5.17   Diastereomers of pentan-3-ol. (H replacement toward or away from the observer)

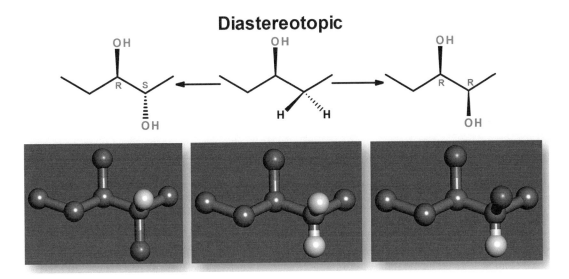

FIGURE 5.18 Diastereomers of pentan-3-ol. (H replacement away from or toward the observer)

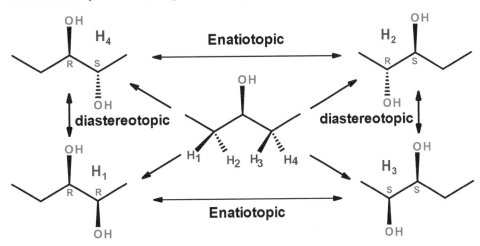

SCHEME 5.4 Enantiomer, diastereomer relationship in pentan-3-ol.

5.7 ENANTIOMERS AND DIASTEREOMERS IN CYCLIC SYSTEMS

Now let us look at the enantiomers and diastereomers in cyclic systems. Here we have a set of compounds, *cis-* and *trans*-2-methylcyclopentanol. Depending on the placement of functional groups, they are called either enantiomers or diastereomers (Figure 5.19).

cis-2-Methylcyclopentanol

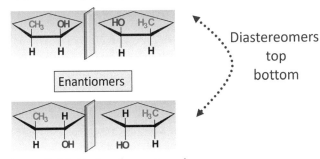

trans-2-Methylcyclopentanol

FIGURE 5.19 cis- and trans-2-methylcyclopentanol.

5.8 MESO COMPOUND

Here we have one more type of compound which is meso compound. Meso compounds have internal reflection, or one part of the molecule is the mirror image of the other part (Figure 5.20).

If you recall what we saw early, meso compounds belong to optically inactive compounds even though they have chiral centers.

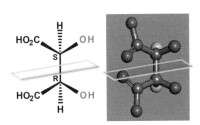

FIGURE 5.20 Meso compound.

5.9   QUESTIONS

1. Draw the structure of the following compounds
   1,4-cis-cyclohexane diol
   Trans 1,4-dibromo pentane
   (R)-Lactic acid
   L-Glucose
   D-Tryptophan
   L-Tyrosine
2. Identify the relationship between A and B, A and C, and B and C (based on topicity) (Figure 5.21).
3. Identify the relationship between both the circled methyl groups in A and the circled methyl and hydrogen in B (based on topicity) (Figure 5.22).
4. If the following compounds undergo a Grignard reaction with MeMgBr, how many products are possible? Comment about the relationship, if more than one compound is possible (Figure 5.23).

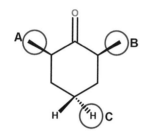

FIGURE 5.21   Relationship among A, B, and C.

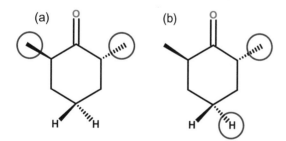

FIGURE 5.22   Relationship between A and B.

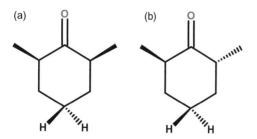

FIGURE 5.23   Predict the products.

BIBLIOGRAPHY

Alan Spivey, Imperial College London, Chemistry I (Organic): Stereochemistry, Fischer Projections, Absolute Configuration and (R)/(S) Notation (https://www.ch.ic.ac.uk/local/organic/tutorial/ACS3.pdf) (accessed, 15-Apr-2024)

Buck, L., & Axel, R. (1991). Odorant receptors and the organization of the olfactory system. *Cell*, 65, 175–187. https://doi.org/10.1016/0092–8674(91)90418-X

Gamini Gunawardena, Utah Valley University, Fischer Projection, in Organic Chemistry (OpenStax) (https://chem.libretexts.org/Ancillary_Materials/Reference/Organic_Chemistry_Glossary/Fischer_Projection) (accessed, 15-Apr-2024)

Ian Hunt, University of Calgary, Chapter 7: Stereochemistry (Fischer Projections) (https://www.chem.ucalgary.ca/courses/351/Carey5th/Ch07/ch7-7.html) (accessed, 15-Apr-2024)

Kalsi, P. S. *Stereochemistry: Conformation and Mechanism*, 9th Ed. New Age International (P) Ltd., 2017, 1–80, 188–209.

Kramer, W. H., & Griesbeck, A. G. (2008). The same and not the same: Chirality, topicity, and memory of chirality. *Journal of Chemical Education*, 85(5), 701. https://doi.org/10.1021/ed085p701

McLaughlin, S., & Margolskee, R. F. (1994). The sense of taste. *American Scientist*, 82, 538–545.

Nasipuri, D. *Stereochemistry of Organic Compounds, Principles and Applications*, 3rd Ed. New Age International Publishers, 2018, 46–54, 102–122.

Smith, M. B., & March, J. *March's Advanced Organic Chemistry Reactions, Mechanisms and Structure*, 5th Ed. John Wiley & Sons, Inc., 2001, 164–166.

Tim Soderberg (University of Minnesota, Morris), Organic Chemistry With a Biological Emphasis by Dr. Dietmar Kennepohl FCIC (Professor of Chemistry, Athabasca University), 5.11: Prochirality (https://chem.libretexts.org/Courses/Athabasca_University/Chemistry_350%3A_Organic_Chemistry_I/Chapter_5%3A_Stereochemistry_at_Tetrahedral_Centres/5.11_Prochirality) (accessed, 15-Apr-2024)

Tim Soderberg (University of Minnesota, Morris), Organic Chemistry With a Biological Emphasis, 3.11: Prochirality (https://chem.libretexts.org/Bookshelves/Organic_Chemistry/Book%3A_Organic_Chemistry_with_a_Biological_Emphasis_(Soderberg)/Chapter_03%3A_Conformations_and_Stereochemistry/3.11%3A_Prochirality) (accessed, 15-Apr-2024)

6 Chiroptical Properties

In this chapter, we will see how we can determine the absolute configuration of stereogenic centers.

Louis Pasteur first observed that a concentrated sodium ammonium tartrate on crystallization (below 28 °C) gave two kinds of crystals. One was right-handed and the other was left-handed. He carefully separated them and checked the optical activity of each set of crystals. He observed that both the solutions gave the same specific rotation but with opposite signs. He was the first to observe the asymmetric arrangement of molecules that are non-superimposable.

6.1 BASICS OF OPTICAL ROTATION

6.1.1 PLANE, CIRCULARLY, AND ELLIPTICALLY POLARIZED LIGHT

Let us recall how plane-polarized lights are produced. We need a light source. We can use two types of light sources either (i) monochromatic (single wavelength) or (ii) polychromatic (multi wavelength) source. If we use the polychromatic source, we need to filter out all other wavelengths except one particular wavelength. It is generally done by a monochromator. It is an optical device that transmits a narrow band of wavelengths of light. This monochromatic light then passes through a polarizer and gets converted to a plane-polarized light (Figure 6.1).

We will quickly recall what is optical rotation? The rotation of the orientation of plane-polarized light by an optically active substance is called optical rotation. If it is rotated clockwise, then it is called dextrorotatory and denoted by (+) or 'd', and if it is rotated anticlockwise, it is called levorotatory and they are denoted by (−) or 'l'. Fischer projection D and optical rotation 'd' are two different notations and they do not have any relationship between them.

A polarimeter is an instrument that is used to measure the amount of rotation of an optically active compound.

Specific rotation $[\alpha]_D$ is defined as the observed rotation when the sample path length (l) is 1 decimeter (10 cm), the sample concentration is 1 gm/mL, and the light used is sodium D line (589 nm) (Figure 6.2).

The sign and amount of optical activity depend on (1) temperature, (2) solvent, (3) wavelength of light, (4) dimensions of the tube (like length, and diameter), (5) nature, and (6) concentration of the substance.

Optical purity of a substance is given in Figure 6.3.

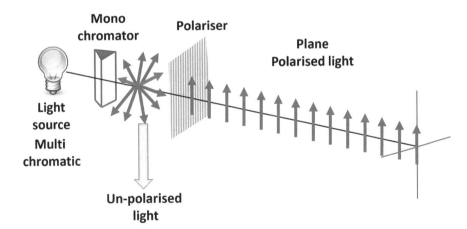

FIGURE 6.1 Polarimeter.

$$[\alpha]_D = \frac{observed\ rotation\ (in\ degrees)}{path\ length\ (l)in\ (dm)\ X\ concentration\ C\ (g/mL)} = \frac{\alpha}{l\ X\ C}$$

FIGURE 6.2 Equation for specific rotation.

$$optical\ purity\ (\%) = \frac{\alpha_{mixture\ of\ enantiomers}}{\alpha_{pure\ enantiomers}}\ X\ 100$$

FIGURE 6.3 Equation for optical purity.

DOI: 10.1201/9781032631165-6

6.1.2 Functioning of Polarimeter

Figure 6.4 depicts the various elements that are part of a polarimeter. It consists of a light source and a monochromator (if the light is polychromatic). A polarizer converts unpolarized monochromatic light to a plane-polarized light. This passes through the sample. Optically active substances are the substances that rotate the orientation of plane-polarized light. Length is the sample path length expressed in deci meter. The angle of rotation alpha is unique to the sample. Here the angle of rotation is found out by the cancellation method. There is a second polarizer called analyzer, kept before the detector which rotates the electrical vector in the opposite direction of rotation or cancels out the angle of rotation. The angle of rotation is compensated by the analyzer before the beam falls on the detector.

6.1.3 Linear and Circularly Polarized Light

Light is a transverse electromagnetic wave. It has both electrical and magnetic components. Since the molecules we study have many electrons, it will be more appropriate for us to focus on electrical components, as to how does it interact with the molecules.

The electrical component is further divided into magnitude and direction, that is, height and angle. When the light is vibrating in a single plane with changing magnitude, it produces linear waves or plane-polarized waves. If light is composed of two plane waves of equal amplitude differing in phase by 90°, then the light is said to be circularly polarized.

Although the light wave consists of two components, we will consider only the electrical vector. Vertically polarized light is formed when the oscillation of electrical vectors on all planes is filtered except a single plane. The green line depicts this. The blue and red components are filtered by the polarizer. Depending on the plane where you observe, the movement of the electrical vector looks different. When the wave moves in a sinusoidal curve, you observe perpendicular to the forward movement of the

light and a line or plane when standing in front of the forward movement of light. In this case, only the amplitude or the magnitude of the electrical vector changes. It goes from +ve to −ve maximum. Figure 6.5 depicts the generation of vertically polarized light.

Light being an electromagnetic wave consists of both electrical and magnetic components. They are mutually perpendicular to each other. A linear polarized light can be formed by the combination of right and left circularly polarized light. A circularly polarized light can be formed by the combination of two mutually perpendicular (y and z axes) linearly polarized (light that has 45° out of phase or one quarter of a wavelength out of phase) light.

Mnemonics: Remember LA (Los Angeles). For linearly polarized light, amplitude changes.

For circularly polarized light, the direction changes and the amplitude remains the same. For a single *frequency*, we can produce a straight line for linearly polarized light and a circle for circularly polarized light. That is why they are given this name. If you look at the overall movement over the entire wavelength, you will notice that the circularly polarized light moves like a helix.

When the wave moves in the sinusoidal movement as well as the linear oscillation, the variation happens only in the magnitude. The rotation or the variation in the angle of forward movement is zero. We end up in circularly polarized light as opposed to linear polarized light.

If we plot all the positions that are covered in the moving direction of the wave in one wavelength, we end up in different representations for linearly polarized light and circularly polarized light.

Let us compare the linear and circularly polarized lights (Table 6.1).

6.2 DICHROISM

Dichroism is a phenomenon in which the absorption of light is different for different directions of polarization. The two major types of dichroism we will now study are

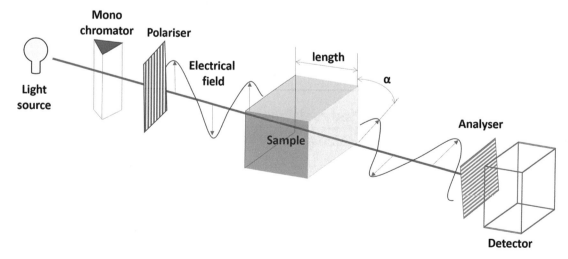

FIGURE 6.4 The various elements of a polarimeter.

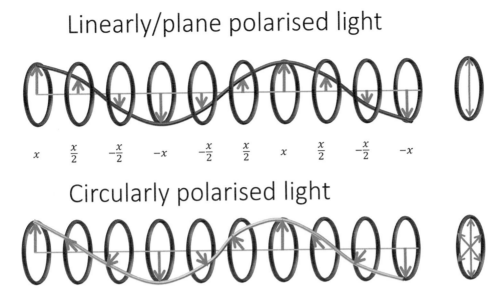

FIGURE 6.5   Depiction of wave nature of linearly and circularly polarized lights.

TABLE 6.1
Comparison between the Linear and Circularly Polarized Lights

Linearly Polarized Light	Circularly Polarized Light
Amplitude/magnitude changes	Angle changes
Sinusoidal (along the direction)	Helical
RCP +LCP (same amplitude)	$LP_{Vertical}+LP_{Horizontal}$ (same amplitude), phase difference $\pm p/2$
The electrical field vector goes from maximum to zero to negative maximum	The electrical field vector remains the same
Refractive index difference gives circular birefringence; absorption coefficient difference gives circular dichroism	

linear dichroism and circular dichroism. This is basically based on the type of light that is used. If linear polarized light is used, that produces linear dichroism, and if circularly polarized light is used, it will produce circular dichroism. The major difference between linear polarized light and circularly polarized light is with respect to the electric vector component only. In linear polarized light, the direction of the electrical vector is constant and the magnitude keeps on changing. In the case of circularly polarized light, the magnitude remains constant, but the direction keeps on changing. There are two types of circularly polarized light: one is right circularly polarized light, and the other one is left circularly polarized light. Both form a helix. The right circularly polarized light forms the right-handed helix, and the left circularly polarized light forms the left-handed helix. The main use of circular dichroism is in the determination of secondary structures of proteins.

6.2.1   CIRCULAR DICHROISM

The circular dichroism is based on the asymmetry of the molecules as a whole. If a molecule is optically active, it may show optical rotatory dispersion and circular dichroism. We already know that the linear polarized light may

be represented as a combination of right and left circularly polarized lights. When the linearly polarized light passes through an optically active medium, the left and the right circularly polarized lights interact differently with the optically active substance. The difference arising out of refractive index changes of the left and right circularly polarized light is called optical rotatory dispersion or ORD. Instead, if there is a difference in the absorption coefficients of the left and right circularly polarized lights, it is called circular dichroism.

The mathematical expression (Beer-Lambert Law) for circular dichroism can be written as follows (Figure 6.6).

A_R and A_L represent the absorption of right and left circularly polarized lights, respectively; ε is the extinction coefficient; $\Delta\varepsilon$ is the molar ellipticity; I is the path length; and c is the concentration.

We can say that circular dichroism occurs due to a macroscopic phenomenon. In other words, it is the property of the molecule as a whole rather than a linear combination of individual chiral blocks.

$$\Delta A(\lambda) = A_L(\lambda) - A_R(\lambda) = [\varepsilon_L(\lambda) - \varepsilon_R(\lambda)]lc = \Delta\varepsilon lc$$

FIGURE 6.6   Beer-Lambert Law for circular dichroism.

$$\theta = 3330\,(\varepsilon_L - \varepsilon_R)$$

FIGURE 6.7 Expression for molecular ellipticity.

Chiral substances show different absorptions of circularly polarized light. This phenomenon is called circular dichroism and is represented in terms of molecular ellipticity (θ) (Figure 6.7)

6.3 OPTICAL ROTATORY DISPERSION

It is a plot of the differences in the refractive indices of the left and right circularly polarized lights while passing through a medium with wavelength. Compounds having similar configuration have similar ORD curves (Table 6.2).

When two lights have different circular polarizations, they will also have different extinction coefficients. In other words, if $\Delta\varepsilon \neq 0$, then $\Delta n \neq 0$.

Circularly polarized light consists of two linearly polarized light components and vice versa.

- ORD is the measurement, as a function of wavelength, of a molecule's ability to rotate the plane of linearly polarized light.
- ORD is the difference in the refractive indices of LCP and RCP.
- CD is based on evaluating the molecule's unequal absorption of right- and left-handed circularly polarized lights.
- CD is the difference in the absorption of LCP and RCP (Figure 6.8).
- In an optically active medium, the refractive indices of the left- and right-handed polarized lights are different, so the plane-polarized light will be rotated through an angle α.

- When n_l and n_r are the indices of the refraction for left- and right-handed polarized lights, respectively, and λ is the wavelength, optical rotation α is given as (radians per unit length) (Figure 6.9).

Optical rotation is given by (Figure 6.10)

- When plane-polarized light passes through an optically active sample, the angular velocities of left and right circularly polarized light differs. This results in the elliptical path for the light and the orientation of the ellipse is called optical rotation.
- The measurement of optical rotation with respect to a particular range of wavelengths is called optical rotatory dispersion.
- If the molecule under study contains chiral chromophores, then one CPL state will be absorbed to a greater extent than the other and the CD signal over the corresponding wavelengths will be non-zero.
- A circular dichroism signal can be positive or negative.
- Birefringent material exhibits a dependence of its refractive indices on the polarization and direction of the light.
- Birefringence is classified as circular, linear, and axial (Figure 6.11)

Factors that affect ORD and CD

- Chiroptical properties depends on (Tables 6.3 and 6.4)
  - → Vibrational/temperature effects
  - → Gauge/origin dependence

TABLE 6.2
Comparison between ORD and CD

ORD	CD
Plane-polarized light	Circularly polarized light
Dispersive phenomena	Absorptive
Plane-polarized light is not converted to elliptical light	Circularly polarized light is converted to elliptical light
Specific rotation is plotted against wavelength	Molecular ellipticity is plotted against wavelength
Refractive indices of left and right polarized light should be different	There is differential absorption of left and right circularly polarized light

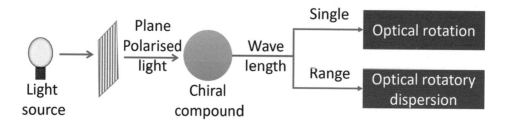

FIGURE 6.8 Relationship between optical rotation and ORD.

$$\alpha = \frac{n_l - n_r}{\lambda}$$

FIGURE 6.9 Expression for angle of rotation.

Optical rotation

$$[\alpha]_\lambda^T = \frac{\alpha\,10^2}{lc}$$
l = cell length in decimetres
c = g/100 mL

$$\phi = \frac{\pi}{\lambda}(n_L - n_R)$$
N = refractive index
λ = wavelength of length
φ = angle of rotation

Molar optical rotation $\quad [\Phi] = [\alpha] \times Mol.wt. \times 10^{-2}$

FIGURE 6.10 Expression for optical rotation and molar optical rotation.

specific ellipticity $= [\varphi] = \dfrac{\varphi'}{c'd'}$

molar ellipticity $= [\theta] = M[\varphi] \times 10^{-2}$

$\varepsilon_l - \varepsilon_r = 0.3032 \times 10^{-3}[\theta]$

$$\theta(\text{rad cm}^{-1}) = \frac{2.303(A_L - A_R)}{4l}$$
θ ellipticity
l path length
A absorption

molar circular dichroism $= \varepsilon_l - \varepsilon_r = \dfrac{k_l - k_r}{c}$

k from $I = I_o 10^{-kd}$

FIGURE 6.11 Expression for specific ellipticity and molar ellipticity.

→ Molecular conformations
→ Electron correlation
→ Condensed-phase and solvent effects

6.3.1 Difference between Circular and Linear Birefringence

6.3.2 Comparison between "Birefringence" and "Optical Activity"

6.4 COTTON EFFECT

The Cotton effect is defined as the characteristic change that is observed in optical rotatory dispersion and/or circular dichroism near the absorption band of a substance. Within a short wavelength region, there will be maximum fluctuation in absorbance. The absolute magnitude of the optical rotation at first varies rapidly with wavelength, crosses zero at absorption maxima, and then moves rapidly in the reverse direction with wavelength. The Cotton effect occurs due to the combination of both (circular birefringence and circular dichroism) effects in the region in which optically active absorption bands are observed.

- It is the change in optical rotation (α) or (φ) with wavelength (λ).
- The absorption changes drastically near the maxima (λ_{max}).
- The difference between n_L–nR (Δn) and ε_{L}-ε_R (Δε) varies with the wavelength and can be either positive or negative.
- Enantiomers show completely opposite Cotton effect curves with the same magnitude/amplitude.

n_R and n_L are the refractive indices of the right and left circularly polarized modes, ε extinction coefficient

TABLE 6.3
Comparison between Circular and Linear Birefringence

Circular Birefringence	Linear Birefringence
Difference between indices of refraction for left- and right-handed circularly polarized light	Difference in two components of the index of refraction relative to the direction of an external field
Faraday effect is a form of **Circular Birefringence** induced by a magnetic field	**Cotton-Mouton effect** is a form of **linear birefringence** induced by a static magnetic field
Does not requires sample to be geometric anisotropic	Requires sample to be **geometric anisotropic**
Optical rotation is a measure	

TABLE 6.4
Comparison between Birefringence and Optical Activity

Birefringence	Optical Activity
bi=two, refringence=refractive index	Rotation of plane-polarized light
Splitting of light rays in to two	**Rotation** of light rays either right or left side
Inherent property of **crystals**	Inherent property of **optically active** compounds (solid/liquid)
Difference between the **extinction coefficients** for right and left circularly polarized components	A difference between the **indices of refraction** for right and left circularly polarized light

When we move from a lower wavelength to higher wavelength, the positive Cotton effect is where the trough comes first, and the peak comes later as shown below. The negative Cotton effect is where the peak comes first, and the trough comes later. The magnitude is the vertical distance between the peak and trough. The sign, magnitude, and shape of the anomalous rotatory dispersion curve are unique for every molecule and can be used to characterize them (Figure 6.12).

6.4.1 Types of Cotton Effect Curves

1. Single Cotton effect curves
   When the Cotton effect curves show only one maximum and one minimum at maximum absorption, they are called single Cotton effect curves (Figure 6.13).

2. Multiple Cotton effect curves
   A rotatory dispersion curve with several extrema is called multiple Cotton effect curve.

6.4.2 Applications of ORD Measurements to Steroidal Ketones

One of the important applications of rotatory dispersion is identifying the location of a carbonyl group in a polycyclic system, especially in natural products. There are a lot of studies focused on the determination of the absolute/relative configuration of steroidal ketones. It is known that many steroidal compounds have a carbonyl group. Due to the presence of C=O (n to π* transition), they are UV active. The most challenging part is to locate the positions of the carbonyl groups in the steroidal ring systems. The following A, B, C, and D carbon skeleton of steroids is extensively studied for the various possible positions of the carbonyl

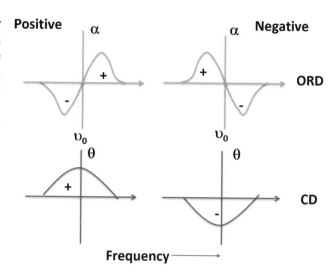

FIGURE 6.13   Types of Cotton effect curves.

group (at 1, 2, 3, 4, 6, 7, 11, 12, 15, 16, and 17). In many steroidal derivatives, the presence of other functional groups (e.g. esters, alcohols, lactones) which do not affect the conformation of the original molecule had very characteristic rotatory dispersion curves and can be used to deduce the absolute and relative configurations (Figure 6.14).

For instance, Figure 6.15 shows the rotatory dispersion curves of the 1- (A), 3- (B), and 7- (C) ketones of the cholestane series.

6.5   AXIAL HALO KETONE RULE

This was developed by Djerassi. To observe this rule, the compound should be properly oriented in such a way that the carbonyl carbon in the chair conformation is pointed upwards. Now on looking through the carbonyl bond if the α-axial halogen is on the right-hand side, it will display a positive Cotton effect, and if it is on the left-hand side, it will produce a negative Cotton effect. This is called axial halo ketone rule.

It is observed that if we introduce an electronegative halogen on the adjacent carbon of the carbonyl (either axial or equatorial), it affects the observed ultra-violet maximum of the ketone chromophore. In fact, a pattern is observed in which an equatorial bromine substituent results in a

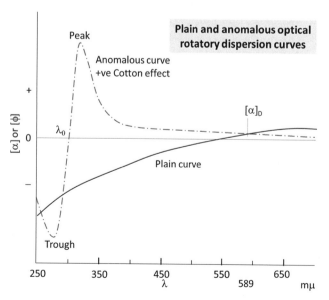

FIGURE 6.12   Cotton effect curves.

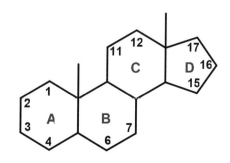

FIGURE 6.14   Numbering of carbons in steroidal ketones.

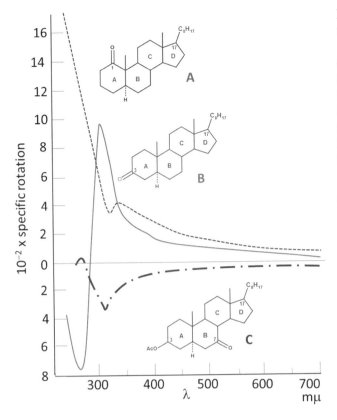

FIGURE 6.15 Examples of Cotton effect curves of cholestane series.

Based on these observations, the "axial haloketone rule" was proposed.

To find out what will be the change in the Cotton effect of the original ketone, we need to view the carbonyl in such a way that the carbonyl oxygen is facing the observer at the top in the chair conformation. If the α-halogen is on the left-hand side when we view it, the compound will show a negative Cotton effect, and if the α-halogen is on the right-hand side, a positive Cotton effect will be observed (Figure 6.16).

6.5.1 Advantages and Limitations of ORD and CD

- Advantages
  - Small volume and low concentration of the sample are sufficient for ORD/CD measurement
  - Can predict the conformations of proteins in dilute solutions
  - Can be used to measure protein denaturation and helix-coil transitions of polypeptides, enzyme interactions with substrates, inhibitors, or coenzymes
- Limitations
  - Cannot predict exact structures that could be determined using X-ray, NMR, and neutron diffraction
  - The absolute stereochemistry of unknown compounds can be deduced only by comparison with the ORD spectra of known compounds
  - Molecular vibrations complicate the interpretation of the ORD spectra

hypsochromic shift (i.e., to a shorter wavelength), and axial bromine results in a bathochromic shift (i.e., to a longer wavelength). Moreover, these α-substituents also affect the sign of the Cotton effect of the parent carbonyl compound.

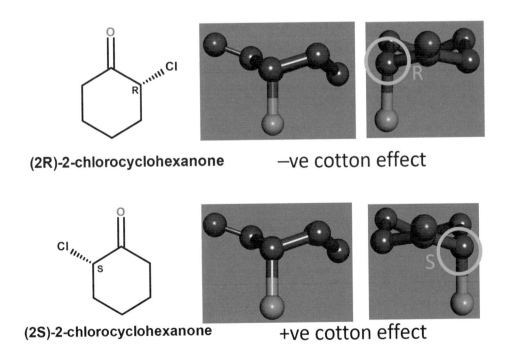

(2R)-2-chlorocyclohexanone −ve cotton effect

(2S)-2-chlorocyclohexanone +ve cotton effect

FIGURE 6.16 Examples of axial halo ketone rule for 2-bromocyclohexanone.

Even though a molecule has various chiral units, if the whole molecule is symmetric then they do not show any circular dichroism. But for dissymmetric molecules, like proteins, one part of the molecule may have different chromophore units compared to the rest of the molecule, so it may show circular dichroism.

In a molecule, the presence of electrical transition dipole, rotational strength, electronic transitions, the composition of chromophores, the nature and the number of base units are all crucial for determining the strength of the CD.

6.5.2 Application of CD

- Chiral molecules of different types and sizes can be studied by CD spectroscopy. In fact, CD is mostly used in the study of large biological molecules
- The secondary structure or conformation of macromolecules can be studied using CD spectroscopy. Since the protein's secondary structure is sensitive to various factors including temperature, environment, or pH, we can use circular dichroism to study how the secondary structure changes with environmental conditions or on interaction with other molecules

6.6 DETERMINATION OF ABSOLUTE CONFIGURATION

The absolute configuration can be determined by two methods. One is a non-chiroptical method, and the other one is a chiroptical method. The non-chiroptical method can be further divided into two major types. One is a physical method, and the other one is a chemical method. In the physical method, we have X-ray structural analysis and NMR spectroscopy. Chemical methods include partial and total syntheses and chemical degradation. In the chiroptical method, we have optical rotatory dispersion and circular dichroism.

6.6.1 Methods Based on Chemical Correlation

Let us look at the correlation. The first one is based on a method that does not affect the configuration of the stereogenic centers.

You think about two parallel processes each reaching the same intermediate with known absolute configuration. Then we can deduce that both the starting materials will have the same absolute configuration at the center under consideration.

You start from a compound with a known absolute configuration. Now perform some chemical reactions such that the chiral center is not disturbed. You will get some intermediate. In the parallel process, you carry out a series of reactions on the compound with unknown configuration to reach the same intermediate. In both processes, we make sure you do not change the stereochemistry of the stereogenic center under consideration. Now you can compare the configuration of both compounds.

In the following example, we are getting (+)-malic acid from (+)-tartaric acid and a compound of unknown stereo configuration. The α-hydroxyl configuration of the unknown compound can be deduced to be (R)- by correlation with (+)-tartaric acid.

6.6.2 Methods Based on Diastereomers

The second one involves diastereomers. This is a relative method. You know diastereomers have more than one chiral center. If the absolute configuration of one of the chiral centers is known, then we can find the relative configuration of the other center. The absolute configuration of quinic acid is established by converting it into a compound of known configuration that is citric acid. (mention what is T) (T is tritium)

Oxidation of quinic acid to 5-dehydroquinic acid followed by treatment with HIO_4 gave the dialdehyde. Oxidation of this dialdehyde using bromine water gave citric acid whose configuration is known. From this, the configuration of α-hydroxyl carbon (attached to the carboxylic acid) can be deduced.

6.6.3 Methods Based on Chiral Center

Here a chemical reaction is carried out on the chiral center with predictable stereochemistry, that is a reaction involving S_N2 substitution. Since S_N2 substitution leads to inversion of configuration, we can very well find out from the final product what was the stereochemistry of the starting

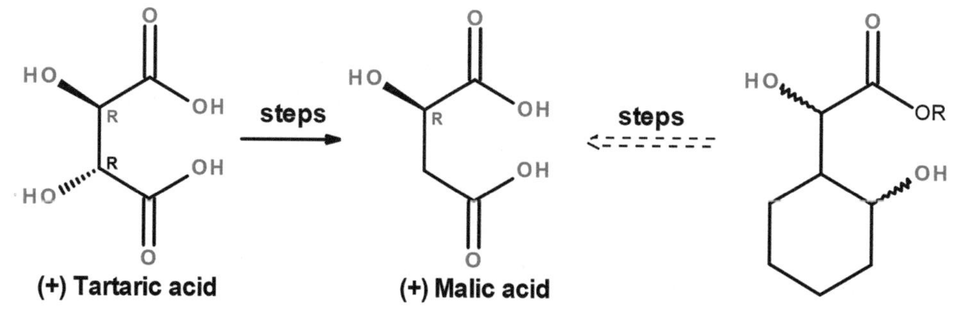

SCHEME 6.1 Example of deduction of unknown configuration using correlation method.

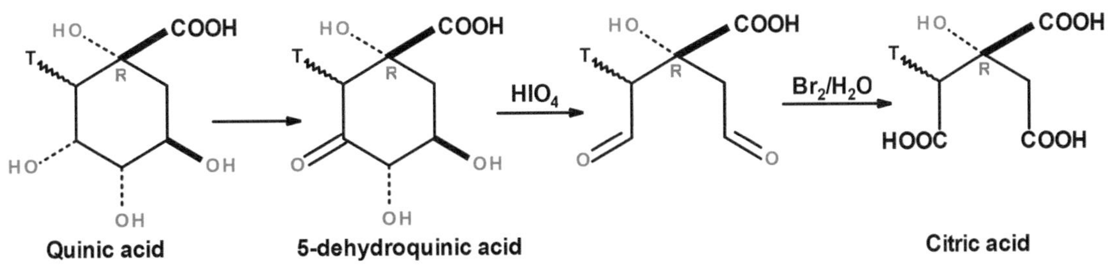

SCHEME 6.2   Example of establishing absolute configuration by conversion into a compound of known configuration.

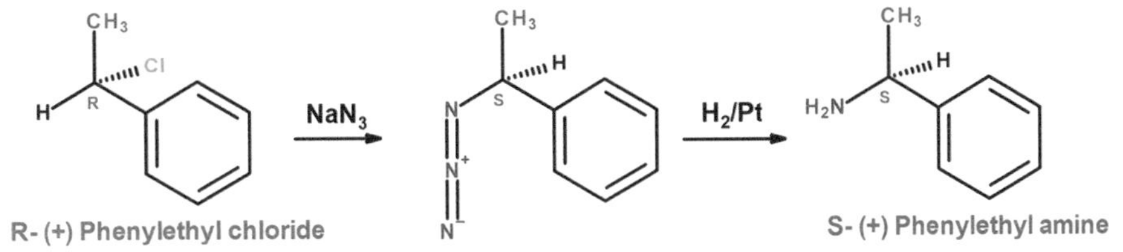

SCHEME 6.3   Example of establishing absolute configuration by carrying out a reaction with predictable stereochemistry.

material. An example is given here. R-phenylethyl chloride is converted to its azide by S_N2 reaction, and this was then reduced using hydrogen and palladium to give the phenethylamine with opposite configuration.

auxiliary is carefully removed to give the pure enantiomer. Many chiral acids can be separated using chiral bases, and vice versa. Chiral acids can also be converted into chiral esters using alcohols and vice versa.

6.6.4   METHODS BASED ON CORRELATION

The next one is based on correlation related to asymmetric synthesis. The major ones are Crams Rule, Felkin-Anh model, and Prelog Rule. We will see more about them in the next chapter.

6.6.5   METHODS BASED ON RESOLUTION

The next one is methods based on resolution. Basically, this is based on separating enantiomers. Since enantiomers will have the same chemical properties, normal separation techniques based on distillation, crystallization, chromatographic separation, and so on do not help us to separate them. What is done to perform separation is to convert the enantiomers into diastereomers by treating them with chiral auxiliaries. Once the enantiomer is converted into a diastereomer, they have different physical and chemical properties. They can be separated. After separation, the chiral

6.7   QUESTIONS

1. If we need to find out the absolute stereochemistry of 2, which of the pathway we need to follow? Why? (Figure 6.17)
2. In the following synthetic scheme the absolute stereochemistry of A and B need to be determined. Can we do it? If so, explain (Figure 6.18).
3. Can we use CD below 200 nm for protein analysis? If not why?
4. Which of the following compounds will show the Cotton effect? (Figure 6.19)
5. For the above set of compounds, can we predict the sign of the Cotton effect, if they show Cotton effect?
6. If you need to predict the sign of Cotton effect, which rule will you follow?
7. Which are the amino acids in proteins that will show CD signals?

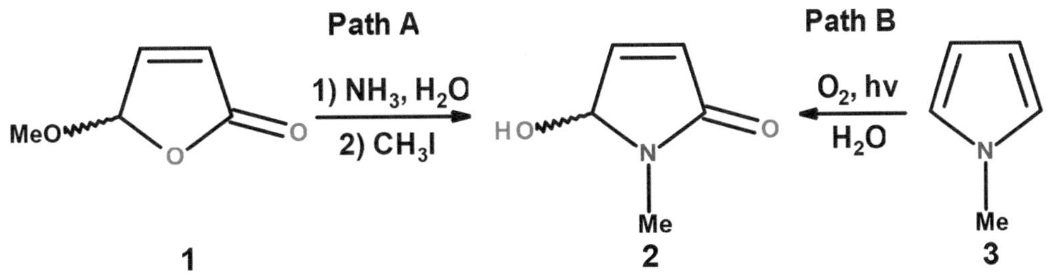

FIGURE 6.17   Example question 1 to find out the absolute stereochemistry.

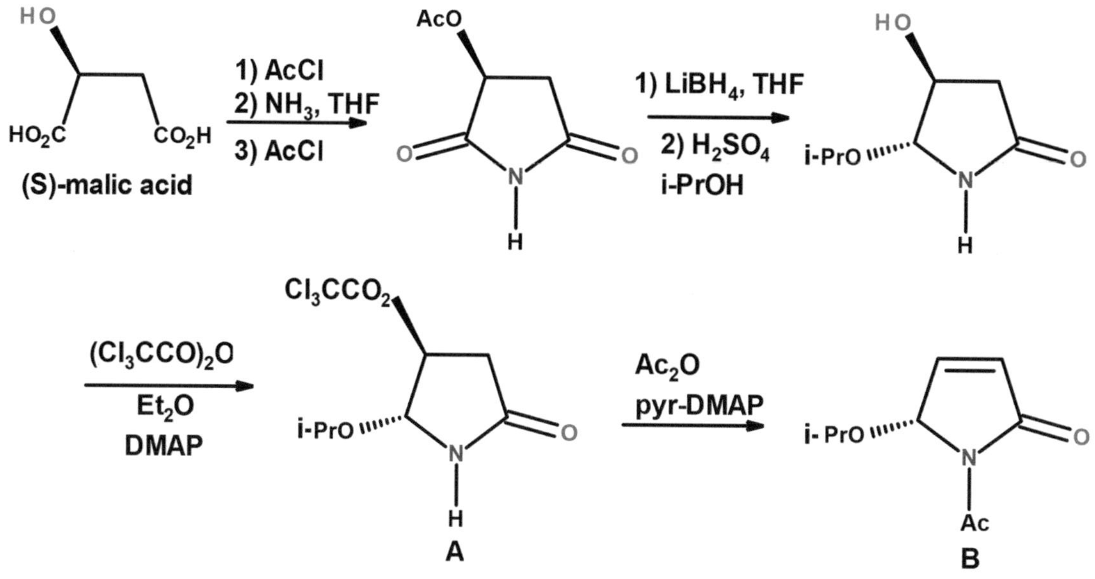

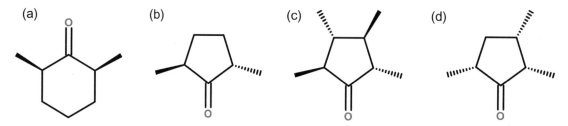

FIGURE 6.18 Example question 2 to find out the absolute stereochemistry.

FIGURE 6.19 Example question 4 to find out the Cotton effect.

BIBLIOGRAPHY

Animated biology With Arpan, https://www.youtube.com/watch?v=aVnncH1lqRg (circular dichroism) (accessed, 15-Apr-2024)

Cuiper, A.D. et al., Chapter 5: Determination of the Absolute Configuration of 3-Pyrrolin-2-ones (https://www.rug.nl/research/portal/files/14526842/c5.pdf) (accessed, 15-Apr-2024)

Finar, I. L. *Organic Chemistry, Volume 2: Stereochemistry and the Chemistry of Natural Products, Chapter 2*, 5th Ed. Pearson Education India, 1956, 96–133.

Hanson, K. R., & Rose, I. A. (1963). The absolute stereochemical course of citric acid biosynthesis. *Proceedings of the National Academy of Sciences of the United States of America*, 50(5), 981–988.

John D. Roberts and Marjorie C. Caserio, California Institute of Technology, 19.9: Optical Rotatory Dispersion and Circular Dichroism, Organic Chemistry (OpenStax) (https://chem.libretexts.org/Bookshelves/Organic_Chemistry/Book%3A_Basic_Principles_of_Organic_Chemistry_(Roberts_and_Caserio)/19%3A_More_on_Stereochemistry/19.09%3A_Optical_Rotatory_Dispersion_and_Circular_Dichroism) (accessed, 15-Apr-2024)

Nasipuri, D. *Stereochemistry of Organic Compounds, Principles and Applications*, 3rd Ed. New Age International Publishers, 2018, 160–165, 455–472, 477–478.

Optical Rotation (https://dasher.wustl.edu/bio5325/lectures/lecture-11.pdf) (accessed, 15-Apr-2024)

Physics Videos by Eugene Khutoryansky, https://www.youtube.com/watch?v=8YkfEft4p-w (Polarization of Light: circularly polarized, linearly polarized, unpolarized light) (accessed, 15-Apr-2024)

van Holde, K. E. et al. *Principles of Physical Biochemistry*, 2nd Ed. Pearson, 2006.

7 Stereochemical Reactions

In this chapter, we will look at stereochemical reactions. We already know the importance of enantiopure drugs for treatments. Let us look at some important terminologies that are used (Table 7.1).

7.1 TYPES OF STEREOCHEMICAL REACTIONS

7.1.1 STEREOSELECTIVE REACTION

Stereoselective reaction refers to reactions in which a single reactant gives a mixture of stereoisomeric products, one of them being present in higher quantities. The creation of a new stereocenter may happen either through a non-stereospecific process or may occur due to a non-stereospecific transformation of an already existing stereocenter. The formation of different products in different ratios occurs due to differences in the steric and electronic effects at the reaction site (the mechanistic pathways) or is responsible for the observed selectivity.

7.1.2 STEREOSPECIFIC REACTIONS

The reactions in which only one particular product is formed are called stereospecific reactions. The stereochemistry of the starting material determines the stereochemistry of the product formed. The S_N2 reaction is stereospecific. You already know S_N2 reaction proceeds via rear-side attack, and the product is obtained with an inverted configuration. For example, if the substrate is an (R)-isomer, a backside nucleophilic attack (S_N2) results in an inversion of configuration, with the formation of the (S)-isomer. This concept also applies to substrates that are *cis* as well as *trans*. We will see more examples of this in the later part of this unit. The Mitsunobu reaction is one of the examples of stereospecific reaction. This involves inversion (S_N2 reactions).

7.1.2.1 Inversion: Mitsunobu Reaction

In this reaction, a chiral alcohol is converted to an ester of opposite configuration. The reagents used are triphenylphosphine (PPh_3) and diethyl azo dicarboxylate. Inversion of the

stereochemistry of the –OH functional group and facile change of functionality via a nucleophilic displacement are some of the salient features of this reaction.

7.1.2.2 Retention

Double inversion leads to retention. When there is anchimeric assistance or neighboring group participation, retention of configuration is possible. Formation of dl-2,3-dibromobutane was observed when threo 3-bromo-2-butanol was treated with HBr, and the meso isomer was produced when the erythro pair underwent reaction with HBr.

Since both products are optically inactive, we cannot differentiate them through the differences in optical rotation. The relationship between the meso isomer and dl pair is that of diastereomeric in nature, and we can expect the dibromides to have different boiling points and different refractive indices. Through comparison with authentic reference compounds, we can identify these products.

It is not just one of the enantiomers, but both of them when reacted separately give the dl pair. The reason for this is that the cyclic intermediate present after the leaving group leaves is symmetrical, so the external nucleophile

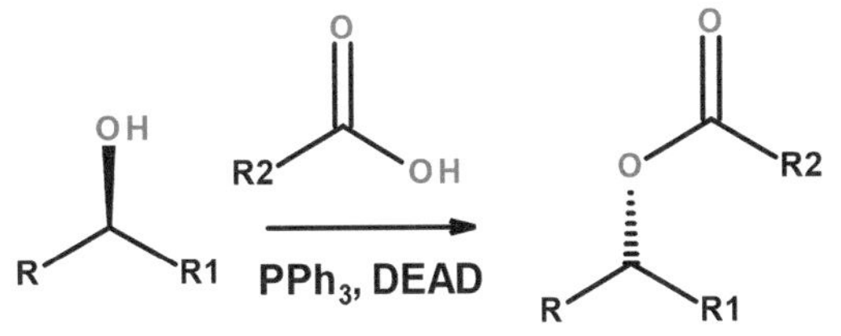

SCHEME 7.1 Mitsunobu reaction.

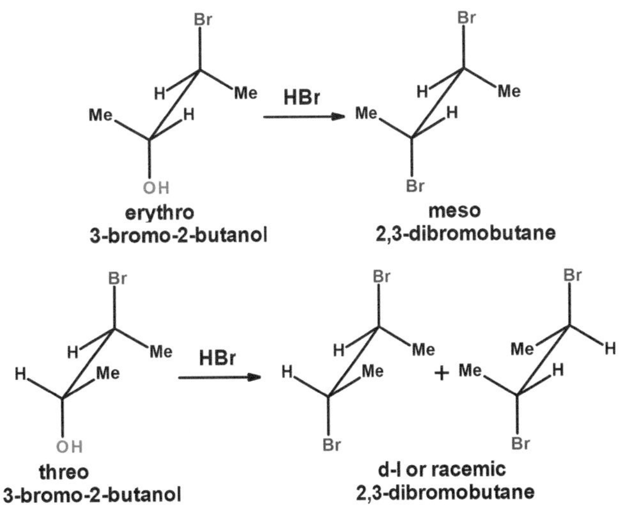

SCHEME 7.2 Stereochemical products.

TABLE 7.1
Stereochemical Terms

Terms	Related to
Stereoselective	Product formation (major and minor compounds)
Stereospecific	Product formation (both isomers giving different products)
Chiral auxiliary	Asymmetric synthesis (induction of stereospecificity or selectivity)
Chiral pool	Asymmetric synthesis (starting material)

DOI: 10.1201/9781032631165-7

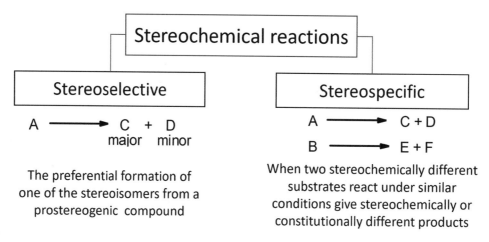

FIGURE 7.1   Stereochemical reactions.

Br⁻ can attack either of the carbon atoms with equal probability leading to a dl pair.

If in a reaction, one stereoisomer of a product is formed preferentially over the other, it is called stereoselective reaction. The reaction does not prevent the formation of two or more stereoisomers, but one predominates (Figure 7.1).

There are two major categories of stereoselective reactions. An **enantioselective reaction** is one in which one enantiomer is formed in preference to the other. From an achiral starting material, by employing a chiral catalyst, an enzyme, or a chiral reagent, it is possible to get one enantiomer in preference to the other.

**HOW TO REMEMBER
STEREOCHEMICAL REACTIONS**

An easy way to remember between stereoselective and stereospecific reactions

When you go to a mall to buy a dress (from Pantaloons) you have different varieties of clothes for men, women, and kids. You have many types to choose from. What you do, you select when you have many options.

When you go to buy slippers, you specifically go to the Bata showroom and get what you want. Here you do not go to Big Bazaar for a choice, and you are very specific to get what you want.

In stereoselective reactions, many products are possible and only one of them will be formed as a major product. In stereospecific reactions, one isomer gives one set of products and another isomer gives another set of products.

A **diastereoselective reaction** is one in which only one of the diastereomers is formed in large excess compared to the other isomers, giving a preferred relative stereochemistry.

7.2   EXAMPLES OF STEREOCHEMICAL REACTIONS

7.2.1   ADDITION OF BROMINE TO ALKENES

Cis or *Z*-but-2-ene gives 2R,3R-isomer and 2S,3S-isomer, and *trans* but-2-ene gives 2S,3R- isomer. One isomer gives one set of isomers, and another compound gives another set of compounds. Look at Scheme 7.3.

Let us first look at the evidences we have.

1. Bromination proceeds with anti-stereochemistry.
2. No rearrangement.
3. The reaction is stereospecific.
4. If water is the solvent, it gets added to the product.

Based on these evidences, we can think of different mechanisms. We rule out radical mechanisms here, because we do not use light or any radical initiators. This cannot be a nucleophilic addition because of electron-electron repulsion between the nucleophile and π electrons. So we can think of electrophilic addition only. One is the carbocation mechanism, and the other is cyclic bromonium ion mechanism. Since there is no rearrangement, carbocation mechanism is highly unlikely.

Let us look at this mechanism in detail. In the reaction mechanism, if we have the initial attack of the bromonium ion produces a carbocation intermediate as given below. Since the carbocation is planar (*sp²* hybridized) now the bromide ion attack the carbocation from the opposite side leading to the product. This is given in the following scheme.

There are a couple of reasons why this is not the actual mechanism:

1. The planar carbocation can be attacked from either above the plane or below the plane.
2. There is also possible C–C single bond rotation of the initial carbocation to give another carbocation.

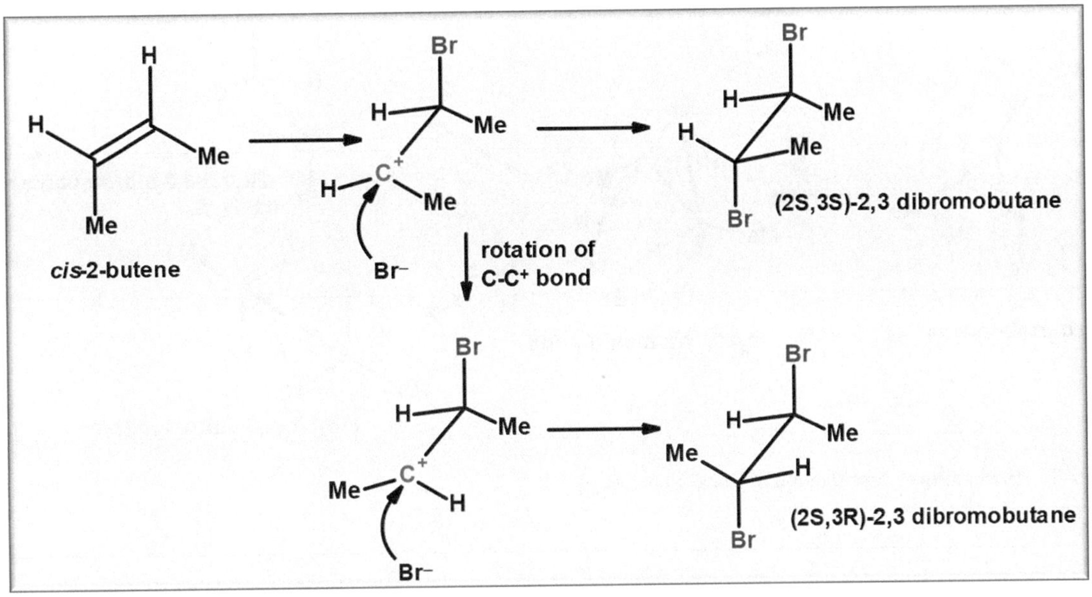

SCHEME 7.3 Mechanism of bromination involving discrete carbocation of cis-2-butene.

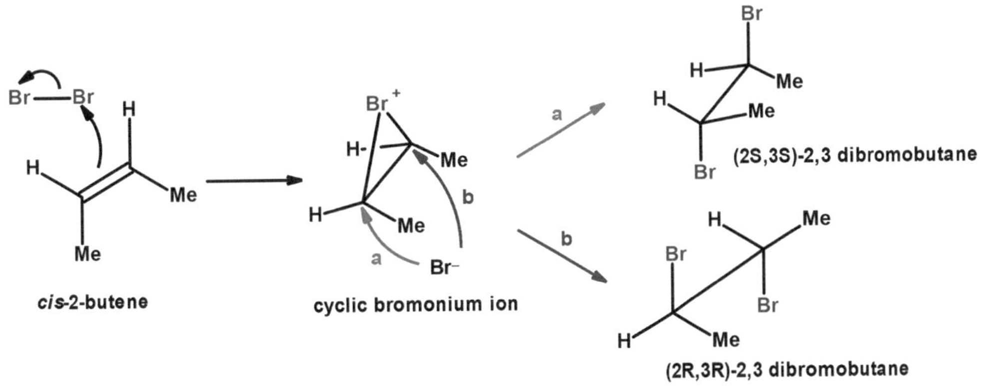

SCHEME 7.4 Mechanism of bromination of *cis*-2-butene.

In other words, an open carbocation mechanism is unlikely to explain the expected product formation. In 1937, to account better for the observed stereochemistry, I. Roberts and G. E. Kimball at Columbia University proposed the following mechanism. Formation of a cyclic bromonium ion intermediate was proposed. This prevents the C–C single bond rotation. The anti-addition now is possible only from the bottom face.

The reaction starts with the addition of bromine to the double bond. The addition can take place from the top face or bottom face of the alkene. Let us consider the attachment of the bromine to the top face. When this happens, the hybridization of the carbon atoms across the double bond also changes from sp^2 to sp^3. In other words, we see a change of structure from planar to tetrahedral. Due to this, the hydrogens and methyl groups are displaced toward the bottom side. The relative positions of the methyl groups do not change. They are still located across from each other,

as they were in the starting olefin. Since the bromine adds across the double bond, we are getting a cyclic bromonium ion intermediate as shown in the above scheme. In the final step, the nucleophilic bromide ion attacks the bromonium ion intermediate from the bottom side. This results in the formation of a new carbon–bromine bond with simultaneous breaking of the old carbon–bromine bond. Since the second attack happens from the face opposite to the first attack it leads to anti-addition. In other words, the first bond is formed from the top face and the second bond is formed from the bottom face. The next question to answer is whether the second attack takes place on the front carbon or at the back carbon. If the attack happens according to the path, (1) we get 2S, 3S-isomer, and if the attack occurs by path, (2) we will get 2R,3R-isomer. We know these two are enantiomers. Since there is no preference for either path (1) or (2), we may get both attacks with equal probability leading to the formation of both the enantiomers in equal

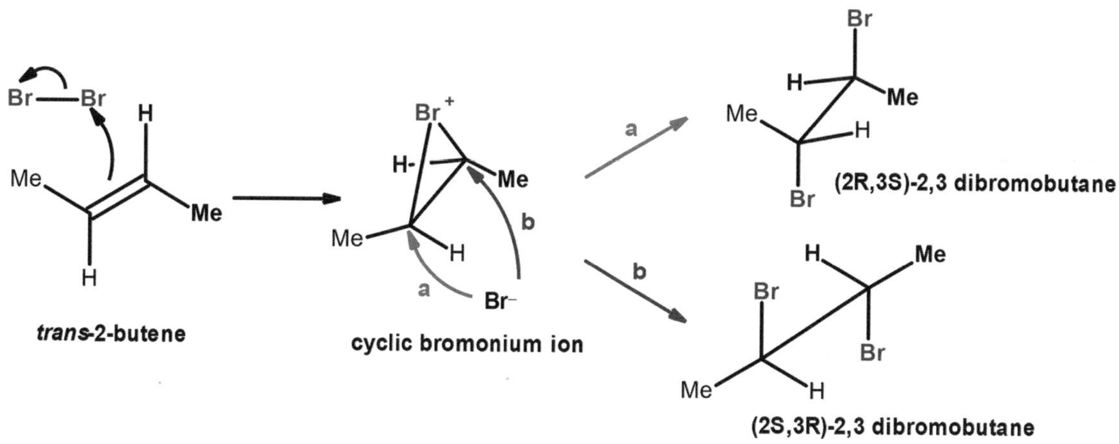

SCHEME 7.5 Mechanism of bromination of trans-2-butene.

amounts, and thus we end up with a racemic modification or dl pair. We can also extend the same set of arguments when the first step is an attack of the bromonium ion happening at the bottom face. We will get the same set of products.

In the same way, the *trans*-but-2-ene gives the meso isomer. The following scheme depicts the product formation.

7.2.2 Epoxidation of Alkene (Syn Addition)

The epoxidation using meta-chloroperoxybenzoic acid (m-CPBA) is a syn addition reaction. In the following example, oxygen is added across the double bond from the same face. For the Z-pent-2-ene isomer here, we see the addition from the less hindered side leading to 2S, 3R-isomer, and the more hindered side attack leads to 2R, 3S-isomer.

In the above example, oxygen is added across the double bond from the same face. For the E-pent-2-ene isomer here, we see the addition from the less hindered side leading to 2S, 3S- isomer, and the more hindered side attack leads to 2R, 3R-isomer. Since both *cis* and *trans* gave one set of isomers (enantiomers), we can say this reaction is an example of stereospecific reaction.

7.2.3 Hydroboration–Oxidation of Alkenes (cis Addition)

In the following example, we will see the hydroboration–oxidation sequence. The conversion of an alkene into an alcohol (hydration of the double bond) is a two-step process. Unlike the earlier example of bromine addition, this process

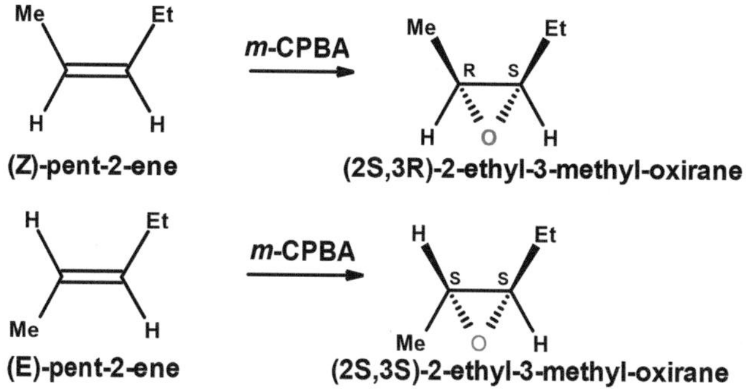

SCHEME 7.6 Mechanism of epoxidation of Z-pent-2-ene.

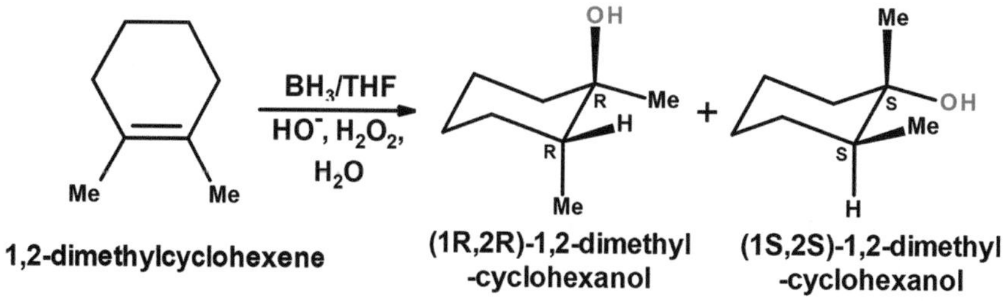

SCHEME 7.7 Hydroboration of 1,2-dimethylcyclohexene.

67

is a *syn* addition of a hydrogen and a hydroxyl group across the double bond (from the same side). Hydroboration–oxidation is occurring according to the anti-Markovnikov principle. The negative part of the addendum, the hydroxide ion ($^-$OH), is getting attached to the less substituted carbon. This reaction is stereospecific. This reaction serves as a complementary regiochemical alternative to other hydration reactions (acid-catalyzed addition and oxymercuration–reduction). Herbert C. Brown received the Nobel Prize in Chemistry in 1979 for his work on boron.

According to the electronegativity principles, the boron–hydrogen bond is polarized toward hydrogen as shown here $B^{\delta+}$—$H^{\delta-}$. Due to this charge polarization, the nucleophilic double bond attacks the electropositive boron atom forming a carbon–boron bond. When there is an unsymmetrical olefin, the least substituted carbon bonds with the boron atom. Since 1,2-dimethylcyclohexene is a symmetrical olefin, we

can write any one of the double-bonded carbons getting bonded to boron.

But no carbocation intermediate is observed. So the stepwise electrophilic addition reaction may not be happening. For this reason, the reaction often is considered to be a four-center concerted addition.

In the stepwise reaction, we can see due to steric crowding, the less substituted carbon bonds with electropositive boron. However, this does not explain the fact that the reactions show no other characteristics of carbocation mechanism. Since the transfer of hydride to carbocation is very fast, we may not be able to see this process stepwise. This could not explain the higher reactivity of alkynes compared to alkenes, which is in stark contrast to normal electrophilic additions of alkenes and alkynes.

The second step of the reaction sequence is the oxidation step where the breaking of the carbon–boron bond happens.

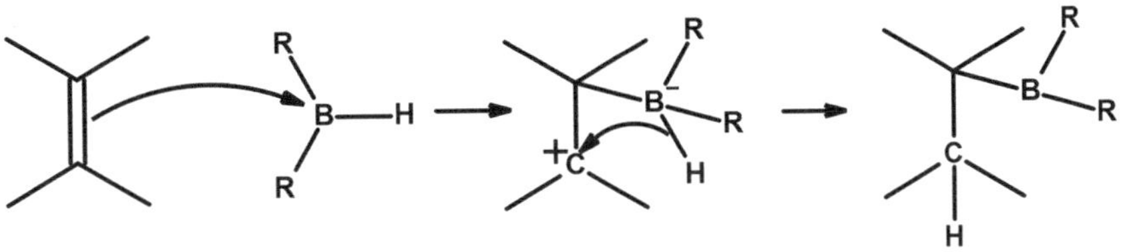

SCHEME 7.8 Hydroboration mechanism of carbon–boron bond.

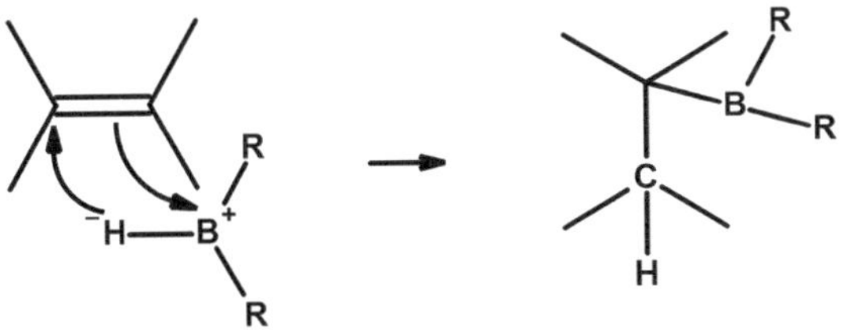

SCHEME 7.9 Hydroboration mechanism four-center concerted addition.

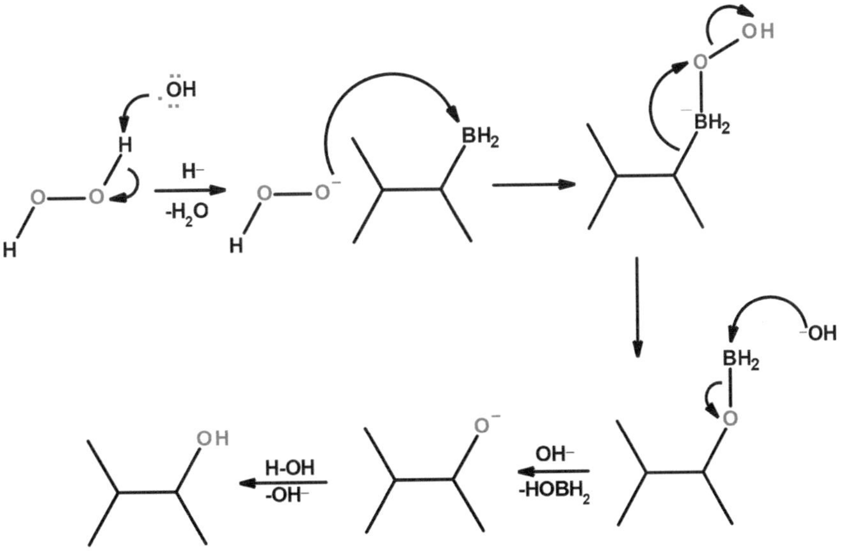

SCHEME 7.10 Hydroboration mechanism oxidation step.

TABLE 7.2

Nature of Addition of Few Reactions

Name of the Reaction	Nature of Addition
Addition of HX (hydrohalogenation)	Anti
Addition of H-OH (hydration)	**Syn** and anti
Addition of HH (hydrogenation)	Metal (**syn**)
Addition of XX (halogenation)	Anti (halonium ion)
Addition of X-OH (halohydrination)	Anti (epoxide opening), (halonium ion opening)
Addition of HO-OH (hydroxylation)	**Syn** (OsO_4), anti (epoxide opening)
Epoxidation	**Syn**
Substitution (S_N2)	**Inversion**
Elimination (E2)	**Anti**

The oxidation reaction involves various steps. The initial step is the formation of the hydroperoxide nucleophile. The deprotonation of hydrogen peroxide by base generates the hydroperoxide nucleophile. In this step, the nucleophilic hydroperoxide anion attacks the boron atom. Subsequently, two changes occur. One is the loss of hydroxide anion, and the other is the migration of substrate unit from boron to oxygen. The hydroxide anion which was liberated in the previous step attacks the rearranged boron (now in the terminus) leading to the breaking of the boron–oxygen (with substrate) bond. Now the protonation of the final intermediate gives the product alcohol.

Table 7.2 is a quick reference for the stereochemical outcome of various addition reactions.

7.3 ACYCLIC AND CYCLIC STEREOSELECTIONS

There are two major types of stereoselections. One involves an acyclic intermediate, and the other involves a cyclic intermediate. Acyclic stereoselection includes reactions of carbonyl and alkenes.

The most challenging aspect of stereoselection is control of the relationship between stereocenters. It is more difficult to achieve in acyclic systems due to the lack of conformational rigidity. It was the pioneering work by Fischer who first reported the stereoselective addition of hydrogen cyanide to aldoses. Scientists are working on identifying the patterns governing acyclic stereocontrol and are trying to model them.

Acyclic stereoselection does not always relate to acyclic systems, because in many cases, acyclic systems undergo stereoselection using highly ordered cyclic transition states, "Acyclic stereocontrol refers to reactions where the reacting moiety is free to undergo rotation relative to a preexisting stereocenter, but adopts a preferred reactive conformation. Reaction can then occur from either one of two diastereotopic faces, as determined by transition state interactions".

The stereoselectivity of nucleophilic addition to a carbonyl adjacent to a stereo center is an example of acyclic stereocontrol (Figure 7.2).

Control of the relationship between two or more stereocenters is readily achieved in cyclic systems in which rigorous conformational constraints are imposed. Some of the cyclic transition states include (1) the relationships between pseudo equatorial and pseudo axial substituents in cyclic Zimmerman–Traxler transition states that determine stereoselectivity in the aldol reaction and (2) the Chelation model.

7.4 ASYMMETRIC INDUCTION MODELS

7.4.1 CRAM'S MODEL

When a prochiral ketone is subjected to nucleophilic addition reaction using organometallic reagents or metal hydride reagents, it can give two diastereomeric products.

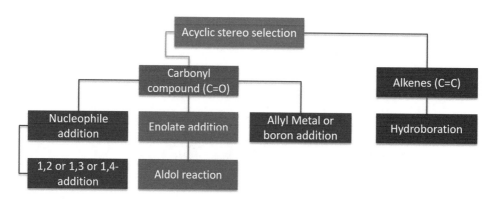

FIGURE 7.2 Acyclic stereocontrol.

Depending on the size of the substituent, we may get one of the isomers in major quantity. Cram's model is an open-chain model. The nucleophile attacks from the side which is opposite to the larger group. In the case of metal-oxygen coordination, the substituent on the metal may be preferentially transferred to the carbonyl carbon. That is from the side adjacent to the smallest substituent.

If you look at the two reactions given below, ketone reduction with lithium aluminum hydride (LiAlH$_4$) gave the threo isomer in large quantity and the Grignard addition to the aldehyde gave the erythro isomer as the major product. In both these reactions, Cram's prediction worked well. Although Cram's model predicts the stereochemical outcome of these reactions successfully, it is very difficult to quantify the steric effect of the substituents.

7.4.2 THE FELKIN AHN MODEL

The Felkin Ahn model invokes the electronic nature of the substituents. In the above reaction, two reactive conformations are involved. They are whether the large group is kept 90° from the carbonyl group or the most electron withdrawing group is placed perpendicular to the carbonyl group. Using the steric repulsion as shown in the examples, the model correctly predicts the formation of the preferred product. In the middle intermediate, there is steric repulsion between the incoming alkyl group and the medium-sized substituent, and in the last one, there is steric repulsion between the larger substituent and the incoming alkyl group. In the first one, the repulsions are minimal. We see more of such reactions in the carbonyl addition reactions section (Figure 7.3).

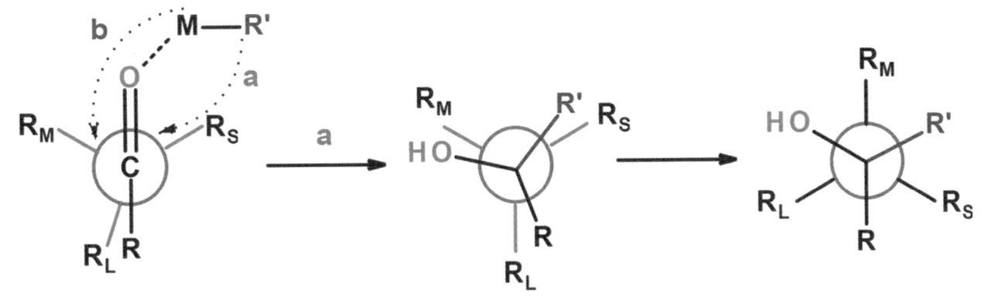

SCHEME 7.11 Cram's model.

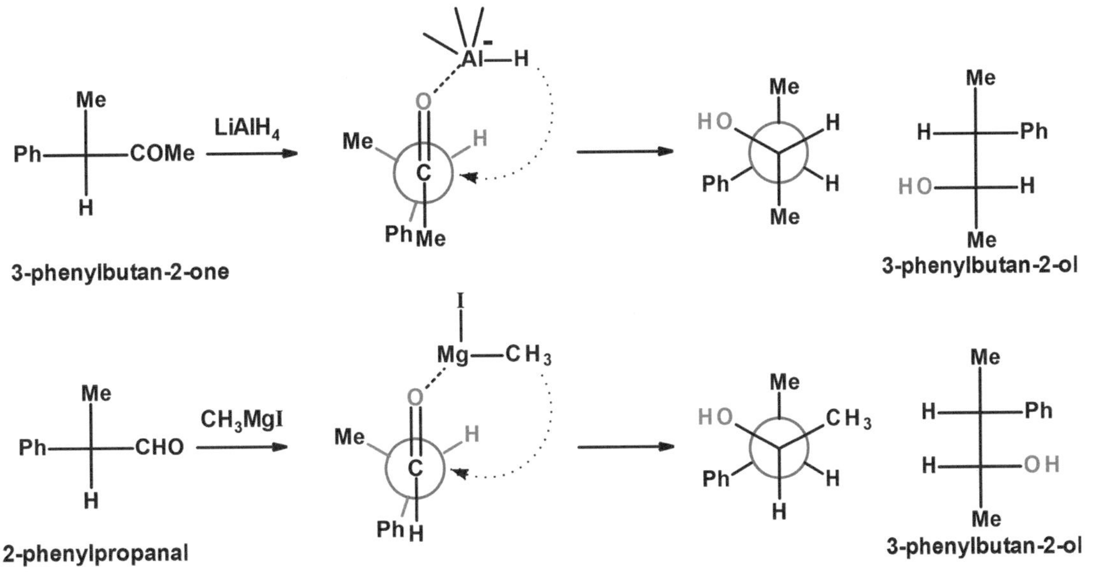

SCHEME 7.12 Examples of Cram's model.

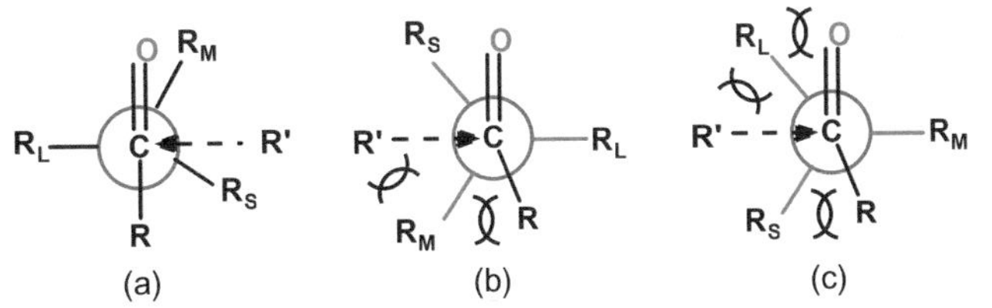

FIGURE 7.3 Felkin Ahn model.

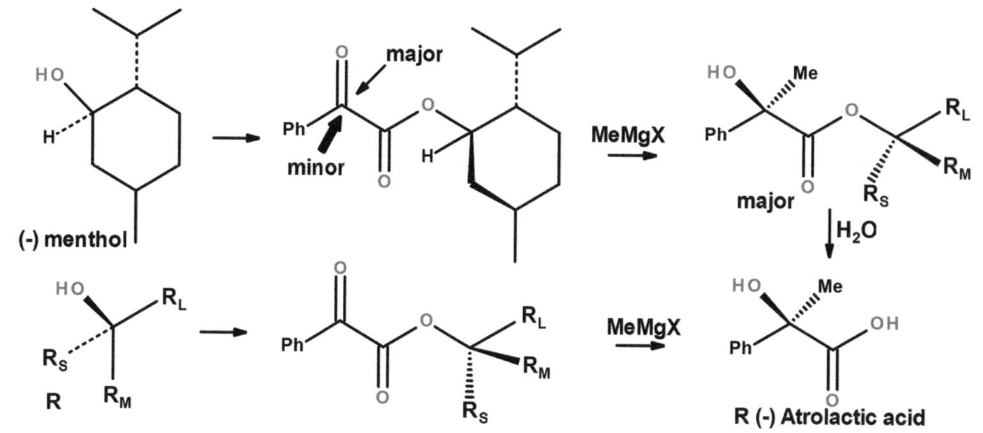

SCHEME 7.13 Prelog rule.

7.4.3 PRELOG RULE

Prelog rule correlates the configuration of chiral alcohols with α-hydroxy acids. In the sequence of reactions, a keto acid is converted into its ester using a chiral alcohol. Now this ester is subjected to Grignard's reaction on the ketone. Hydrolysis of the resulting hydroxy ester produces Alpha hydroxy acids. Although the chiral induction in this reaction is moderate, still it can help ascertain the configuration.

7.5 QUESTIONS

1. Which of the product(s) will be formed in the following reaction? (Figure 7.4)

2. In the following reaction, both *syn* addition and *anti*-addition occur and give the following products. Comment about the reaction mechanism (Figure 7.5).
3. Why in the absence of any chiral environment (catalyst, reagent, ligand, and solvent), achiral compounds cannot be converted to chiral compounds?
4. Why free radical addition of halogens to alkanes lead to racemic modification?
5. What is the relationship between transition state energy and stereoselectivity of reactions?

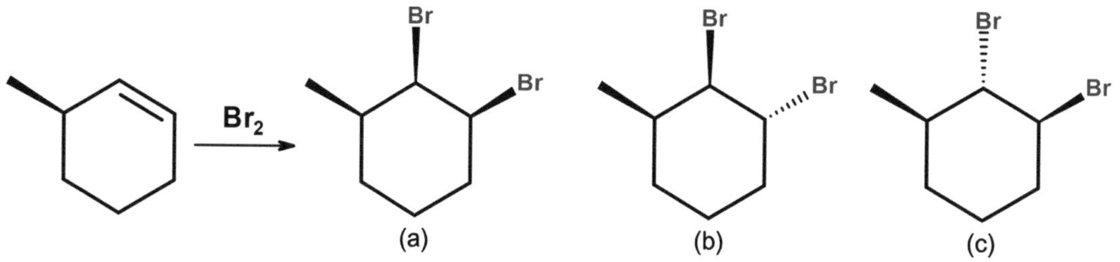

FIGURE 7.4 Question 1.

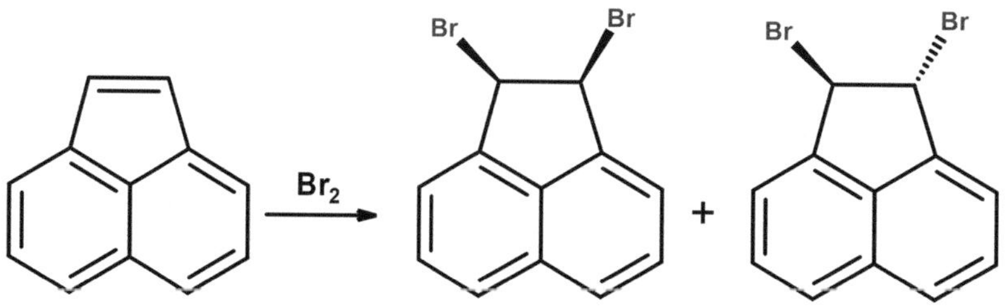

FIGURE 7.5 Question 2.

BIBLIOGRAPHY

Bode, ETH Zürich, Key Concepts in Stereoselective Synthesis, 2015, (https://www.ethz.ch/content/dam/ethz/special-interest/chab/organic-chemistry/bode-group-dam/documents/open-source-lecture-notes/key-concepts-in-stereoselective-synthesis-2015.pdf) (accessed, 15-Apr-2024)

Farmer, S. 8.7: Oxidation of Alkenes - Epoxidation and Hydroxylation in Organic Chemistry (OpenStax) (https://chem.libretexts.org/Bookshelves/Organic_Chemistry/Organic_Chemistry_(Morsch_et_al.)/08%3A_Alkenes-_Reactions_and_Synthesis/8.07%3A_Oxidation_of_Alkenes_-_Epoxidation_and_Hydroxylation) (accessed, 15-Apr-2024).

John D. Robert and Marjorie C. Caserio, 11.6: Addition of Boron Hydrides to Alkenes. Organoboranes, in Organic Chemistry (OpenStax) (https://chem.libretexts.org/Bookshelves/Organic_Chemistry/Book%3A_Basic_Principles_of_Organic_Chemistry_(Roberts_and_Caserio)/11%3A_Alkenes_and_Alkynes_II_-_Oxidation_and_Reduction_Reactions._Acidity_of_Alkynes/11.6%3A_Addition_of_Boron_Hydrides_to_Alkenes._Organoboranes) (accessed, 15-Apr-2024)

Kalsi, P. S. Stereochemistry: Conformation and Mechanism, 9th Ed. New Age International (P) Ltd., 2017, 412–456.

Morrison, R. T., & Boyd, R. N. Organic Chemistry, Chapter 7, 6th Ed. Prentice-Hall of India Pvt-Ltd, 2002, 225–235, 239–245.

O'Brien, A. G. (2011). Tetrahedron report number: 957, Recent advances in acyclic stereocontrol. Tetrahedron, 67, 9639–9667.

Smith, M. B., & March, J. March's Advanced Organic Chemistry Reactions, Mechanisms and Structure, 5th Ed. John Wiley & Sons, Inc., 2001, 147–150, 166–167, 404–407.

8 Aromaticity and Polyaromatic Compounds

In this chapter, the topics that will be covered include the criteria for aromaticity, ring current, and Musulin–Frost diagrams for cyclic systems.

You are already familiar with aromatic compounds. In the earlier part of the nineteenth century, the term aromatic compounds was given to molecules that possessed aroma (odor). They are more stable than their linear counterparts, their bond lengths are intermediate between the C-C (single bond) and the C=C (double bond), they have π-electron ring current and they undergo substitution rather than addition reactions.

When we talk about involvement in life processes, we have both beneficial and toxic effects associated with aromatic compounds. 2,4-Diaminotoluene and 2,4-diaminoethylbenzene are examples of mutagenic compounds. Genotoxic compounds include benzo(a)pyrene and 20-methylcholanthrene, and teratogenic compounds include acridines and 9-methylacridine. Polycyclic aromatic hydrocarbons (PAHs) are formed from anthropogenic activities (Figure 8.1).

8.1 IMPORTANT FEATURES OF AROMATIC COMPOUNDS

When we talk about energy, aromatic compounds have greater thermodynamic stability than their analogous isoelectronic linear polyenes (responsible for resonance energies). With respect to magnetism, aromatic compounds have ring current. The planar structure is responsible for delocalization. In addition to that aromatic compounds have uniform bond length and bond angle. Finally for the reactivity aspect, aromatic compounds undergo substitution rather than addition reactions.

We will now look at what are the criteria that a compound should possess to show aromatic characters. There are four common rules.

8.2 CRITERIA FOR AROMATICITY

1. Conjugation: Each atom must be either sp^2 or sp hybridized
2. Cyclic: All the atoms in the ring must be involved in the π system (*i.e.* no sp^3 atoms)
3. Planar: Good overlap/interaction between the adjacent "p" orbitals
4. Follow Hückel's rule of $(4n+2)$ π electrons

Let me talk about something which you are very familiar with. Imagine you are having a tasty meal or experiencing the beautiful smell of the roses in the garden or the lovely breeze you encounter after the drizzle. You can feel it, but can you measure it? Aromaticity per se is exactly similar to our feeling.

Aromaticity is not a property of a molecule that we can measure but a concept.

In this unit, we will try to understand aromaticity in terms of energy and magnetic criteria. When we talk about energy or stabilization, we are referring to ground-state thermodynamic properties. When we talk about reactivity criteria for aromaticity, we are talking not only about the ground state but also about the transition state and other intermediates that may be formed.

When we study about Clar rule, polyacenes the correlation between reactivity criteria and aromaticity will be discussed.

8.3 ENERGETIC MEASURE OF AROMATICITY

Let us focus on some energy studies. We will start with the unusually low reactivity of benzene toward addition. Although halogen addition across the C=C is difficult, benzene can be forced to undergo an addition reaction with hydrogen that is a reduction to give cyclohexane.

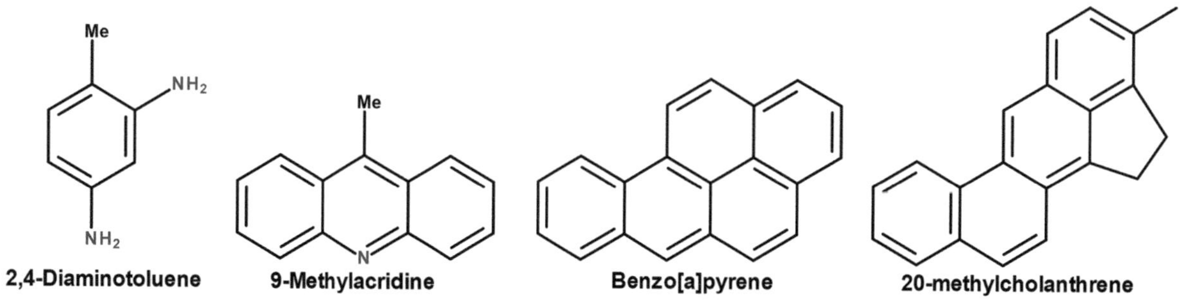

FIGURE 8.1 Examples of polycyclic aromatic hydrocarbons.

DOI: 10.1201/9781032631165-8

We can study the heats of reactions like combustion and hydrogenation to derive an empirical value for resonance energy. Let us compare catalytic reduction and the energy required for the same. Since benzene on complete reduction gives cyclohexane, we will have the energy of cyclohexane as our reference. We will do a comparison in energy associated with the hydrogenation of various cyclohexane systems.

Now you will have an important question in your mind, why do we need to relate reduction to stability?

Let me answer. You Think like this. Resonance stabilization occurs due to the movement of π electrons and this is called delocalization. Now, to study resonance stabilization, we should know about the movement of π electrons or delocalization. If you stop the movement of π electrons or delocalization, you can find out resonance. How to stop the movement of π electrons, you already know you cannot confine electrons between a single π bond in a conjugated system, the best way to do this is to remove the π electrons, which is actually reducing the bond. In other words, we can say, heat of hydrogenation of alkenes that is reduction is a measure of the stability of carbon–carbon double bonds. We can find out this energy by two methods. One is *theoretical calculation* and the second one is *experimental observation*. If both values are the same, then there is no resonance stabilization, if they are different then the **difference** may be due to **resonance stabilization** (Figure 8.2).

Since benzene is a cyclic six carbon unit, we will take cyclic six carbon compounds having one, two, or three double bonds. Cyclohexene which has one double bond requires –28.6 kcal/mol of energy for its reduction. When we go to cyclohexadiene where the double bonds are not in conjugation requires –57.4 kcal/mol of energy for complete reduction. On the other hand, cyclohexadiene where the double bonds are in conjugation requires –55.4 kcal/mol of energy for complete reduction. In this case, the difference between experimental values and theoretical values is called the ***resonance stabilization energy***, which in this case is –1.8 kcal/mol. When we take benzene, which is having three double bonds the predicted energy for the reduction is –85.8 kcal/mol. However, the experimental value is –49.8 kcal/mol. So, the difference –36 kcal/mol is the resonance stabilization energy.

8.4 AROMATIC RING CURRENT

A compound with the ability to sustain an induced ring current is called diatropic. Let us see how this affects the protons in the aromatic ring. In the presence of an external magnetic field the electrons in the aromatic ring (as in an NMR instrument), experiencing diamagnetic ring current, induces a field of its own. From the following figure, we can see the protons are falling on the induced field created by the diamagnetic ring current. Due to this, those protons experience some additional magnetic field than the applied external magnetic field. This results in the protons getting moved downfield (to higher δ) compared with where they would be if electron density were the only factor. This can

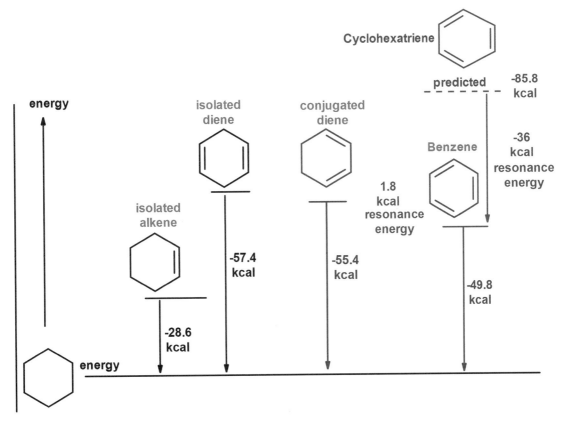

FIGURE 8.2 Potential energies of hydrogenation of various cyclohexane systems.

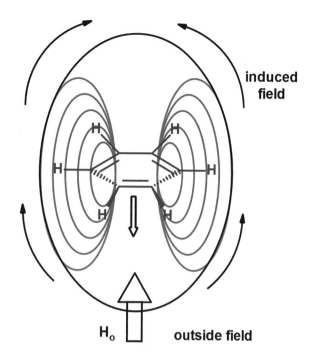

FIGURE 8.3 Depiction of the magnetic field felt by aromatic protons.

be clearly seen in the proton resonances of alkene and aromatic protons. Ordinary alkene hydrogen resonate at 5–6 δ, whereas the hydrogens of benzene rings are resonating around 7–8 δ (Figure 8.3).

There are two instances where the protons will be diatropic.

As mentioned earlier, the ring protons experience more magnetic field (deshielding, because the induced magnetic field and the external magnetic field are in the same direction, an additive effect). This leads to a downfield shift of the protons attached to the ring compared with the normal alkene region, we can conclude that the molecule is diatropic and hence aromatic.

The protons above and within the ring experience less magnetic field (less shielding, because the induced magnetic field and the external magnetic field are in the opposite direction, a subtractive effect). In addition, if the compound has protons above or within the ring, then if the compound is diatropic, these will be shifted upfield. We will see this effect in other annulenes.

The following Figure 8.4 represents the various bonding and anti-bonding arrangements before and after overlap. There are zero nodes in the lowest energy bonding orbital and three nodes in the highest energy anti-bonding orbital.

8.5 MUSULIN–FROST CIRCLE

What is Musulin–Frost circle? It is a simple representation of bonding and anti-bonding MOs of cyclic π systems. This elegantly reproduces the mathematical solution to the wave equation in the form of a geometric representation. *A circle… is inscribed with a polygon with one vertex pointing down; the vertices represent energy levels with the appropriate energies. Depending on the number of π electrons in the system we can easily arrange them using this simple representation* (Figure 8.5).

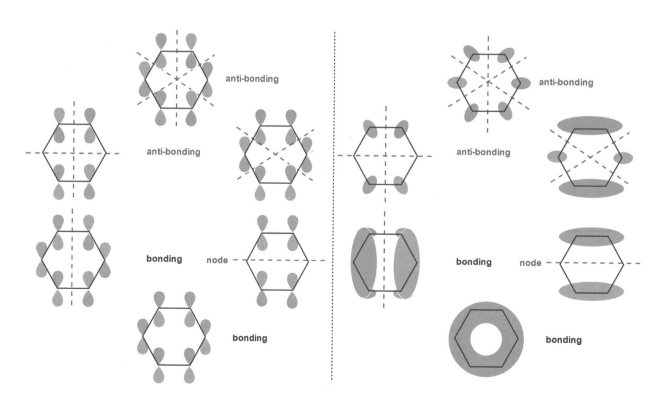

FIGURE 8.4 Various bonding and anti-bonding arrangements before and after overlap for benzene.

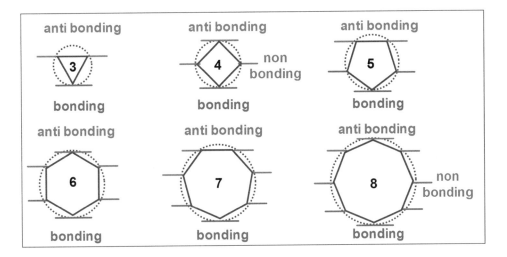

FIGURE 8.5 Depiction of various Musulin–Frost circles for cyclic π systems.

We have studied resonance stabilization energies. There is one problem. The cyclohexatriene is a hypothetical molecule and does not exist. So the resonance stabilization we have calculated may not be accurate. So to overcome this problem, we can resort to other types of calculations.

8.6 DEVIATIONS OF AROMATIC CRITERIA

There are a few deviations of aromatic criteria for strained and other complex systems. Aromatic stabilization energies (ASEs) of strained and more complicated systems are difficult to evaluate. In the case of Annulenes (aromatic and anti-aromatic), the ring current effects cannot explain the upfield and downfield chemical shifts of protons located inside and outside of the aromatic rings.

The six inner hydrogens of aromatic [18]-annulene, resonate at −3.0 ppm vs δ=9.28 for the outer protons. This relationship is inverted dramatically in the anti-aromatic [18] annulene dianion, $C_{18}H_{18}^{2-}$, where δ=20.8 and 29.5 (inner) vs. −1.1 (outer).

8.7 NON-AROMATIC AND ANTI-AROMATIC COMPOUNDS

Non-aromatic compounds

- Lack of planarity or disruption of delocalization
- They may contain 4n or (4n+2) π electrons

Anti-aromatic compounds are

- Planar
- Cyclic systems
- Conjugated systems with an even number of pairs of electrons (4n π electrons)
- Highly unstable (destabilized by a closed loop of electrons)

In contrast to aromatic systems, paramagnetic ring current is observed in anti-aromatic systems. Due to this, protons on the outside of the ring experience a lesser applied magnetic field compared with the applied magnetic field hence they are shifted upfield while the inner protons are shifted downfield. This is in sharp contrast to a diamagnetic ring current, which causes shifts in the opposite directions in aromatic systems. Compounds that sustain a paramagnetic ring current are called paratropic; and are prevalent in four- and eight-electron systems.

The following Table 8.1 gives a glimpse of a comparison between aromatic and anti-aromatic protons (Table 8.2).

Let us take one of the examples with 4n π electrons. We will take the example of cyclooctatetraene (COT) (8 π electrons). Using Musulin–Frost circle we can assign the 8 π electrons in various orbitals. It is shown below (Figure 8.6).

Out of the 8 π electrons, 6 π electrons are in the bonding orbitals and two unpaired electrons are in the non-bonding orbitals. We already know, molecules with unpaired electrons are unstable and reactive. So we can conclude that cyclooctatetraene with 8 π electrons is unstable and hence, is not aromatic. Since it is not aromatic, it did not gain in stability by being planar. So the molecule in fact exists as tub shape rather than a planar.

TABLE 8.1
Comparison between Aromatic and Anti-Aromatic Protons

	Type	Outside Ring Protons	Above or Inside Protons	Ring Current
Aromatic	Diatropic	Downfield	Upfield	Diamagnetic
Anti-aromatic	Paratropic	Upfield	Downfield	Paramagnetic

TABLE 8.2

Comparison of Cyclic π Electron Systems

System	π e⁻	Cation	Neutral	Anion
Cyclopropene	2	**Aromatic (+)**	Non-aromatic	Anti-aromatic
Cyclobutadiene	4	**Aromatic (di +)ᵃ**	Anti-aromatic	**Aromatic (di −)ᵇ**
Cyclopentadiene	4	Anti-aromatic	Non-aromatic	**Aromatic**
Benzene	6		**Aromatic**	**Aromatic**ᶜ
Cycloheptatriene	6	**Aromatic**	Non-aromatic	Anti-aromatic
Cyclooctatetraene	8		Non-aromatic (non-planar)	

ᵃ4 e⁻

ᵇ6 e⁻

ᶜ8 e⁻

The two extra electrons are in sp² orbital perpendicular to p orbital

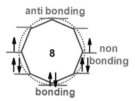

FIGURE 8.6   Musulin–Frost diagram for cyclooctatetraene.

8.8   POLYCYCLIC AROMATIC HYDROCARBONS

Let us study about a different class of organic compounds called PAHs, Clar rule, polyacenes, Baird's rule, and annulene.

Three or more fused unsubstituted benzene rings are called PAHs. These compounds are mostly hazardous, but some of them are useful to mankind.

Let us look at some of the fusions that are commonly present in PAH.

8.8.1   ORTHO FUSION (LINEAR AND ANGULAR)

"Polycyclic compounds in which two rings have two, and only two, atoms in common. Such compounds have n common faces and 2n common atoms." Two adjacent rings that have only **two atoms** and **one bond** in common are said to be *ortho*-fused.

8.8.2   ORTHO AND PERI FUSED (CLUSTER)

Polycyclic compounds in which one ring contains two, and only two, atoms in common with each of two or more rings of a contiguous series of rings. Such compounds have n common faces and less than 2n common atoms (Figure 8.7).

For your ease of visualization, in the above structure, the common atoms between two rings are highlighted in different colors.

8.8.3   USEFUL APPLICATIONS OF PAH

PAH is used mainly in the electronics industry, especially in the field of molecular electronics. Due to the ability to tune their properties like adjusting the bandgap and supramolecular assemblies, these compounds find application in the molecular engineering field.

In fact, the field of molecular materials got an excellent boost, thanks to the isolation of graphene by Geim and Novoselov (Nobel Prize winners, 2010). Graphene has excellent electronic, thermal, and mechanical properties and is used in energy storage to optoelectronics.

8.9   POLYACENES

These compounds are similar to polycyclic aromatic hydrocarbons. They are obtained by the linear annelation of

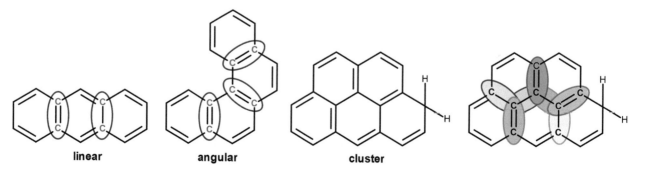

linear   angular   cluster

FIGURE 8.7   Some of the fusions that are commonly present in PAH.

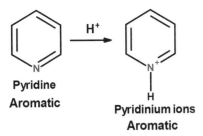

SCHEME 8.1 Protonation of pyridine.

benzene rings, general formula $C_{4n+2}H_{2n+4}$. They are highly reactive benzenoid hydrocarbons, which are aromatic. The aromatic stability gradually decreases and the reactivity increases rapidly with the size. They have distinctive structural geometries with useful electronic features.

8.9.1 Unique Properties of Polyacenes

They are paramagnetic in the ground state. Some of the higher members are anti-ferromagnetic (AFM) (nonacene). They possess a diradical configuration for open-shell singlet state when the number of rings is more than six.

8.10 6-MEMBERED AROMATIC HETEROCYCLES

Let us now look into some heteroaromatic systems. The first one we can consider is nitrogen. Pyridine is aromatic with three π bonds or 6 π electrons. Although nitrogen has its lone pair of electrons, they are present in the sp^2 orbital that is perpendicular to the π system (which forms the aromatic system). This is confirmed by the resonance energy of pyridine and its reactivity. Similar to aromatic systems, pyridine undergoes substitution and not addition reactions.

Unlike benzene where none of the carbon has a lone pair of electrons, pyridine has a nitrogen with a lone pair. Due to this, pyridine acts like a mild base. However, derivatives such as pyridinium ion are still aromatic (Figure 8.8).

Where oxygen or sulfur is the heteroatom we have 4H-pyran or 4H-thiopyran, they have two lone pairs on the heteroatom. Similar to pyridine, the lone pair of electrons are not involved in π conjugation and hence it is not aromatic. Only when the hetero atom forms the cation, as shown for pyrylium ion, the 6 π electrons are present in the ring system and it becomes aromatic. Thus, pyran is not aromatic, but the pyrylium ion is (Figure 8.9).

8.11 OTHER MONOCYCLIC RING SYSTEMS (6 π e⁻)

Here we see some examples of monocyclic systems. we can compare benzene, pyrrole, furan, and thiophene. Benzene is more stable compared to five-membered heterocycles (Figure 8.10).

Cyclopentadienyl anion and tropylium cation both have 6 π electrons and hence they are aromatic.

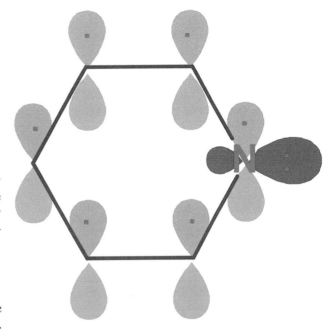

FIGURE 8.8 Depiction of lone pair of electrons on pyridine nitrogen.

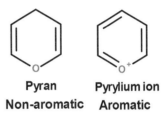

FIGURE 8.9 Structure for pyran and pyrylium ion.

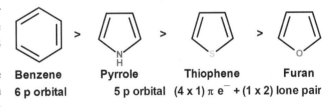

FIGURE 8.10 Examples of five-membered heterocycles.

8.12 BICYCLIC RING SYSTEMS

Here we can see 8, 10, and 12 π electron systems. Ten π electrons follow Hückel's rule and hence it is aromatic, whereas the other two systems are not (Figure 8.11).

8.13 ANNULENE

Annulenes are compounds that are unsubstituted, monocyclic, and with alternating single and double bonds. According to this definition, we can say benzene is [6] annulene and cyclooctatetraene is [8] annulene. We already knew benzene which follows $(4n+2)$ π electrons is aromatic

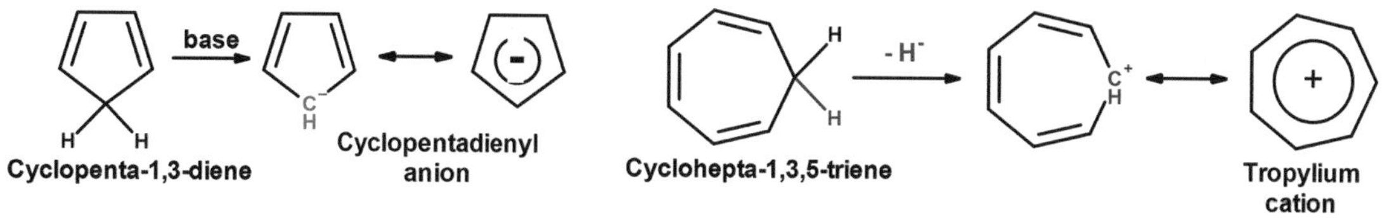

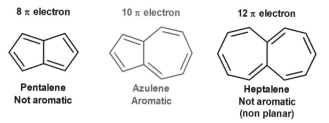

SCHEME 8.2 Formation of cyclopentadienyl and tropylium cations.

8 π electron   **10 π electron**   **12 π electron**

Pentalene   **Azulene**   **Heptalene**
Not aromatic   **Aromatic**   **Not aromatic**
                                    (non planar)

FIGURE 8.11 Examples of bicyclic ring systems.

and cyclooctatetraene which follows 4n-π electrons is not aromatic.

Let us look at the next higher annulenes, [10] annulene. There are three geometrically possible isomers of [10] annulene: the all-*cis*, the mono-*trans*, and the *cis–trans–cis–cis–trans*. If Hückel's rule applies, they should be planar. But it is far from obvious that the molecules would adopt a planar shape, since they must overcome considerable strain to do so. For a regular decagon, the angles would have to be 144°, considerably larger than the 120° required for sp^2 hybridization. Some of this strain would also be present in the twist structure, but this kind of strain is eliminated in naphthalene-like structure since all the angles are 120°. However, it was pointed out by Mislow that the hydrogens in the one and six positions should interfere with each other and force the molecule out of planarity (Figure 8.12).

Next, we will study about 12, 14, and 16 annulenes. 12-annulene has 4n π electrons, 14- annulene has (4n+2) π electrons and 16- annulene has 4n π electrons. Accordingly, 12- annulene is not aromatic, 14- annulene is aromatic, and 16- annulene is not aromatic. The three inner protons of [12] annulene resonate at δ 7.8 ppm and the outer protons resonate at δ 5.9 ppm. The three inner protons of [14]

annulene resonate at δ 0.0 ppm and the outer protons resonate at δ 7.6 ppm. The three inner protons of [16] annulene resonate at δ 10.4 ppm and the outer protons resonate at δ 5.4 ppm (Figure 8.13).

Here we see some more examples of annulenes. There are clearly two sets of protons, inner and outer. They resonate at different frequencies.

8.14   HOMOAROMATIC COMPOUNDS

When cyclooctatetraene is protonated using concentrated H_2SO_4, one of the double bonds is protonated and we get homotropylium cation. In this species, an aromatic sextet is spread over seven carbons, as in the tropylium cation. The eighth carbon is an sp^3 carbon and hence this cannot take part in the aromaticity. The presence of diatropic ring current can be deduced from the NMR spectra. The proton H_b is found at δ 0.3; H_a at δ 5.1; H1 and H7 resonate at δ 6.4, and H2–H6 at δ 8.5. The homoaromatic compound can be defined as a compound that contains one or more sp^3-hybridized carbon atoms in an otherwise conjugated cyclic system. For the orbitals to overlap most effectively, the sp^3 atoms are forced to lie almost vertically above the plane of the aromatic atoms. As shown in the figure, H_b is directly above the aromatic sextet, and so is shifted far upfield in the NMR. All homoaromatic compounds so far discovered are ions.

8.15   BAIRD'S RULE

Although **Hückel's rule** applies to a variety of benzenoid systems, Clar rule was introduced to polycyclic systems. But all these rules deal with ground-state singlet species only. There is no rule to explain the stability or reactivity of

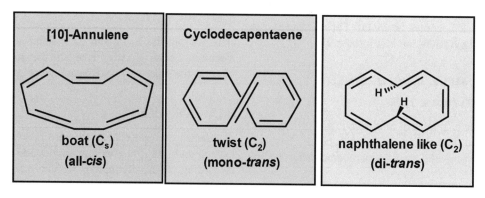

FIGURE 8.12 Possible isomers of [10] annulene.

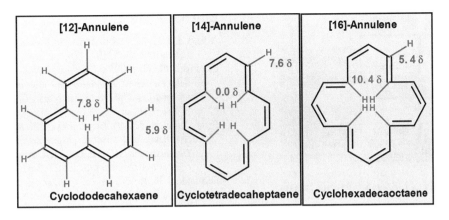

FIGURE 8.13 NMR resonances of 12, 14, and 16 annulenes.

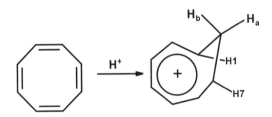

SCHEME 8.3 Formation of homoaromatic compound.

excited systems. Baird extended the aromaticity concepts to the lowest triplet states.

Baird's rule estimates whether the **lowest triplet state** of a planar cyclic structure will have aromatic properties or not. Baird's rule looks at the **excited state**, where the electron-count patterns for aromaticity and antiaromaticity are reversed. A 4n π electron count makes a ring system anti-aromatic in the ground state by **Hückel's rule**, whereas the same 4n π electrons in the triplet state will be aromatic.

Using this rule, we can predict and explain the photochemical reactions of benzenoid systems.

8.16 MÖBIUS AROMATICITY

The concept of "Möbius aromaticity" was conceived by Helbronner in 1964 when he suggested that large cyclic 14 annulenes might be stabilized if the *p*-orbitals were twisted gradually around a Möbius strip. This concept is illustrated by the diagrams labeled Hückel, which is a destabilized 4n system, in contrast to the Möbius model, which is a stabilized 4n system. Zimmerman generalized this idea and applied the "Hückel- Möbius concept" to the analysis of ground-state systems, such as barrelene (Figures 8.14).

8.16.1 CRITERIA FOR AROMATICITY

The following Figure 8.15 depicts the ground-state and triplet-state aromaticity criteria for both aromantic and anti-aromatic systems (Figure 8.15).

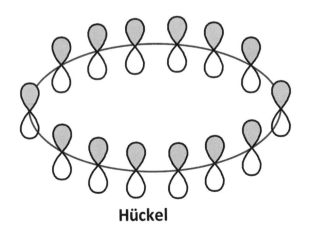

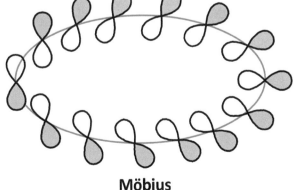

Hückel **Möbius**

FIGURE 8.14 Hückel and Möbius cyclic 14 annulenes '*p*' orbitals.

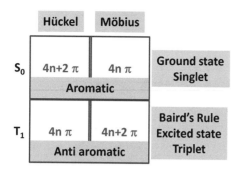

FIGURE 8.15 Hückel and Möbius aromaticity criteria.

8.17 QUESTIONS

1. For the following four compounds identify which of them belong to (a) aromatic (b) non-aromatic (c) anti-aromatic.
2. In the above four compounds, how many of them contain a conjugated π electron system?
3. What are the limitations of aromatic stabilization energies?
4. Why PAH usage should be controlled?
5. Why there are more than one Clar structure in many polycyclic aromatic compounds?
6. Whether the presence of electron-donating groups in the PAH will affect the Clar structure?
7. Is it possible to predict the approximate chemical field shift of inner and outer protons of 10–18 Annulenes?

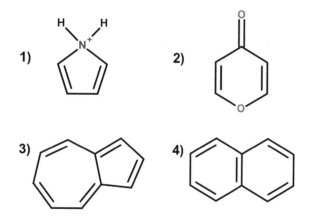

FIGURE 8.16 Question 1.

BIBLIOGRAPHY

Aihara, J.-I. (1982). Aromaticity and diatropicity. *Pure and Applied Chemistry*, 54, 1115–1128.

Cyranski, M. K. (2005). Energetic aspects of cyclic pi-electron delocalization: evaluation of the methods of estimating aromatic stabilization energies. *Chemical Review*, 105, 3773–3811.

Feng, X., et al. (2009). Large polycyclic aromatic hydrocarbons: Synthesis and discotic organization. *Pure and Applied Chemistry*, 81(12), 2203–2224. https://publications.iupac.org/pac/pdf/2009/pdf/8112x2203.pdf

Fernandes, P. R. N., et al. (2009). Evaluation of polycyclic aromatic hydrocarbons in asphalt binder using matrix solid-phase dispersion and gas chromatography. *Journal of Chromatographic Science*, 47, 789.

Herndon, W. C., & Ellzey, M. L. (1974). Resonance theory. V. Resonance energies of benzenoid and nonbenzenoid π systems. *Journal of the American Chemical Society*, 96(21), 6631–6642. https://doi.org/10.1021/ja00828a015

Ian Hunt, University of Calgary, Chapter 11: Arenes and Aromaticity (https://www.chem.ucalgary.ca/courses/350/Carey5th/Ch11/ch11-qu.html) (accessed, 15-Apr-2024)

ICHI AIHARA resonance energy and experimental stability for many compounds (8-10) (https://pdfs.semanticscholar.org/4f86/fbf6588ec3d0e9199d2d66235a9ccb6ddbc7.pdf) (accessed, 15-Apr-2024)

Lawal, A. T. (2017). Polycyclic aromatic hydrocarbons. A review. *Cogent Environmental Science*, 3, 1339841. https://doi.org/10.1080/23311843.2017.1339841

Nikolić, S., Trinajstić, N., & Klein, D. J. (1990). The conjugated-circuit model. *Computers & Chemistry*, 14(4), 313–322. https://doi.org/10.1016/0097-8485(90)80038-4

Polycyclic Aromatic Hydrocarbons (PAHs), CDC Environmental health, Nov. 2009, (https://www.epa.gov/sites/production/files/2014-03/documents/pahs_factsheet_cdc_2013.pdf) (accessed, 15-Apr-2024)

Randić, M. (1976). Conjugated circuits and resonance energies of benzenoid hydrocarbons. *Chemical Physics Letters*, 38(1), 68–70. https://doi.org/10.1016/0009-2614(76)80257-6

Solà, M. (2013). Forty years of Clar's aromatic π-sextet rule. *Frontiers in Chemistry*, 1(22), 22. https://doi.org/10.3389/fchem.2013.00022

Szűcs, R., Bouit, P.-A., Nyulászi, L., & Hissler, M. (2017). P-containing polycyclic aromatic hydrocarbons. *ChemPhysChem*, 18(19), 2618–2630. https://doi.org/10.1002/cphc.201700438.

Toxicological Profile for Polycyclic Aromatic Hydrocarbons, U.S. Department of Health and Human Services, Aug. 1995 (https://www.atsdr.cdc.gov/toxprofiles/tp69.pdf) (accessed, 15-Apr-2024)

9 Aromatic Electrophilic and Nucleophilic Substitutions

In this chapter, we will see a brief overview of aromatic electrophilic substitution reactions. We will divide this into three major parts. What, which and how? In "**What**" category, we will study the **product and its ease of formation**. That is the study of **thermodynamics and kinetic control of the reaction**. Under "**Which**," we will study **the factors that affect product formation,** and in "**how**" category, we will see how the products are formed, that is, the **type of intermediates**.

9.1 TYPES OF ELECTROPHILIC SUBSTITUTIONS

9.1.1 CHARGE TRANSFER OR π COMPLEX

When we talk about electrophilic addition on aromatic systems, there are two types of reactions possible. The first one is a π complex, and the second one is σ complex. In the case of π complex, we can expect that there will be an interaction between delocalized π electrons and the electrophile leading to π complex formation. One such example is toluene which forms a complex with HCl at –78 °C. When we say π complex, there is no true bond that is formed between the ring carbon and the electrophile, but they are placed above the benzene ring at right angles to the plane of the benzene ring. These complexes are also called *charge transfer complexes.*

When we analyze this reaction mixture, the solution is a non-conductor of electricity. When the reaction occurs, there is no color change, and there is negligible change in UV. All these observations lead us to conclude that in this complex, there is no disturbance of π electron cloud. In other words, there is no ionic species formed in this reaction.

Although the electrophile resides above the π electron cloud (close to 90°), the molecule is not perfectly 90° symmetrical with respect to the planar aromatic ring (C_6 axis), and it is tilted by 6.1° as shown below.

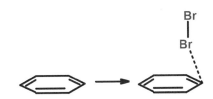

SCHEME 9.1   Charge transfer complex.

9.1.2 σ COMPLEX OR WHELAND INTERMEDIATE

On the other hand, if a Lewis acid like $AlCl_3$ is used, we get a different type of intermediate called Wheland intermediate or the σ (sigma) complex. How do we know that there are two different types of complexes?

If we look at the reaction of benzene with DCl in the presence of $AlCl_3$ at –78 °C, the following things are observed. When we analyze this reaction mixture, the solution conducts electricity, and when the reaction occurs, there is a green color development as the reaction progresses. If we measure the UV of the reaction medium before and after the reaction, we observe a change in UV.

The orientation and reactivity effects are explained on the basis of resonance and field effects of each group on the stability of the intermediate arenium ion. Since in most substitution reactions, the overall rate is highly dependent upon the nature of the substituent, we can say that the formation of the intermediate arenium ion is the rate-determining step as shown below.

Some of the reactions are irreversible, and the others are usually stopped well before equilibrium is reached. Therefore, which of the three possible intermediates is formed is dependent not on the thermodynamic stability of the products, but on the activation energy necessary to form each of the three intermediates.

An aromatic system with stabilization getting converted into high energy σ complex loses its aromaticity. It will be a difficult process having high activation energies ($\Delta G_1^{\ddagger}$) as shown below. But we do see aromatic substitution at room temperature. What is the driving force?

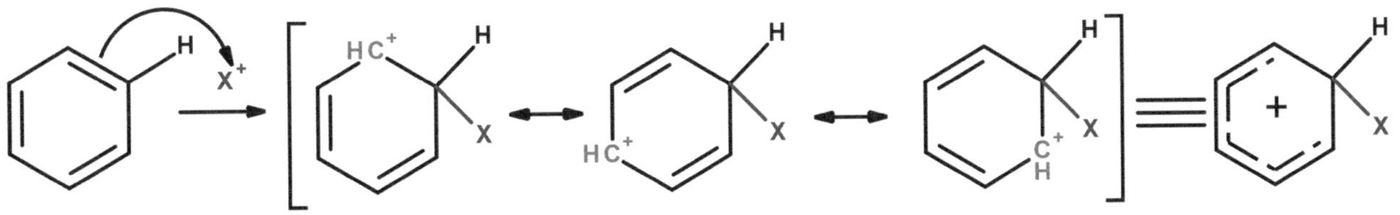

SCHEME 9.2   Formation of arenium ion.

DOI: 10.1201/9781032631165-9

1. The energy released by the new bond with electrophile
2. The sigma complex itself is not high in energy due to the delocalization of positive charge.

We are still not clear how the reaction proceeds. Because the arenium ion intermediate or the Wheland intermediate can undergo a couple of reactions, (1) it can go back to the starting material, (2) it can react with the nucleophile to give a product, and (3) it can form isomeric products due to resonance.

If it goes back to the starting material, then no reaction. If it reacts with a nucleophile, we will get additional product. But that is highly unlikely because the resulting addition compound (as shown below) will lose its aromaticity. Loss of proton is the most probable reaction, and it gives the substituted product.

The next question we must answer is "How do we know the existence of arenium ion intermediates?". We need to investigate experimental evidences to prove the formation of the arenium ion intermediate. (1) No isotope effect is observed in aromatic electrophilic substitution, which means the C–H bond breaking is not the rate-determining step and (2) there are instances where researchers had isolated the arenium ion intermediates or the Wheland intermediate. One such example is given below.

Based on those evidences, we can conclude the existence of arenium ion. Overall, the aromatic electrophilic substitution occurs through three steps. The first one is the addition of the electrophile to the aromatic system. This is a very slow process or the rate-determining step. The activation energy for this process is very high because the molecule loses its aromaticity. It is an endergonic step. It can either form the π charge transfer complex or it can form the Wheland intermediate. In the second step, the Wheland intermediate can be stabilized by resonance structures. In the last step loss of proton occurs to give the final product. This is a very fast process because the product now regains its aromaticity. This step is an exergonic process.

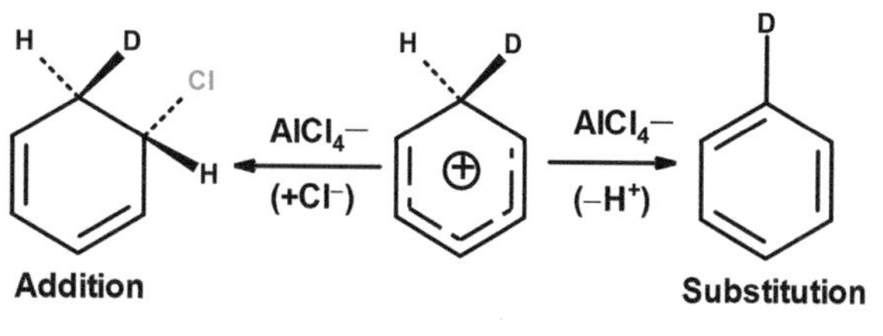

9.1.3 Examples of Electrophilic Substitution

The following scheme gives a brief overview of the various important electrophilic substitutions we study elaborately. They are (1) halogenation, (2) nitration, (3) sulfonation, (4) Friedel–Crafts alkylation, and (5) Friedel–Crafts acylation.

9.2 NITRATION

The following scheme depicts a typical nitration reaction of benzene. The reaction is carried out by the nitrating mixture which consists of conc: nitric acid (HNO_3) and conc: sulfuric acid (H_2SO_4) in catalytic amounts and water is produced as the by-product.

You will be wondering about the role of sulfuric acid. It absorbs water. But it is used only in catalytic amount, so it cannot remove all the water. What will happen if water is not removed? Does it add to nitrobenzene? Probably not. Nitration without the catalytic amount of sulfuric acid is slow. Benzene does not react with sulfuric acid. This leads to the conclusion that maybe sulfuric acid reacts with HNO_3. How do you prove it? The four-fold freezing point depression of HNO_3 in H_2SO_4 was close enough to show that there are four ions involved. One H_3O^+, one $^+NO_2$, and two HSO_4^-.

There are various evidences to show that $^+NO_2$ is actually involved in the nitration of aromatic systems:

1. The $^+NO_2$ ion could be isolated in solution and as salts.
2. The Raman spectrum of the reaction mixture shows absorption at 1,400 cm^{-1} corresponding to a linear triatomic species.
3. In addition to sulfuric acid, other acids ($HClO_4$) can also produce the $^+NO_2$ ion.
4. Finally, nitration does not take place with nitric acid alone.

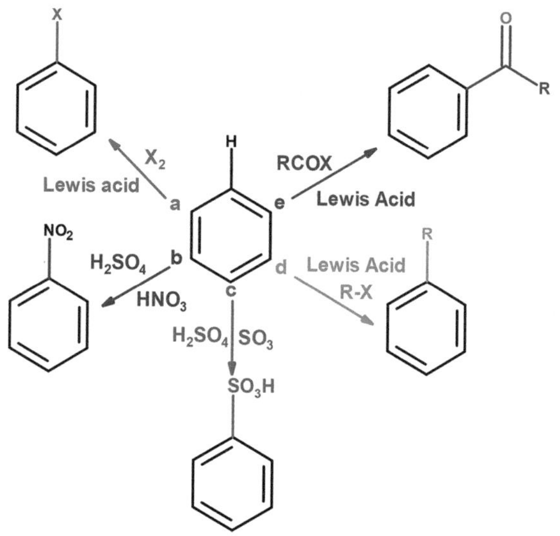

SCHEME 9.3   Addition or substitution reactions in benzene.

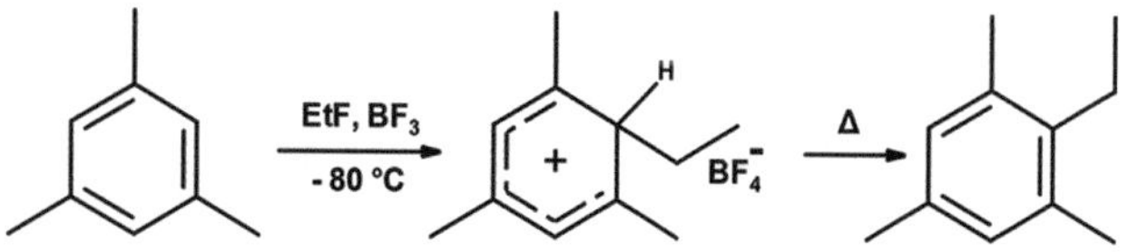

SCHEME 9.4   Alkylation of aromatic system.

SCHEME 9.5   Examples of electrophilic substitution.

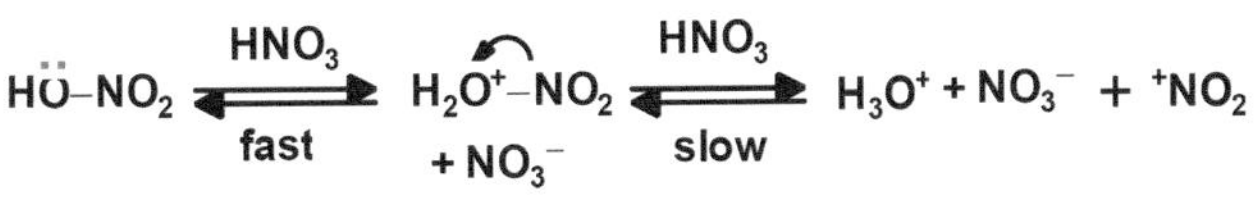

$$HO-NO_2 \underset{fast}{\rightleftharpoons} \overset{HNO_3}{} H_2O^+-NO_2 \underset{slow}{\rightleftharpoons} \overset{HNO_3}{} H_3O^+ + NO_3^- + {}^+NO_2$$
$$+ NO_3^-$$

SCHEME 9.7 Formation of nitronium ion in the presence of nitric acid.

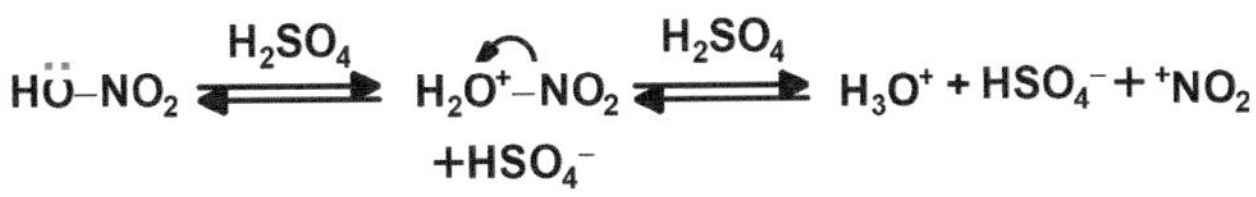

$$HO-NO_2 \rightleftharpoons \overset{H_2SO_4}{} H_2O^+-NO_2 \rightleftharpoons \overset{H_2SO_4}{} H_3O^+ + HSO_4^- + {}^+NO_2$$
$$+HSO_4^-$$

SCHEME 9.6 Formation of nitronium ion in the presence of sulfuric acid.

The generation of nitronium ion can be written as shown below

In the above-mentioned process, based on the rate at which the intermediates are produced, we can identify the rate-determining step.

1. Protonation is a fast process. So, it is not the rate-determining step.
2. Formation of cation is slow, and it is rate-determining step.

With this information, we can write the rate law. The reaction depends both on the concentration of the aromatic substrate and the nitronium ion.

$$Rate = k[Ar-H][^+NO_2]$$

However, this ideal equation does not apply to all the cases. Some of the limitations include (1) low solubility of Ar–H in nitrating mixture and (2) in nitrating mixture $[^+NO_2] = [HNO_3]$, but in other mixtures used for nitration (acetyl nitrate, $CH_3C(O)ONO_2$, nitric acid and mercury(II) nitrate or sodium nitrite, and 0.5 mol% $Pd_2(dba)_3$), the nitronium ion concentration is not related to the concentration of nitric acid $[^+NO_2] \neq [HNO_3]$.

9.2.1 MECHANISM OF NITRATION REACTION

To explain the product formation, we can think of two different possibilities. One involves a single transition state which is a concerted pathway. Here, the formation of the C–NO$_2$ bond and the breaking of C–H bond occur simultaneously.

The next one is a non-concerted pathway or two-step pathway involving the σ complex. This can further be divided into two possibilities. In one instance, the C–NO$_2$ bond formation is the rate-determining step. In the second instance, the C–H bond cleavage is the rate-determining step. How do we decide what is the probable path?

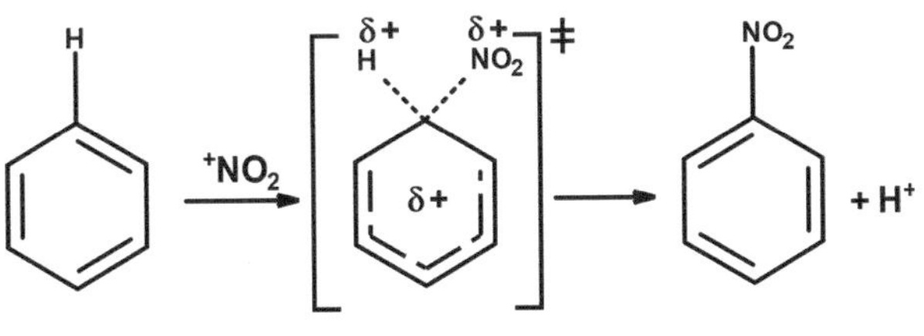

As we mentioned earlier, there is no kinetic isotopic effect. So, the C–H bond breaking is not the rate-determining step. That means, the concerted mechanism or the C–H bond breaking is not the acceptable pathway based on experimental evidences.

We can draw the reaction coordinate diagram for this transformation involving two transition states. The first may be the approach of the nitronium ion toward the aromatic system (TS$_1$) to give the intermediate sigma complex. This undergoes loss of proton (TS$_2$) to give the product. Refer to Figure 9.1 for details. According to Hammond's postulate (discussed in Chapter 10), *species resembling in energy are resembling in structure* also.

9.2.2 SPECIAL CASE OF NITRATION

Phenol undergoes nitration with dilute HNO$_3$. But under the reaction conditions, nitronium ion concentration is not high. But the actual cation is nitrosonium ion (^+NO) formed from the nitrous acid which nitrosates the arene. This Nitroso derivative gets oxidized to nitro derivative under atmospheric conditions (Figure 9.2).

9.3 HALOGENATION

We have two types of halogenations: one is a *free radical addition*, and the other is the *electrophilic substitution*. The free radical reaction occurs photochemically. The electrophilic substitution occurs in the presence of Lewis acid (AlCl$_3$) catalyst.

When a neutral halogen molecule interacts with the benzene, the reaction does not occur because there is no interaction. When there is a chloride ion, the benzene π electron cloud repels it, so here also no reaction takes place. However, the π electron cloud can interact with an electrophile. Now the question is what should be the strength of the electrophile weak or strong? When the π electron cloud and electrophile interaction takes place, the aromaticity will be disturbed or lost. A weak electrophile cannot do

SCHEME 9.8 Nitration of benzene.

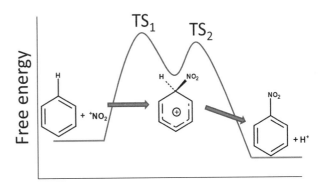

FIGURE 9.1 Potential energy diagram charge transfer complex.

$$HNO_2 + 2\,HNO_3 \rightleftharpoons H_3O^+ + 2\,{}^-NO_3 + {}^+NO$$

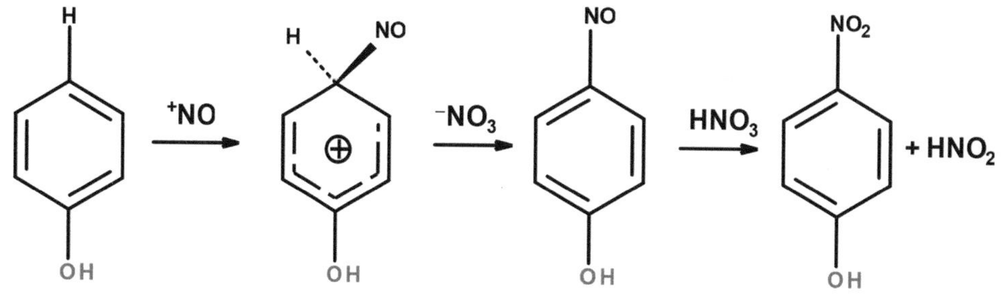

FIGURE 9.2 Nitration of phenol.

this. We need a strong electrophile. So, to facilitate the electrophile - π electron cloud interaction we need a Lewis acid that can polarize the neutral molecule in such a way to create a powerful electrophile.

Let us look at the mechanism by which the electrophilic halogenation occurs. Benzene needs to form a π complex. The Lewis acid (AlX₃/FeX₃) interacts with the Br–Br bond and polarizespolarizes. Since the Lewis acid pulls the electron, the halogen bond polarizespolarizes, and one end becomes a little electron deficient.

First, it forms a π-complex, and then it breaks down to give a σ-complex. Then the electron-rich Lewis acid complex removes the proton from the σ complex and gives the final product.

9.4 SULFONATION

Unlike the previous cases where an electrophile is involved in the substitution reaction, in the case of sulfonation reaction, SO_3, which is a neutral molecule, acts as an electrophile. The highly polarized sulfur-centered sulfonium moiety is the electrophile. When sulfuric acid is used, it produces SO_3 and this is a reversible reaction. When oleum is used in sulfonation, the SO_3^- group is protonated first before the C–H bond is broken.

Do not get confused between the neutral sulfur trioxide electrophile and depiction of sulfonic group as minus charge (as given in the following scheme). Due to the presence of strong electron-withdrawing oxygen, which pulls the electrons toward itself, making the oxygen negative at the same time making the central sulfur atom highly electron deficient. SO_3 adds to the benzene ring to give the sulfonium ion.

$$2\,H_2SO_4 \rightleftharpoons H_3O^+ + 2\,HSO_4^- + SO_3$$

SCHEME 9.10 Formation of sulfur trioxide from sulfuric acid.

When the sulfur trioxide is added to the benzene ring, one of the S=O (double bond) polarizes toward oxygen giving a negative sulfonic acid anion. Loss of proton from this intermediate leads to regaining of aromaticity. This anion gets protonated to the final product (it gets protonated in the acid medium).

9.5 FRIEDEL–CRAFTS ALKYLATION

Similar to the halogenation, the electrophilic carbon of alkyl halide is not a strong electrophile to affect aromatic electrophilic substitution. But alkyl halide in the presence of a Lewis acid catalyst, namely, $AlCl_3$, can do this job, and this reaction is called Friedel–Crafts reaction (FCR). In fact, at low temperatures, it is possible to even isolate the RX-Lewis acid complex. In the *tert*-butyl bromide case, the alkylation is carried out by the *tert*-butyl carbocation, as part of an ion pair. In this reaction, a polarized complex as shown below is formed which is similar to the complex formed in the halogenation.

The observed rate law is given by the following equation. Since the Lewis acid is also involved in the rate law, we can compare the effectiveness of Lewis acids as well.

$$Rate = k[Ar–H][RX][MX_3]$$

The following is the effectiveness of Lewis acid catalysts that are generally employed in FCR.

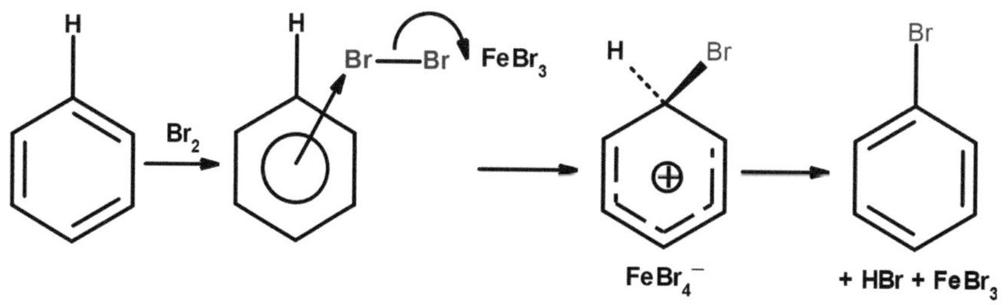

SCHEME 9.9 Halogenation of benzene.

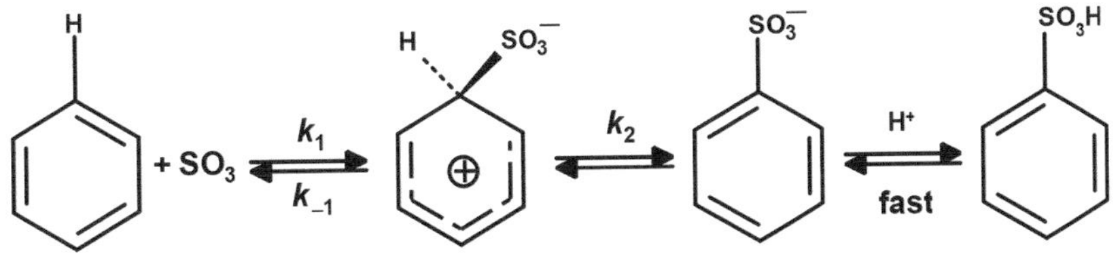

SCHEME 9.11   Sulfonation of benzene.

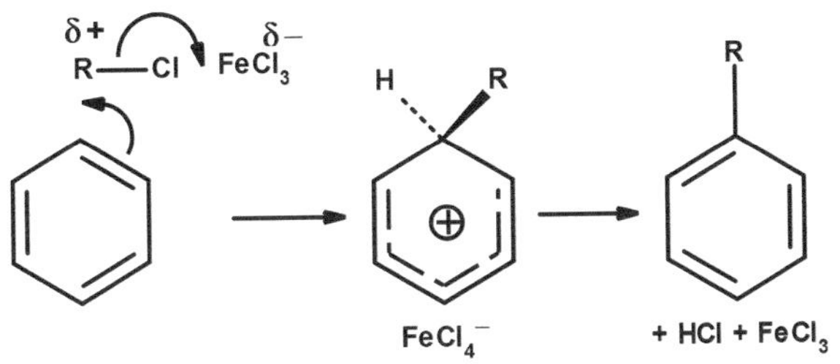

SCHEME 9.12   Friedel–Crafts alkylation of benzene.

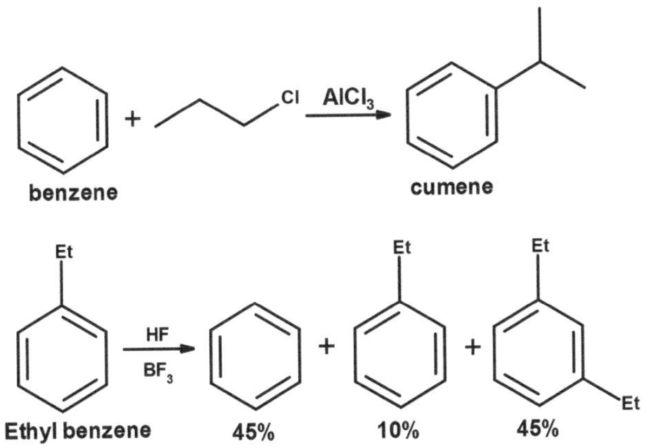

SCHEME 9.13   Examples of rearrangements in Friedel–Crafts alkylation.

$$AlCl_3 > FeCl_3 > BF_3 > TiCl_3 > ZnCl_2 > SnCl_4$$

9.5.1   LIMITATIONS OF FRIEDEL–CRAFTS ALKYLATION

Although FCR is an important reaction for preparing various alkyl benzenes, a major limitation is rearrangements. Whenever the reacting electrophile can undergo rearrangement, we are bound to get more than one product. Moreover, carrying out selective mono-alkylation is also a challenge, because after the first alkylation, the aromatic ring becomes more activated, hence it can further undergo electrophilic substitution leading to polyalkylation. So, to effect mono-alkylation first we have to carry out acylation and then the resulting carbonyl compound is subjected to Wolff–Kishner reduction or Clemmensen reduction to give the monoalkyl derivative.

As shown in the above alkylation reaction of benzene with propyl chloride, the major product obtained is isopropyl benzene (cumene) and not n-propyl benzene.

This is because the propyl cation undergoes rearrangement. Similarly, in the following reaction, dealkylation of ethyl benzene occurs in the presence of Lewis acid.

Due to this reason, introduction of a primary alkyl chain in FCR is always a challenge. What researchers do is to use the acid chloride for acylation and later reduce the carbonyl. This is done purely to avoid both ortho/para isomeric products and rearranged products.

9.6   FRIEDEL–CRAFTS ACYLATION

FC acylation follows a similar rate law like alkylation. Although the electrophile is the carbonyl carbon, does it exist as an ion pair or polarized complex is difficult to predict. Both species are given in the scheme. In either case, we get the same acylated product.

Similar to the alkylation the rate law for acylation also is affected by the concentration of arene, acyl halide, and the Lewis acid catalyst. When we compare both acylation and alkylation there are many similarities, but there are a few differences as well. Unlike alkylation, the Lewis acid essentially complexes with the acyl oxygen, thus preventing polyacylation, due to which Friedel–Crafts acylation requires more than 1 equivalent of Lewis acid.

Various acylating agents that are used in Friedel–Crafts acylation include acid chlorides, acid anhydrides, carboxylic acid, carbon monoxide, and HCl (Gattermann–Koch reaction), nitrile and HCl (Houben–Hoesch reaction).

9.7   GATTERMANN–KOCH REACTION

Formylation is carried out by a mixture of CO and HCl gas along with $AlCl_3$. One significant characteristic of this process is that the reaction is favored toward the product. The formylated product complexes with $AlCl_3$, thereby preventing the reverse process. Of course, to effect this reaction we need to use more than 1 eq. of the Lewis acid.

9.8   ELECTROPHILIC SUBSTITUTION OF SUBSTITUTED BENZENES: ORIENTATION

Now we will look into mono-substituted benzene derivatives. This in turn may lead to three different products: 1,2-, 1,3- and 1,4- products. The orientation and reactivity effects are explained on the basis of resonance and field effects of each group on the stability of the intermediate arenium ion.

SCHEME 9.14   Friedel–Crafts acylation of benzene.

SCHEME 9.15   Orientation in electrophilic substitutions of substituted benzenes.

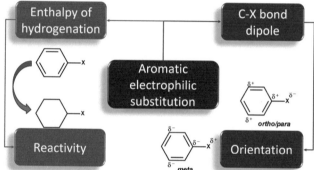

FIGURE 9.3   Bond dipole and orientation.

We can explain the electrophilic substitution using the enthalpy of hydrogenation and bond dipole. The enthalpy of hydrogenation of the ring determines the relative reactivities of AR–X, and the direction of the AR–X bond dipole determines the orientation that is ortho, para, or meta substitution.

Predicting the product formation

- Based on Field effect
  → Z that has an electron-donating field effect (+I) will stabilize all three ions
  → Z that has an electron-withdrawing field effect (−I) will destabilize all three ions
- Based on the resonance effect (Figure 9.3)

We know the −NH$_2$ is highly an activating group and the −NO$_2$ is a highly deactivating group toward aromatic electrophilic substitution. This also follows the linear free energy relationship of Hammett's equation (Chapter 10) (Figure 9.4).

Enthalpies of hydrogenation give a good picture of the electron densities of the aromatic ring. Since there is a correlation between the experimental enthalpy of hydrogenation and the order of the reactivities of the substituents, we can easily understand which species stabilizes the carbocation and how fast they undergo electrophilic substitution.

With the simple electron donating or withdrawing group we can arrange the order of reactivities. There are limitations. Under different conditions, the same substrate reacts at different rates with different electrophiles. If we compare toluene and benzene, depending on the electrophile, solvent, and temperature the rate varies. So, by using the enthalpy of hydrogenation, we can predict the reactivity in different conditions. We need to use different Hammett plots.

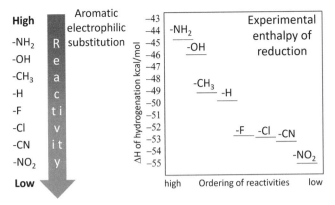

FIGURE 9.4 Relationship between the experimental enthalpy of hydrogenation and the order of the reactivities of the substituents.

Here we will try to see the orientation with respect to electron density on the substituent carbon. By looking at ^{13}C NMR chemical shift, we can predict the polarization of the C-X bond. Because ^{13}C NMR chemical shift is a good measure of the shielding or presence of electron density at the particular carbon. The ^{13}C NMR chemical shift of the *ipso* carbon of all the *meta* directing or deactivating groups is more than 128.36 ppm (Figure 9.5).

We already know halogens are an exception. For all deactivating groups, the bond dipole has a negative end on the *ipso* carbon ($C^{\delta-}$-$X^{\delta+}$). But for halogens, it is the opposite. That is the reason the halogens are *ortho-para* directing even though they are electron withdrawing.

9.9 APPLICATION OF EAS

Many useful aromatic compounds are prepared by this reaction. For example, many colorful organic dyes can be prepared by this method (precursor to azo dyes).

9.10 AROMATIC NUCLEOPHILIC SUBSTITUTION

Unlike electrophilic substitution, the diversity of aromatic nucleophilic substitution is less. Because there are some inherent issues related to an electron-rich nucleophile attacking an electron-rich π system of the aromatic substrates. The first issue is the repulsion between π electron cloud and the incoming nucleophile; the other issue is when you add two more electrons to the stable six π electrons aromatic system you are introducing destabilization. Unless some strong electron-withdrawing group is present in the aromatic ring to reduce the available electron cloud on the ring system, there is no incentive for the aromatic system to undergo nucleophilic substitution. How can we make the nucleophilic substitution work? We have to use strong electron-withdrawing substituents like –NO$_2$ group on the aromatic system to carry out nucleophilic substitution.

9.10.1 AROMATIC NUCLEOPHILIC SUBSTITUTION (S$_N$AR MECHANISM)

In aromatic nucleophilic substitution, the nucleophile displaces a good leaving group, such as a halide, on an aromatic ring.

When we say nucleophilic substitution, what comes to your mind is S$_N$1 and S$_N$2 reaction mechanisms. Unfortunately, they do not operate on aromatic systems. Do you know why? If we take the S$_N$2 mechanism, the nucleophile has to attack the reaction center from the rear side with the simultaneous loss of the leaving group. The nucleophile can only come perpendicular (i.e., 90° angle) to the leaving group and not from the opposite side (180° angle). So S$_N$2 attack is not possible. If we think about the S$_N$1 mechanism, we need to have a highly unstable phenyl cation. The π electron cannot stabilize the cation. Because

X	p-^{13}CNMR	directing
-NO$_2$	134.71	meta
-CHO	134.43	meta
-CN	132.84	meta
-H	128.36	-
-Cl	126.43	o/p
-CH$_3$	125.38	o/p
-F	124.16	o/p
-OH	121.09	o/p
-NH$_2$	118.39	o/p

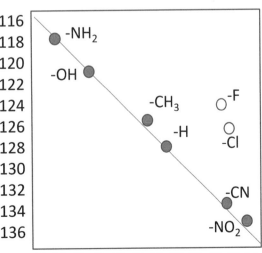

FIGURE 9.5 Relationship between ^{13}C NMR values of ipso carbon and orientation.

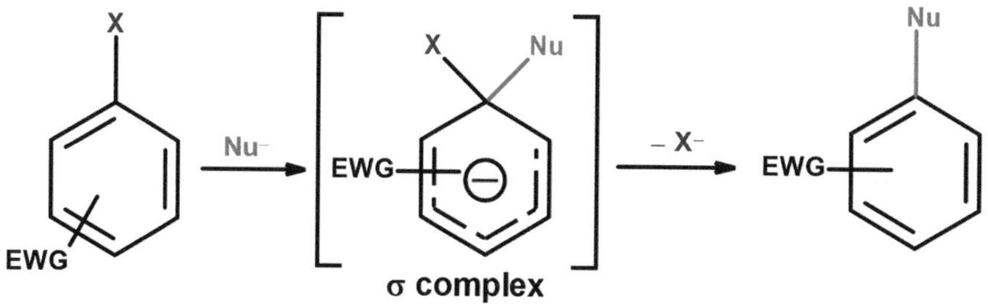

SCHEME 9.16   Aromatic nucleophilic substitution.

FIGURE 9.6   Why S_N2 or S_N1 mechanism does not operate in aromatic system.

the π orbital and sp^2 orbitals are orthogonal to each other. So, we need a different type of mechanism to explain the product formation (Figure 9.6).

We can think the reaction may occur in two steps. The first step involves an attack of the nucleophile Y⁻ at the carbon bearing the leaving group. This will lead to an intermediate carbanion on the *ortho* carbon. During this step, the aromaticity of the ring is destroyed, and the reaction site where the leaving group was present has to change the hybridization from sp^2 to sp^3, that is, planar to tetrahedral to accommodate the nucleophile. Due to these reasons, mono-substituted arenes with a leaving group like haloarenes do not undergo nucleophilic substitution reaction.

9.11   TYPES OF AROMATIC NUCLEOPHILIC SUBSTITUTION MECHANISM

The first one is addition–elimination. This happens when EWG are present (e.g., –NO₂ and –COOH) and the leaving group is *ortho* or *para* to the EWG. The second one is elimination–addition. This happens when EDG are present

(e.g., –CH₃ and –OCH₃), and this requires a strong leaving group and harsh reaction conditions.

9.11.1   ADDITION–ELIMINATION MECHANISM

Here the nucleophile attacks the carbon on which the leaving group is present. Simultaneously the EWG pulls the electrons away from the aromatic ring. This leads to the formation of an enolate-like intermediate. The middle one is in the following structure.

Few things do not take place as in conventional reactions. First protonation does not take place as in Michael's addition, because that would result in a loss of aromaticity. Second, the leaving group does not abstract a proton from the solvent.

What actually happens is that during the charge reversal, the leaving group goes away with its bonding pair of electrons. This also restores the aromaticity.

We know *para*-chloronitrobenzene when treated with 1N sodium hydroxide at 150 °C gives *para*-nitrophenol. When there is only one electron-withdrawing group (–NO₂), the nucleophilic substitution occurs at a very high temperature and in the presence of a very strong base. Instead, when two nitro groups are present in the same aromatic ring, this substitution occurs with a thousand times lower concentration of NaOH and at 100 °C. But when we have the three nitro groups, the reaction occurs with simple water itself at 40 °C.

9.11.2   CHICHIBABIN REACTION

Another example is pyridine undergoing nucleophilic substitution with amide (⁻NH₂) nucleophile. Instead of an external electron-withdrawing group (–NO₂) pulling the electron outside of the aromatic ring, it is also possible for an internal nitrogen (pyridine) to pull the electron toward itself,

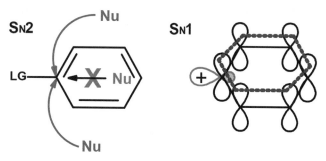

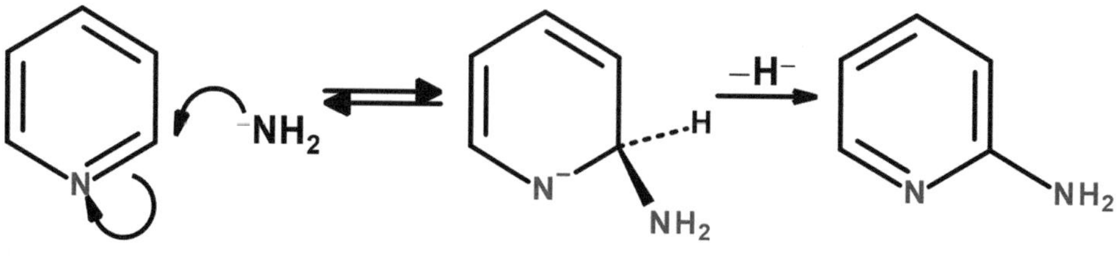

SCHEME 9.17   Chichibabin reaction.

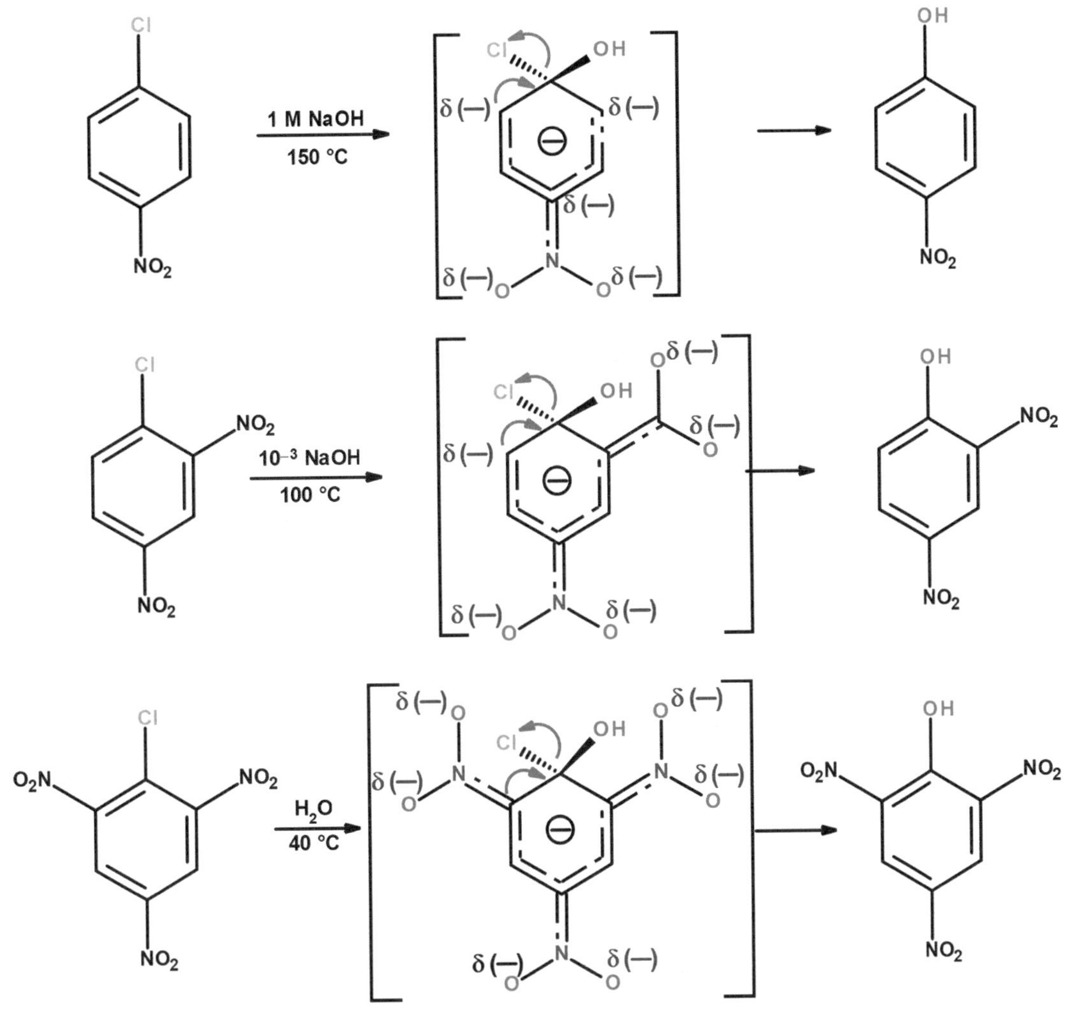

SCHEME 9.18 Nucleophilic displacement of chloride group by hydroxyls.

thereby facilitating the attack by the nucleophile. Pyridine when treated with sodamide in N,N-dimethylaniline as solvent gives 2-aminopyridine. This is called the Chichibabin reaction. The mechanism is given below.

Let us look at the conversion of *para*-chloronitrobenzene to *para*-chlorophenol. Here chloride is the leaving group, and the hydroxide is the nucleophile. In the first step, the nucleophile adds to the aromatic system. Since we are adding two electrons to the system, there is a building up of partial negative charge at the *ortho* and *para* carbon atoms.

Now the nitro group pulls the electron cloud toward itself and it helps in the distribution of the electron cloud. The halogen now goes away with its bonded pair of electrons, this now restores the aromaticity, that is, 6 electrons to the aromatic system. However, the conditions are pretty harsh, 150 °C.

In this mechanism, there is some doubt about whether the hydroxide nucleophile came from the sodium hydroxide or the solvent water. The role of the base in these reactions is to form the active nucleophile.

The situation becomes better when there are two EWGs. When we look from the mechanical aspect, there is only one *ortho* carbon that has δ−. However, the overall negative charge is distributed nicely over both the two nitro groups. Loss of the leaving group is similar to the previous case.

In the case of trinitro derivative, the reaction occurs even without the base. This implies excellent negative charge distribution across all the nitro groups. As a result, the aromatic ring is highly deactivated. This helps in the attack of the nucleophile as well as the departure of the leaving group.

How do you know the mechanism we just proposed is an acceptable process. The German Scientist Jakob Meisenheimer in 1902 reported the isolation of the 1:1 adduct formed between an arene-carrying electron-withdrawing groups (the nitro group) and an alkoxy nucleophile.

9.11.3 MEISENHEIMER COMPLEX

In the case of electrophilic substitution, we saw Wheland intermediate or σ complex as an intermediate. Here we have the Meisenheimer complex which is the product of the reaction of either the methylaryl ether with potassium ethoxide or the ethylaryl ether and potassium methoxide.

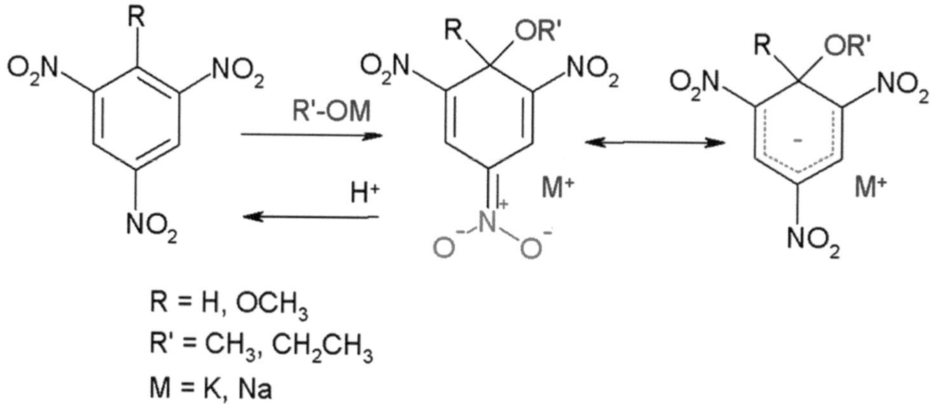

R = H, OCH$_3$
R' = CH$_3$, CH$_2$CH$_3$
M = K, Na

SCHEME 9.19 Meisenheimer complex.

Isolation of the above complex also confirms the rate law. Unlike the aliphatic S$_N$2 reaction, where the attack of the nucleophile takes place from the rear side, in the case of aromatic nucleophilic substitution, the attack presumably takes place to the right or left side of the *ipso* substituent.

$$Rate = k[ArX][Y^-]$$

A cyclohexadienyl derivative formed as Lewis adduct from a nucleophile (Lewis base) and an aromatic or heteroaromatic compound, is called Jackson–Meisenheimer adduct. In literature, many people use the term "Meisenheimer complex" to the typical Meisenheimer alkoxide adducts of nitro-substituted aromatic ethers.

H$^-$ is a very poor leaving group compared to H$^+$. Due to this, loss of H$^-$ is very difficult in a nucleophilic substitution reaction. Unlike electrophilic substitution, where, the loss of H$^+$ from the same carbon, the nucleophilic substitution requires a better leaving group than H$^-$. In the case of aromatic *nucleophilic* substitution, the better leaving groups include Cl$^-$, Br$^-$, N$_2$, SO$_3^{2-}$, and $^-$NR$_2$.

There is a very special case where nitrogen is displaced from a diazonium ion. When neutral N$_2$ is liberated from diazonium ion (ArN$_2^+$), it will result in Ar$^+$ which is a highly unstable species. It is so reactive that, it reacts with any nucleophile that is present in the solution. In fact, decomposition of aryl diazonium salt in the presence of nucleophiles is the only reaction where aryl cations are formed in solution. The driving force is the formation of nitrogen gas.

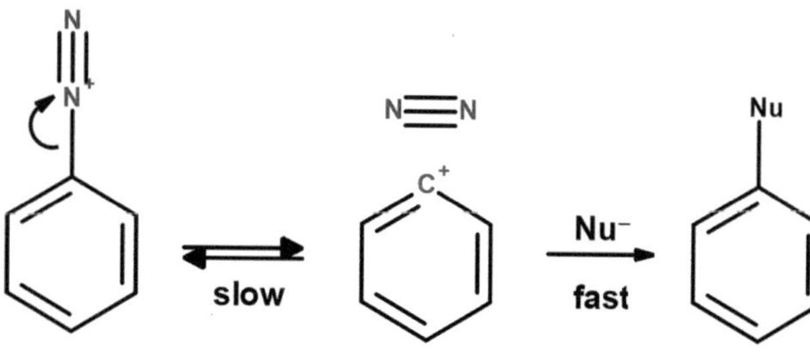

Several nucleophiles that are otherwise difficult to be introduced can be introduced into an aromatic system easily using diazonium salt. For example, it is not possible to introduce fluorine atom into an aromatic nucleus by simple nucleophilic substitution with F$_2$. The introduction of fluorine into aromatic system is possible only by the diazonium ion salt route.

9.11.4 SANGER REACTION

The most important application of aromatic nucleophilic substitution is the famous Nobel Prize winning Sanger reaction. Sanger's reagent, which is 1-fluoro-2,4-dintrobenzene reacts with N-terminal amino acid in polypeptides chain and helps in protein sequencing. Sanger used this reaction to sequence insulin.

9.11.5 ELIMINATION–ADDITION MECHANISM

Let us look at how substrates with EDG groups undergo nucleophilic substitution. To explain this reaction, the elimination–addition mechanism is proposed. Unlike the previous case, here we are getting isomers.

If we take *para*-chlorotoluene and treat it with sodium hydroxide at 350 °C, which is an extremely high temperature and strong basic condition, the substitution of the halide with hydroxide ion takes place giving two isomeric products. When bromobenzene is treated with sodamide we get aniline. When we treat *ortho*-chloroanisole with sodamide, we get *meta*-methoxyaniline.

Let us look at the mechanism. The initial step is the elimination. The base abstracts the *ortho* hydrogen, giving a carbanion. This leads to a benzyne intermediate. Now the hydroxide attacks the benzyne to give various products. If you look carefully, you need an *ortho* hydrogen for the benzyne formation.

You may be wondering, when the triple bond benzyne is formed, is there a change in hybridization from *sp*^{2} to *sp* in those two carbons?

The benzyne is similar to alkyne. Normal alkynes are linear, whereas the triple bond in benzyne is present in a

SCHEME 9.20 Decomposition of aryl diazonium salt.

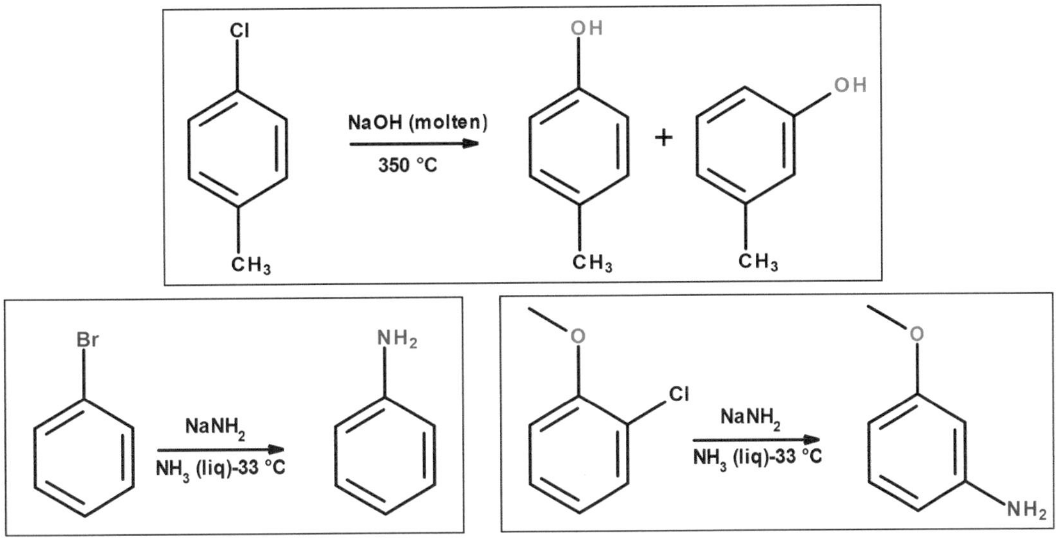

SCHEME 9.21 Examples of elimination–addition mechanism.

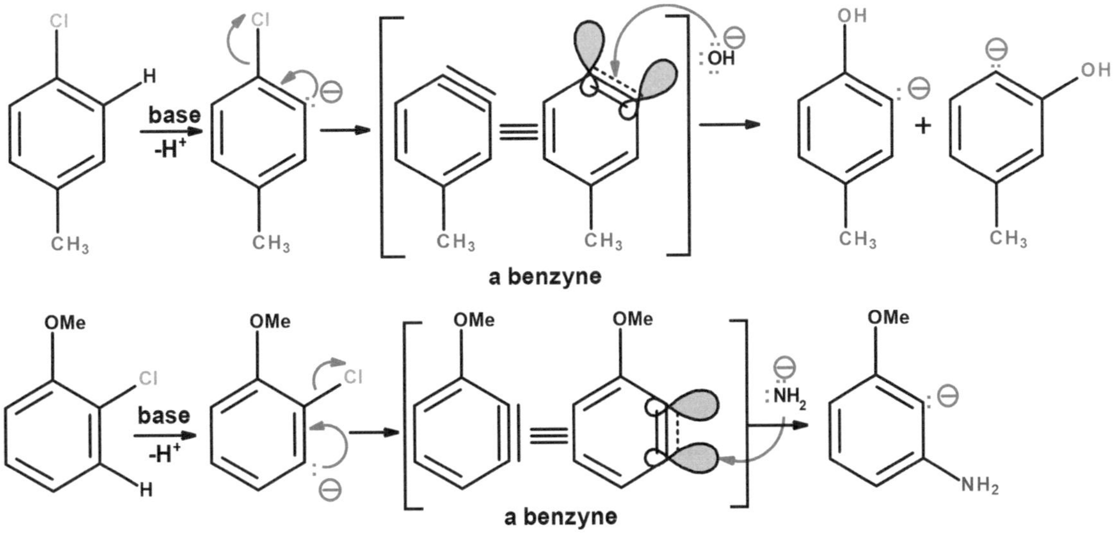

SCHEME 9.22 Benzyne formation.

cyclic system, this introduces some strain. Due to this strain, the orbital overlap is weak. In reality, what we see is a hybridization of orbitals of the adjacent C–C bonds that is neither purely sp^2 nor sp but it is in between sp^2 and sp.

The *ortho*-halogeno-compounds give *meta*-amines, rearrangement being essentially quantitative. However, the *meta*-halides give the *meta* product with no rearranged products. The formation of benzyne can be explained based on the relative acidities of the hydrogens. Because, the base can abstract the most acidic hydrogen. The acidities of benzenoid hydrogens depend on the inductive effects of substituents. If the substituent is electron withdrawing, then the acidity order is $o > m > p$, and if the substituent is electron donating, then the acidity order is $p > m > o$. From this, we can understand the benzyne formation.

The *meta*-substituted haloarene reacts quite differently. Since it has two *ortho* protons, we can anticipate to get all the three isomers. One product where the proton adjacent to the R group is replaced by the nucleophile, the second one where the proton *para* to the R group is replaced by the nucleophile, third product where the leaving group is replaced by the nucleophile. But except for the last one, the other two reactions do not occur.

The substituents we are dealing with are basically electron donors. They can stabilize the intermediate through resonance or hyperconjugation. If we take the *path a* mechanism, when the R group is electron donating, the *ortho* carbon is δ^- and the *meta* carbon is not. So the nucleophile can attack *meta* position easily compared to *ortho*. In *path b*, we can apply the same logic. But *para* carbon

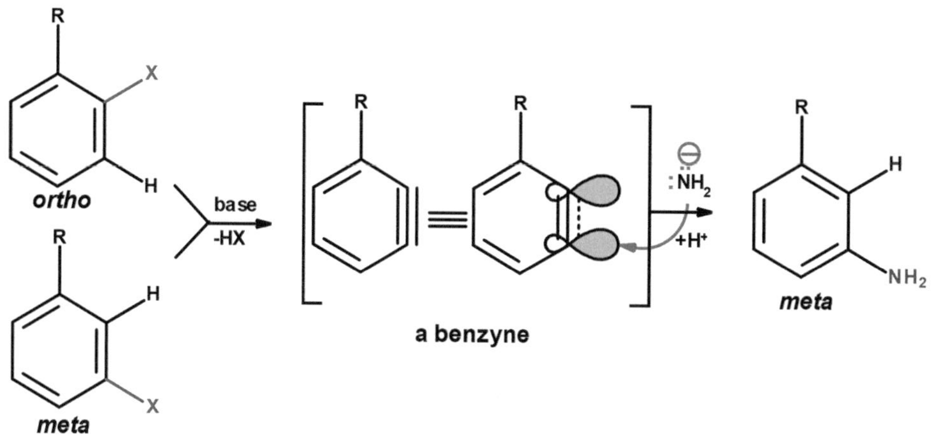

SCHEME 9.23   Mechanism for replacement of ortho or meta halogen by nucleophile.

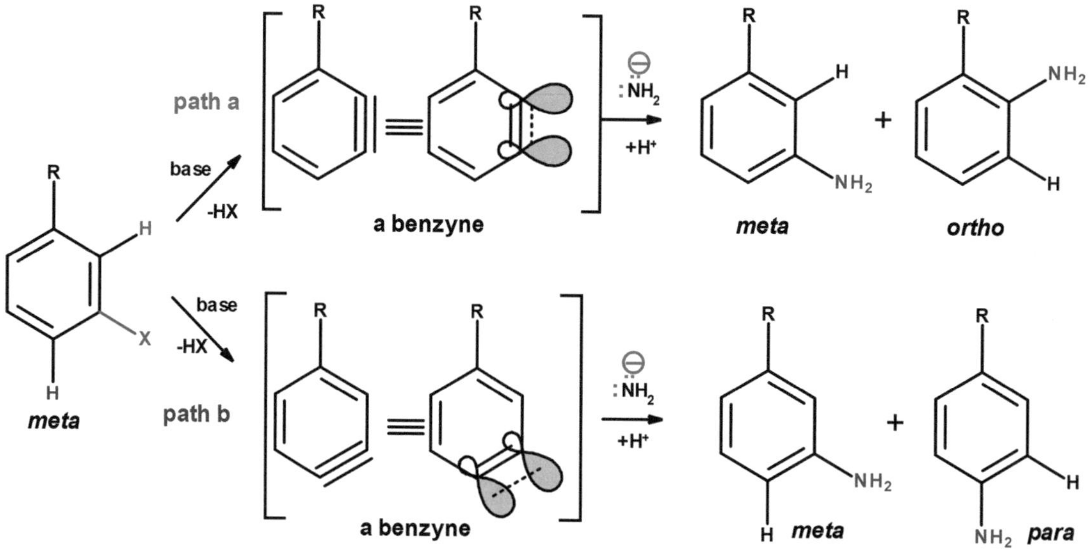

SCHEME 9.24   Mechanism for replacement of meta halogen by nucleophile.

is not electron-rich like the *ortho* carbon, so we may get a very small amount of the *para* isomer, but they are always formed in minor amounts. Whether we follow *path a* or *path b* we get the *meta* product.

When the group R is electron withdrawing in nature by inductive effect, a 3-substituted benzyne should give predominantly the *meta*-substituted aniline and a 4-substituted benzyne may give the *para*-substituted aniline as the major product. Where R is weakly electron donating by its inductive effect 3- and 4- substituted benzynes may give preferentially the *o*- and *m*-substituted anilines, respectively; we are not considering the steric effect of the substituent which may influence the formation of the *ortho* product.

In the following example, we have the leaving group *para* to the EDG. This reaction leads to roughly 1:1 mixture of two products namely the normal and the rearranged products. To sum up, we can say, *ortho* and *meta* isomers

give *meta* products in major amounts and the *para* isomer gives a mixture of *meta* and *para* products.

To recap what we saw, the *ortho*- and the *para*-halides are unique ones because they follow different pathways and give different intermediates, 3- and 4-substituted benzynes, respectively.

9.11.5.1   Evidence for Benzyne Mechanism

The ^{14}C labeled chlorobenzene when treated with potassium amide gave two isomeric products. In one of the products, the amino group was found on the quaternary carbon with ^{14}C label and in the other on the carbon without ^{14}C labeling.

The ^{13}C NMR values and IR values support the formation of benzyne. In addition, the dimeric and trimeric products formed in small quantities in the reaction also point to the formation of benzyne intermediate.

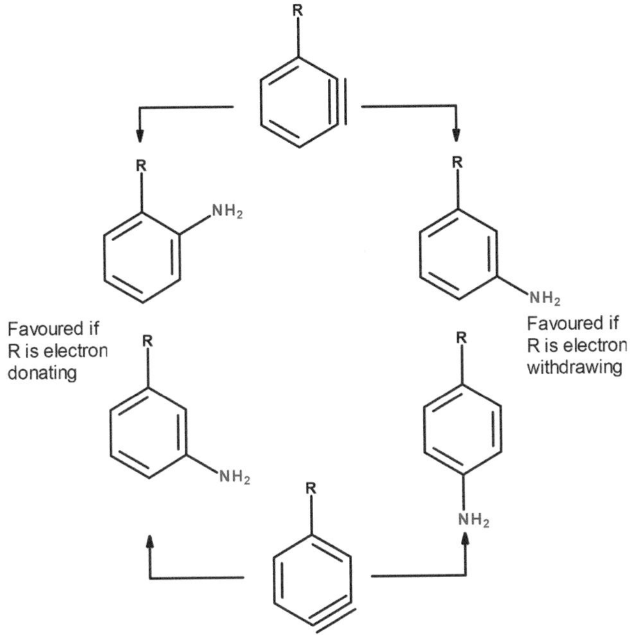

SCHEME 9.25 When substituent is EDG or EWG.

Due to the presence of triple bond in the ring, there is a significant deviation from the preferred 180° bond angle that leads to a low-lying LUMO. Low LUMO explains extreme reactivity toward nucleophiles and dienes.

Moreover, the Diels–Alder reaction of Furan with benzyne gives an adduct; in the second reaction, we have 2+2 cycloaddition of benzyne and an olefin leading to a fused cyclobutane derivative.

9.12 IPSO, CINE, AND TELE SUBSTITUTIONS

Let us look at *ipso* substitution, *cine* substitution, *tele* substitution, vicarious substitution, and ANRORC (Addition of the Nucleophile Ring Opening Ring Closure) reaction.

Why study *ipso* attack or *cine* substitution? Is it because of academic interest or do we have any industrial applications? Although it started more out of academic curiosity, it was later on found to have some useful applications. *Ipso* attack is now extensively used in organometallic chemistry or arene metal complexes.

There are some unexpected products formed in the bromination reaction of *tert*-butyl benzene.

In the above reaction, all the three substituted bromo derivatives are formed. But in addition to that, a very minor

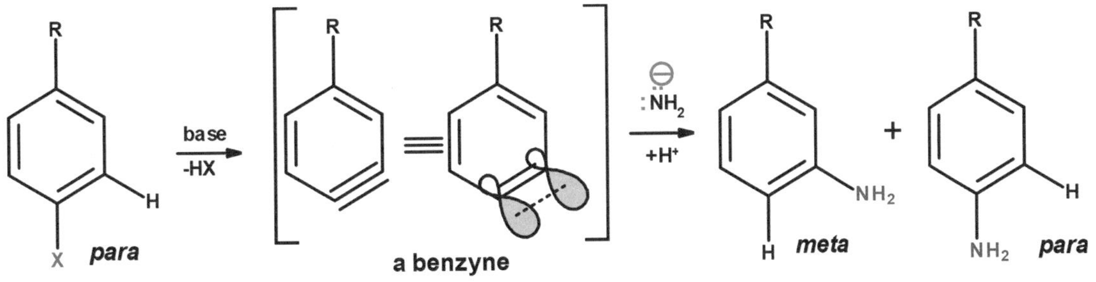

SCHEME 9.26 Mechanism for replacement of para substituent by nucleophile.

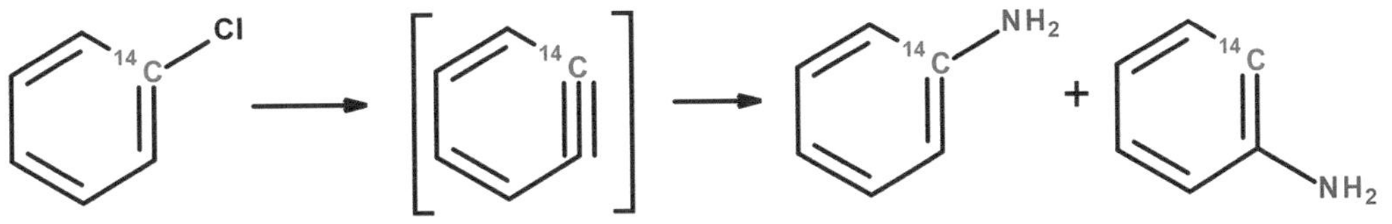

SCHEME 9.27 Evidence for benzyne mechanism.

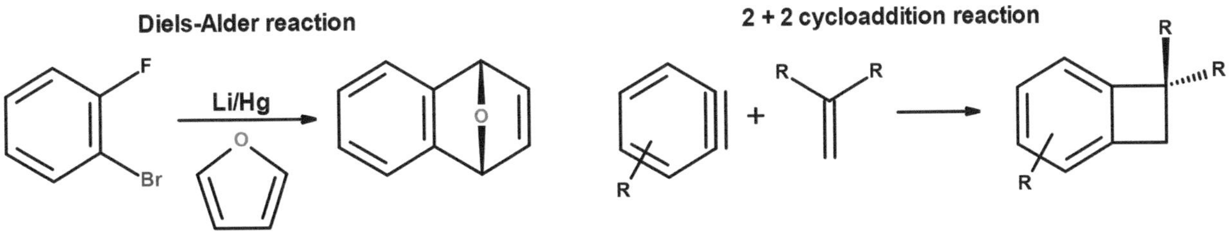

SCHEME 9.28 Diels–Alder reaction and 2+2 cycloaddition of benzyne.

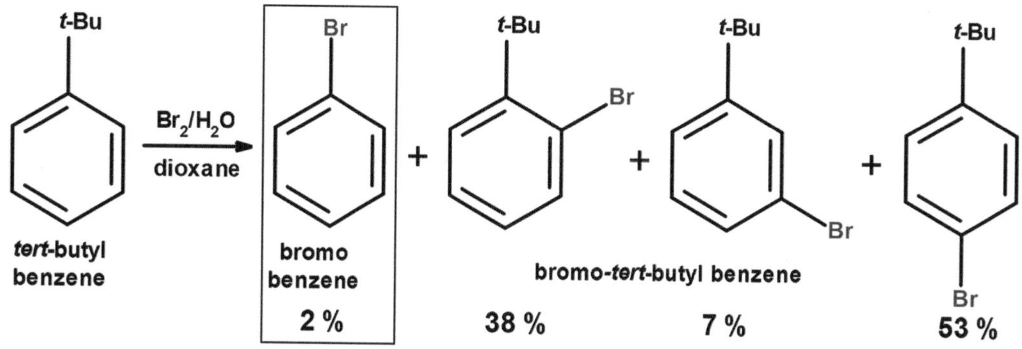

SCHEME 9.29   Bromination reaction of *tert*-butyl benzene.

amount of dealkylated bromo derivative is also formed. So far whatever mechanism we have studied cannot be used to explain the product formation. There is no strong nucleophile, so the benzyne type intermediate is ruled out. If we assume, electrophilic substitution, bromo derivative formation can be explained, but we cannot account for the loss of *tert*-butyl group. This substitution is called *ipso* attack.

Let us look at a brief overview of various substitution reactions of aromatic systems. We have nucleophilic, electrophilic, and free radical substitutions. All these reactions are influenced by the substituents already present in the aromatic system. The substituents exert two kinds of effects one is to dictate regioselectivity or we call that as directing effect. We have *ortho-para* directing and *meta* directing groups. When we talk about which product is formed and at what rate, we label the substituents as *activating* and *deactivating* groups.

9.13   TYPES OF AROMATIC SUBSTITUTION REACTIONS

Replacement of hydrogen on benzene by other atoms or groups is called substitution. Different prefixes are used when other atoms or groups are replaced. We can broadly classify them into two major categories. When hydrogen is replaced on a mono-substituted benzene we will get *ortho*, *meta*, or *para* isomer. In a mono-substituted benzene if the atom or group replaced is not hydrogen, then we have *ipso* attack, *cine*, *tele*, substitutions. If the carbanions that bear leaving groups at the nucleophilic center replaces hydrogen from the aromatic ring, it is called vicarious substitution. ANRORC is a special case of ring opening ring closure reaction.

If the substitution happens at a benzylic position, it is called *meso* substitution and for naphthalenes, we have *peri* substitution. Figure 9.7 depicts the various types of substitutions in the aromatic system.

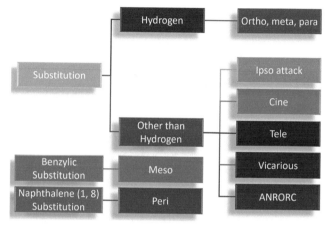

FIGURE 9.7   Various types of substitutions in the aromatic system.

9.13.1   Ipso Attack

In 1971, Pec term "*ipso*" to refer a substituted position in an aromatic ring. According to IUPAC gold book, ipso attack is defined as "the attachment of an entering group to a position in an aromatic compound already carrying a substituent group (other than hydrogen). The entering group may displace that substituent group but may itself also be expelled or migrate to a different position in a subsequent step" (Figure 9.8).

We have four types of substitutions taking place. Reactions 2, 3, and 4 are related to *ortho*, *meta*, and *para* substitution of haloarenes, respectively. But the first one is unprecedented. In the first one, we have the *ipso* attack where the incoming group replaces the halide. If we take the simple nitration of bromobenzene, which gives nitrobenzene as one of the products, we have difficulty in explaining nitronium ion replacing bromide.

Here we see two examples of *ipso* attack (Scheme 9.30). In both the reactions, the electrophile is proton (H$^+$). In the

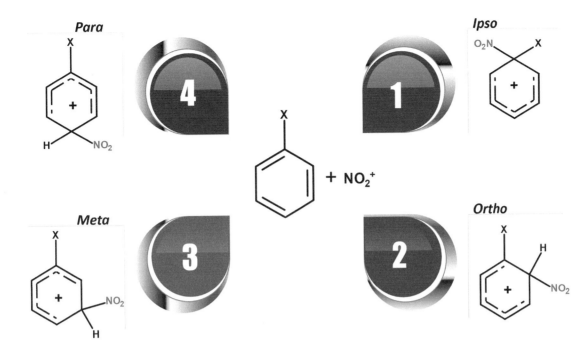

FIGURE 9.8 Different types of electrophilic substitutions of haloarenes.

first reaction, the leaving group is a trimethylsilyl group. In the second reaction, the leaving group is sulfonic acid group. Desilylation means the removal of silyl group. When we say *Protodesilylation* what we mean is the replacement or removal of silyl group by proton. To put the terminology into simple words, we say the first part does the second part. What is the first part proto, that is, proton, what is the second part, desilylation. Proton, that is, the acid removes the silyl group.

As we saw earlier, in sulfonation reaction, the first step is reversible. In the presence of a strong acid, the sulfonic acid group is lost. In such a scenario, a question may come to your mind, is it even possible to make sulfonic acids?

In the electrophilic sulfonation case, we use oleum, which is highly acidic, under that condition the polar sulfur trioxide electrophile is highly reactive. Although the initial step is reversible, we still get the final product, but the yields are not very high. In the *ipso* attack, we are using water. So essentially, we are talking about two different reaction conditions.

In fact, the use of sulfonic acid group as a temporary blocking group in organic synthesis is very common.

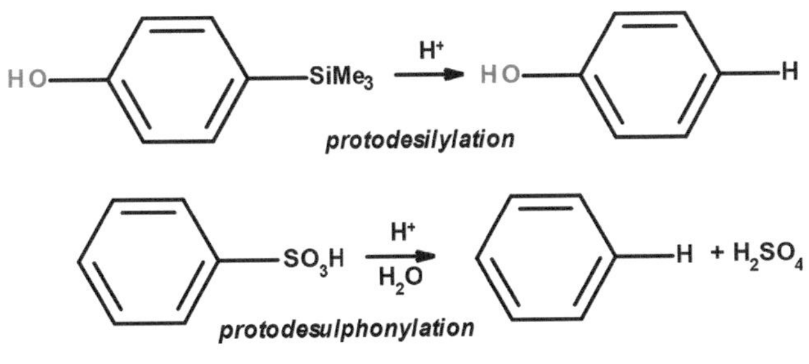

Since *ipso* attack can occur for both electrophilic as well as nucleophilic substitution, there will always be a leaving group. If it is electrophilic substitution, then alkyl groups are good leaving groups. The leaving group ability is similar to carbocation stability, a *tert* carbocation is a better leaving group than secondary or primary. There are instances where bromonium or iodonium ion can also be a good leaving group but not chloronium ions. In the case of nucleophilic *ipso* attack, halides are good leaving groups.

Nitrodealkylation and *nitrodehalogenation* are some examples of electrophilic *ipso* attack. Since the *ipso* attack requires the loss of a group other than a proton, the leaving group should leave as a cation in electrophilic substitutions. In fact, this is clearly seen in the nitration of p-diisopropylbenzene, the *ipso* attack is the major pathway. In the case of nitration of *p*-bromoanisole, around 30% of *ipso* attack takes places, and here the nitro group replaces bromine.

To study about *ipso* attack, the nitration reaction is very useful. The products obtained are so varied that they cover up to five different scenarios as given below (Figure 9.9).

Path a: The arenium ion can lose $^+NO_2$ and go back to the starting compounds. That means we do not see any reaction and the starting material is recovered. The problem is that this change cannot be identified or measured.

Path b: The arenium ion can lose the cation Z^+, in which case this is simply an aromatic substitution with a leaving group other than H. Some examples include *hydro-de-alkylation* occurring in Friedel–Crafts reaction. We had seen this as an unwanted side reaction in FCR alkylations. Loss of *tertiary* carbo cation is prominent compared to secondary or primary carbocation loss. Another example is the formation of arenes from Grignard reagents. The *hydro-de-metallation* occurs in the presence of acidic

SCHEME 9.30 Protodesilylation and protodesulfonation.

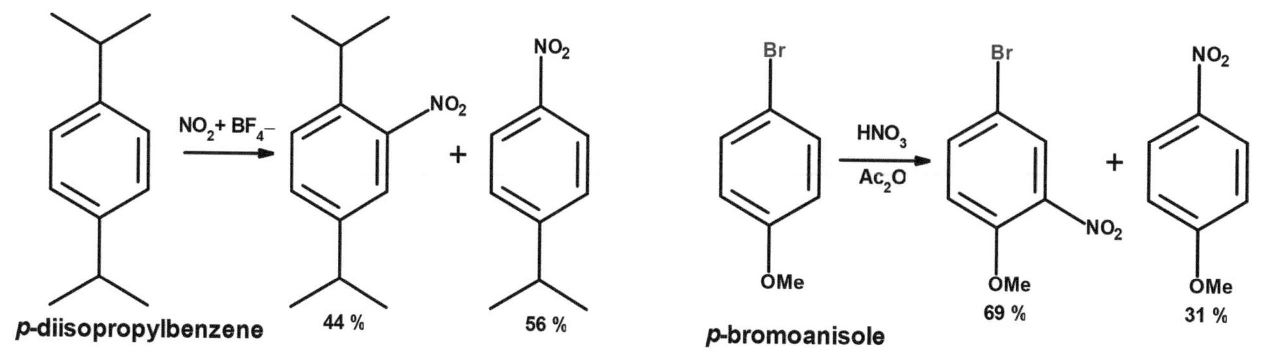

SCHEME 9.31   Nitrodealkylation and nitrodehalogenation.

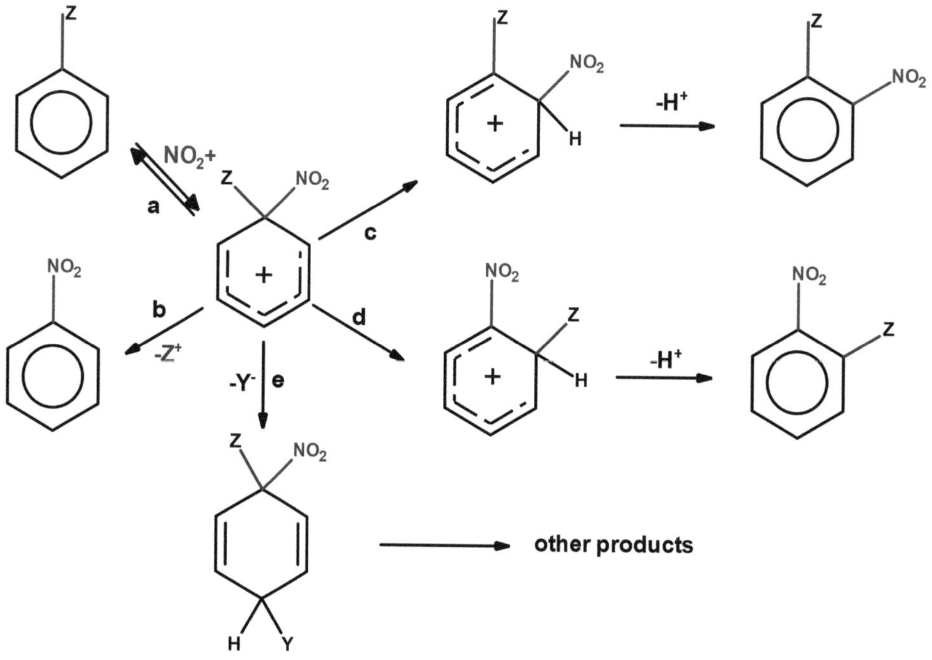

FIGURE 9.9   Different possible paths for nitration.

substances. This again is an unwanted side reaction. This is one of the reasons, we use dry solvents (no moisture) for Grignard reagent preparation.

Path c: The electrophilic group (in this case ⁺NO₂) can undergo a 1,2-shift, followed by loss of proton. If the migration takes place to the second position, it is normal *ortho*-substitution. We do not have any tools to find out whether the *ortho*-substitution occurred by 1,2-shift or direct replacement by the electrophile.

Path d: There are similarities and differences between the previous path and this path. Similarity is 1,2-shift. The difference is which group migrates. In the previous case, it was the electrophile that migrated, and in this case, it is the *ipso* substituent that migrates. The *ipso* substituent (Z) can undergo 1,2-migration, which also produces the *ortho*

product. This pathway is very minor, at least when the electrophile is ⁺NO₂.

Path e: In the reaction medium there are nucleophiles present. Now these nucleophiles can attack the *para* position. In some cases, the products of such an attack (cyclohexadienes) have been isolated but further reactions are also possible.

9.13.1.1   Possible Reactions of ipso Intermediate

The *ipso* intermediate can undergo various reaction pathways, some of which are discussed here.

    a. Loss of the electrophile or loss of ipso substituent
    b. Nucleophilic capture
    c. Substituent migration

9.13.1.2 Loss of ipso Substituent

We already saw the bromination of *tert*-butylbenzene where the loss of the *ipso* substituent occurs, that is, the *tert*-butyl group is lost. This reaction occurs only if the leaving cation is stable.

9.13.1.3 Nucleophilic Capture

Due to their cationic nature, Wheland intermediates can be attacked by the nucleophile if they exist in the reaction mechanism. Thus, during the nitration of *ortho*-xylene in acetic anhydride, we get the nitro derivatives. The formation of the major product of the reaction, 4-acetoxy-o-xylene, was due to the nucleophilic attack of the solvent on the Wheland intermediate.

It is even proved later by the isolation of two epimeric intermediates, which when lose nitrous acid give the acetoxy derivative.

9.13.1.4 Side Chain Modification

In the chlorination of hexamethylbenzene, the major product is a side chain modified product which is the chloromethyl derivatives. This reaction is second order in both chlorine and substrate and the rate is unaffected by light. This indicates that the reaction does not involve free radical species, rather it occurs with an initial heterolytic electrophilic attack on the ring.

The rearrangement step involves, the most likely formation of a methylene derivative and intramolecular transfer of the chlorine atom.

9.13.1.5 Substituent Migration

Migration of either the *ipso* substituent or the electrophile may also happen. Nitration of 2,6-dibromo-4-methylphenol gives a product by replacement of one bromide by the nitro group. Although the reaction looks straightforward electrophilic substitution, it was shown to involve a 1,3- migration of nitro group. Initially, the nitro group does not attack the bromo carbon but attacks the methyl carbon. After this, a 1,3-shift of the nitro group occurs from the methyl-attached carbon to the bromo carbon, bromine is lost to give the final nitro product. The 4-nitro dienone is the intermediate.

9.13.2 Cine Substitution

A substitution reaction in which the incoming group takes up a **position adjacent** to that occupied by the leaving group is called cine substitution. If you recall, when we studied the benzyne intermediate, we have seen similar examples.

9.13.3 Tele Substitution

A substitution reaction in which the incoming group takes up a position **more than one atom away** from the atom to which the leaving group is attached is called tele substitution.

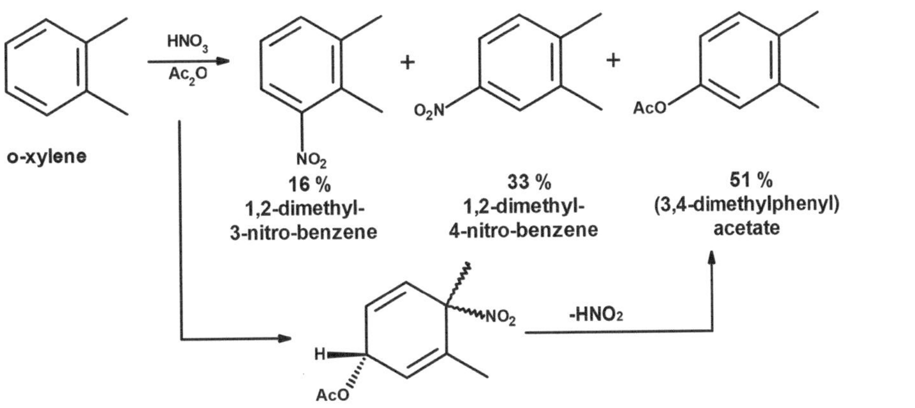

SCHEME 9.32 Nitration of o-xylene.

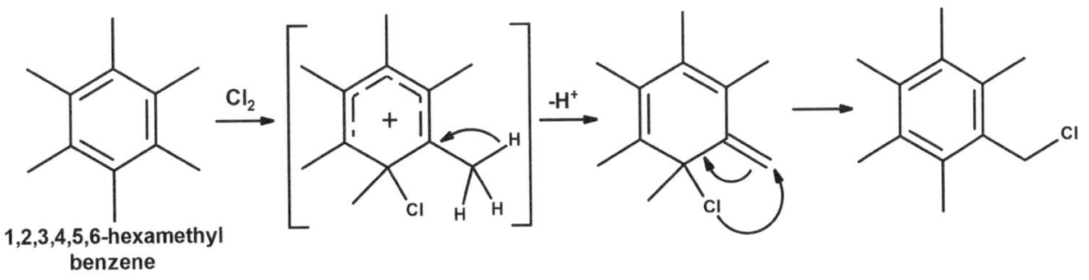

SCHEME 9.33 chlorination of hexamethylbenzene.

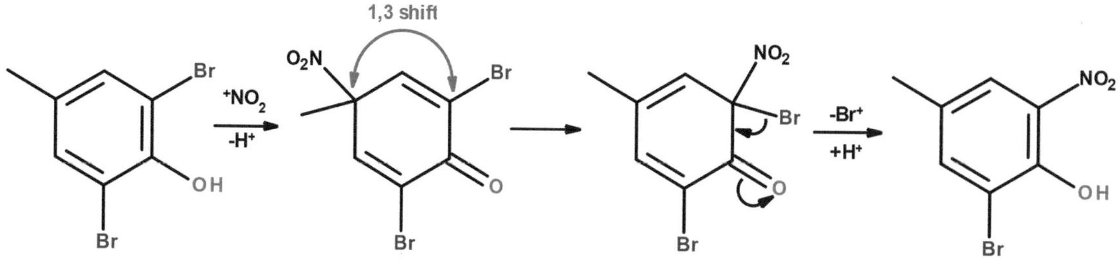

SCHEME 9.34   Example for migration of either the ipso substituent or the electrophile.

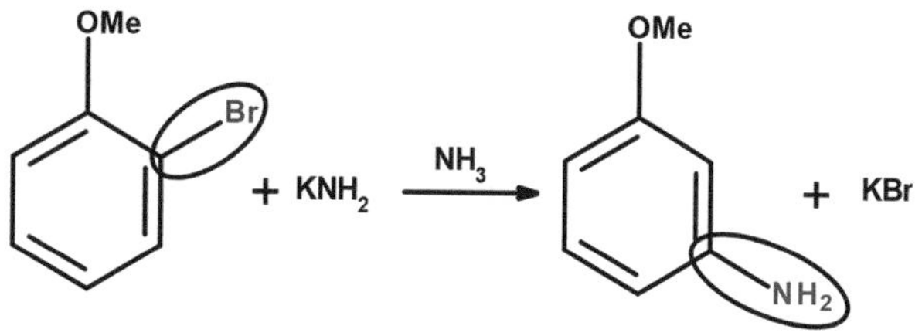

SCHEME 9.35   Example for cine substitution.

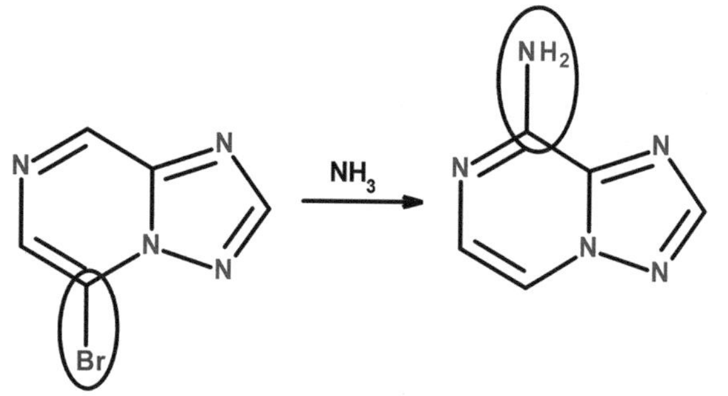

9.13.4   Vicarious Substitution

A nucleophile (carbanions that bear leaving groups at the nucleophilic center) replaces a hydrogen atom on the aromatic ring and not leaving groups such as halogen substituents which are ordinarily encountered in S_NAr.

Nitroarenes react with carbanions that are usually generated from active methylene compounds. The nitrobenzylic carbanions that are formed in the reaction are highly colored (blue, red, etc.). Acidic work-up leads to the final products.

9.13.5   Meso Substitution

Substitution of benzylic hydrogen by other groups is called mess substitution. It is observed in compounds such as calixarenes (formed by the condensation between *para*-substituted phenols and aldehydes) and acridines.

9.13.6   Peri Substitution

Substitution occurs at eight positions of naphthalenes.

An overview of various approaches to substituted aromatics are given here.

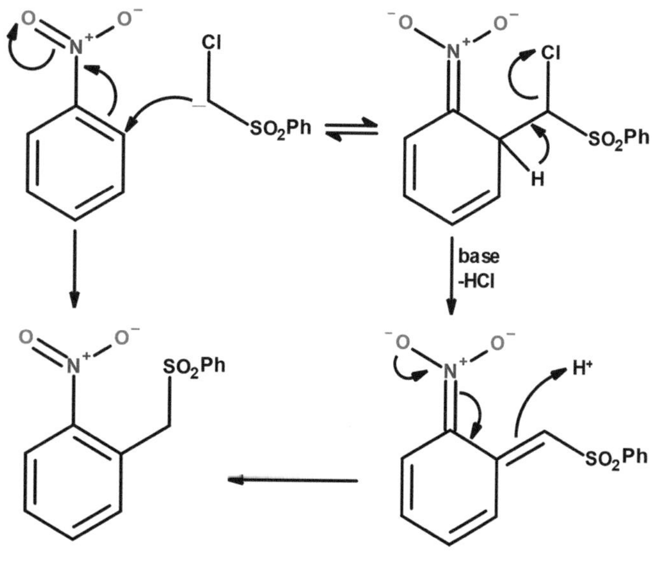

SCHEME 9.36   Example for tele substitution.

SCHEME 9.37   Example for vicarious substitution.

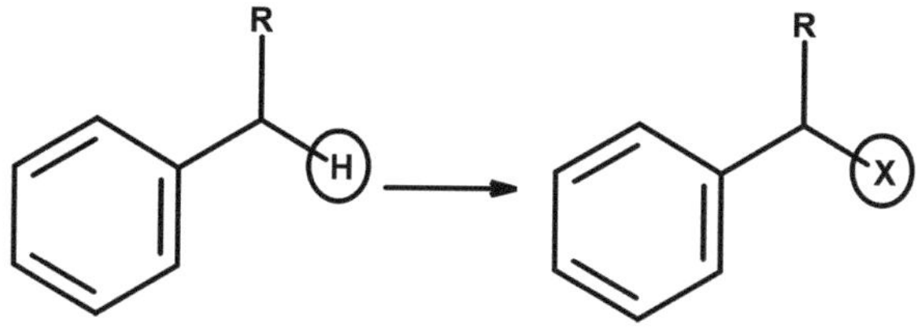

SCHEME 9.38   Example for meso substitution.

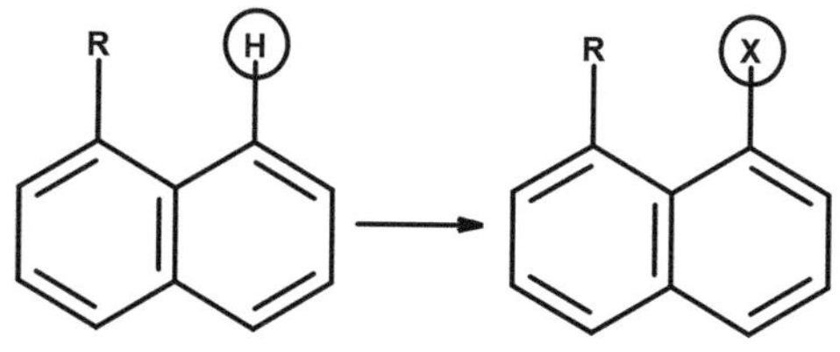

SCHEME 9.39   Example for peri substitution.

9.13.7 σ^H Adducts in Nucleophilic Substitution

The transformation of aromatic nitro compounds to carboxylic acids through the use of alcoholic potassium cyanide was first described in 1871 by von Richter.

When halonitrobenzenes undergoes nucleophilic substitution, there are two possibilities. One is on the carbon carrying the halogen, that is, formation of σ^X adduct. Another one is addition to the *ortho* carbon having hydrogen to form the σ^H adducts. Experimental evidence has shown that the formation of σ^X adducts is slower than the departure of the X anion from the adducts. In other words, the addition is an irreversible process. We extend the same to σ^H adducts as well. Logically, the actual competition lies between σ^X adducts and σ^H adducts only, whichever forms faster decides the product formation.

Initial fast and reversible addition in the position *ortho* to the nitro group gives σ^H adducts. Since hydride anion is a very poor leaving group it cannot depart in a spontaneous reaction. Since the addition is a reversible process, the σ^H dissociates to the starting material. This is followed by slower addition at the *para* position to form σ^X adducts. Since halide is a good leaving group, it departs as a chloride

ion from the σ^X adduct giving the products of the S_NAr substitution. This general picture implies that σ^H adducts are initial products of the interaction of nitroarenes with nucleophiles, if we can find efficient ways for fast conversion of the σ^H adducts we can develop a general process of nucleophilic substitution of hydrogen. Moreover, in the reaction between nucleophiles and halonitrobenzenes, the nucleophilic substitution of hydrogen should be the primary, fast process whereas the conventional S_NAr of halogen should be a slower secondary process.

von Richter initially assumed that cyanide ion, from which the carboxyl group is derived, displaced the nitro group directly from its position on the aromatic nucleus. But later he concluded that the carboxyl function must take up a position on the aromatic ring *ortho* to that vacated by the nitro group, which is cine substitution.

Initially, the cyanide nucleophile adds to the *ortho* position of the nitro group. The negative oxygen attacks the electropositive cyanide carbon and forms a cyclic derivative. This opens up to give an amide which leads to a pyrazolone derivative which on hydrolysis loses nitrogen to give the carboxylic acid.

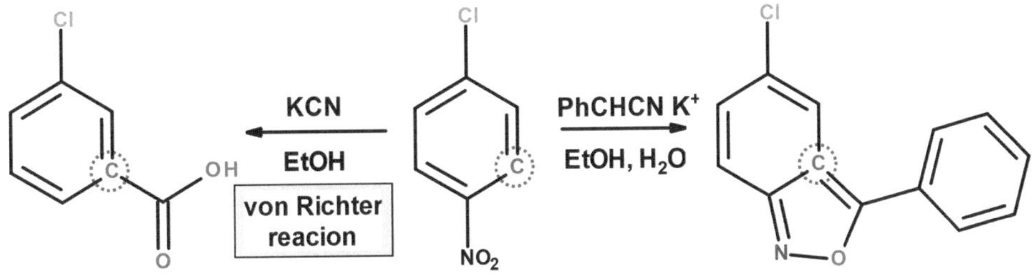

SCHEME 9.40 Example for von Richter reaction.

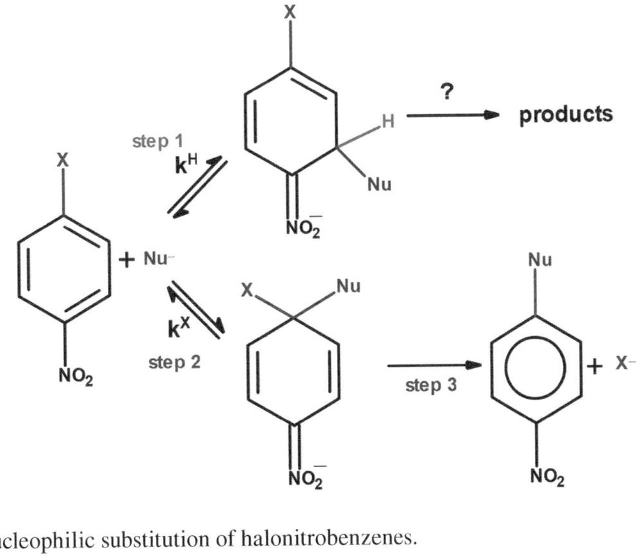

SCHEME 9.41 Example of nucleophilic substitution of halonitrobenzenes.

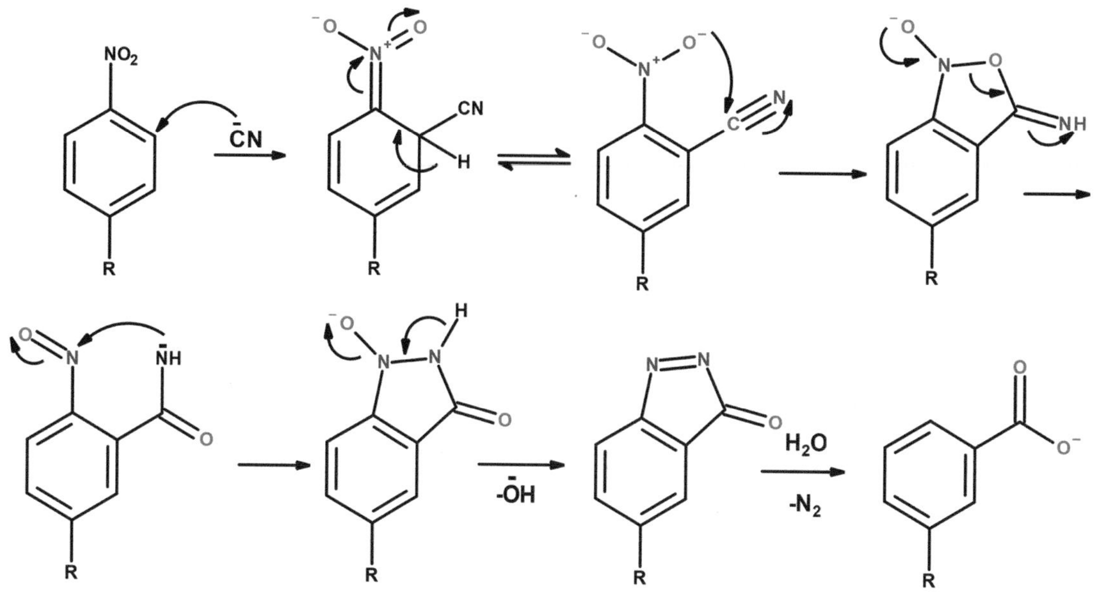

SCHEME 9.42   Mechanism for von Richter reaction.

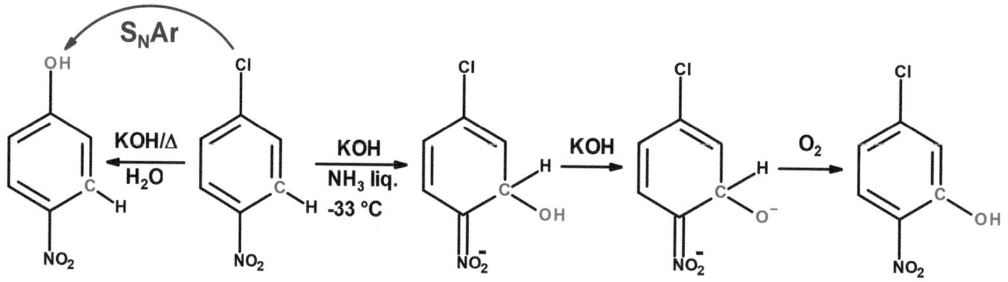

SCHEME 9.43   Example of hydroxide introduction into halo-nitrobenzene.

9.13.8   S$_N$Ar vs Oxidative Nucleophilic Substitution

Here we see two examples of hydroxide introduction into halo-nitrobenzene. The first one is the aromatic nucleophilic substitution where the chloride is replaced by the hydroxide, the ipso attack. The other example is the oxidative addition.

9.14   QUESTIONS

1. Predict the product for the following transformation (Figure 9.10).
2. Why the following transformation B is not observed but A is possible (Figure 9.11).
3. In the following reaction, one of the products is not formed. Identify which one is not formed and give the reason (Figure 9.12).
4. Write the mechanism of nitration in the presence of perchloric acid.
5. Which one of the substrates undergoes reaction with NaOCH$_3$ (Figure 9.13).

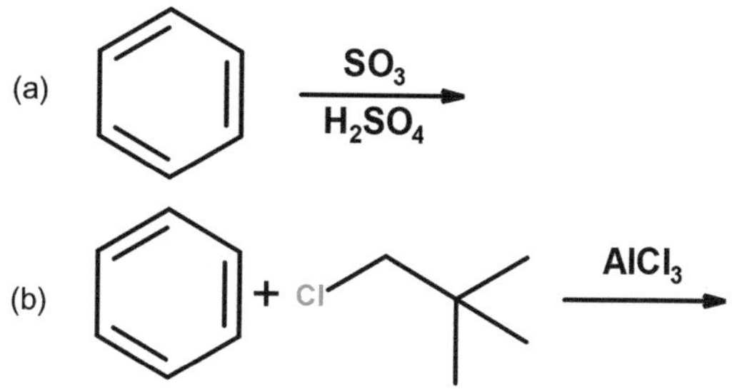

FIGURE 9.10   Question 1.

6. Can you plan a synthetic route to make 2,4-dinitrobenzonitrile from benzene. (hint see the previous question)
7. Can you propose a mechanism for the following transformation? Identify the reagents also (Figure 9.14).
8. Propose a mechanism for the formation of A (Figure 9.15).

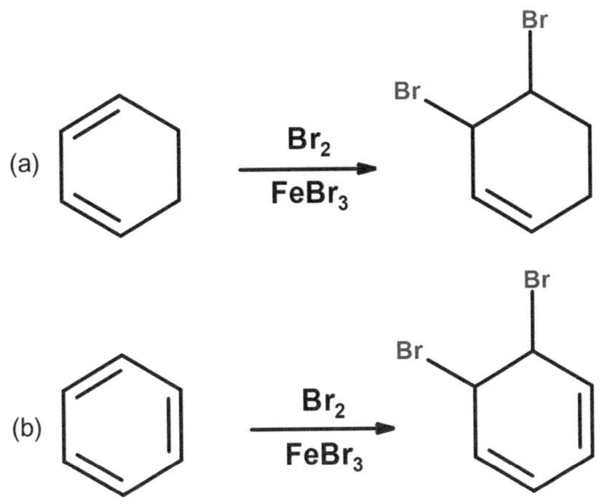

FIGURE 9.11 Question 2.

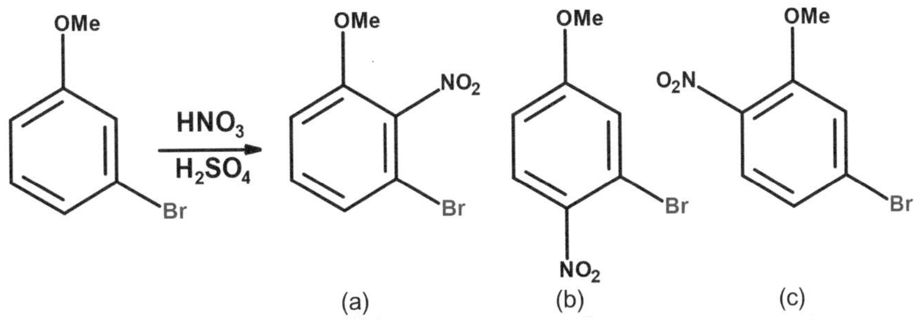

FIGURE 9.12 Question 3.

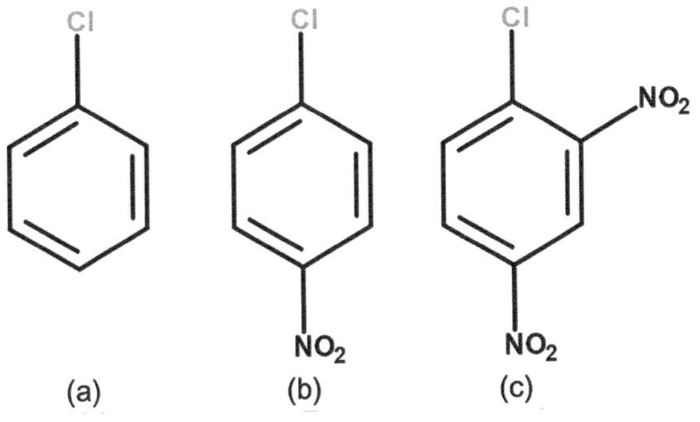

FIGURE 9.13 Question 5.

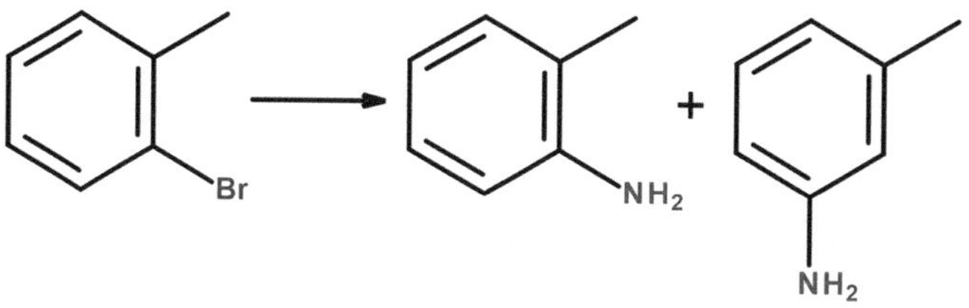

FIGURE 9.14 Question 7.

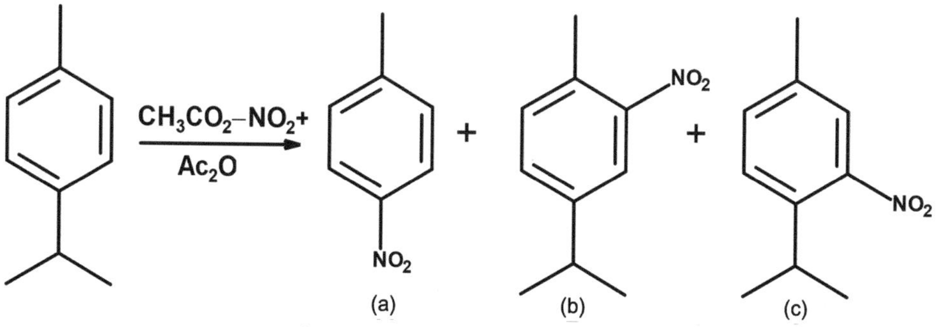

FIGURE 9.15 Question 8.

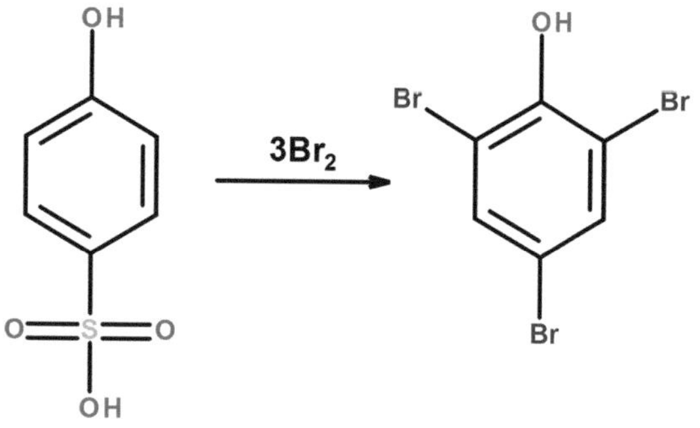

FIGURE 9.16 Question 9.

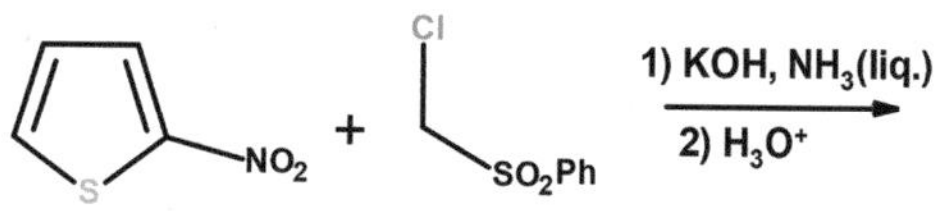

FIGURE 9.17 Question 10.

9. Propose a mechanism for the formation of the product (Figure 9.16).
10. Identify the product formed in the following transformation (Figure 9.17)

BIBLIOGRAPHY

Auwers, K. (1902). Ueber das Nitroketon und das Chinol des Dibrom-*p*-kresols. *Chemische Berichte*, 35, 455.

Baciocchi, E., et al. (1965). The side-chain halogenation of methylbenzenes via electrophilic nuclear attack. IV. Product analysis, kinetics, and mechanism of the chlorination of some hexasubstituted benzenes. *Journal of the American Chemical Society*, 87, 3953.

Benesi, H. A., & Hildebrand, J. H. (1949). A spectrophotometric investigation of the interaction of iodine with aromatic hydrocarbons. *Journal of the American Chemical Society*, 71, 2703–2707.

Blackstock, D. J., et al. (1970). Diene intermediates in the acetoxylation of o-xylene. *Journal of the Chemical Society D: Chemical Communications*, 641a.

Blackstock, D. J., et al. (1970). Isomeric diene intermediates and an acetate rearrangement in aromatic acetoxylation. *Tetrahedron Letters*, 2793.

Bruice, P. Y. *Organic Chemistry*, 4th Ed. Prentice Hall, 2003, 607.

Clayden, J., Greeves, N., & Warren, S. *Organic Chemistry, Chapter 22, Electrophilic Aromatic Substitution*, 2nd Ed. Oxford University Press, 2014, 547, 589.

David Feldman et al. (1991). Nucleophilic aromatic substitution by hydroxide ion under phase-transfer catalysis conditions: Fluorine displacement in polyfluorobenzene derivatives. *The Journal of Organic Chemistry*, 56, 7350–7354. (https://chem.ch.huji.ac.il/rabinovitz/refs/154.pdf) (accessed, 15-Apr-2024)

de la Mare, P. B. D., & Harvey J. T. (1957). The kinetics and mechanisms of aromatic halogen substitution. Part III. Partial rate factors for the acid-catalysed bromination of *tert.*-butylbenzene by hypobromous acid. *Journal of Chemical Society*, 1957, 131.

Farmer, S., et al., 16.7: Nucleophilic Aromatic Substitution (https://chem.libretexts.org/Bookshelves/Organic_Chemistry/Map%3A_Organic_Chemistry_(McMurry)/Chapter_16%3A_Chemistry_of_Benzene_-_Electrophilic_Aromatic_Substitution/16.07_Nucleophilic_Aromatic_Substitution) (accessed, 15-Apr-2024).

Feldman, D., et al. (1991). Nucleophilic aromatic substitution by hydroxide ion under phase-transfer catalysis conditions: Fluorine displacement in polyfluorobenzene derivatives. *The Journal of Organic Chemistry*, 56, 7350–7354. (https://chem.ch.huji.ac.il/rabinovitz/refs/154.pdf) (accessed, 15-Apr-2024).

George A. Olah et al. (1982). Recent aspects of nitration: New preparative methods and mechanistic studies (A Review). *Proceedings of the National Academy of Sciences of the United States of America*, 79, 4487–4494 (https://www. pnas.org/content/pnas/79/14/4487.full.pdf) (accessed, 15-Apr-2024).

Gordon, J. L. M. (1993). University of Canterbury, IPSO ATTACK OF AROMATIC SYSTEMS (https://ir.canterbury.ac.nz/bit-stream/handle/10092/7859/gordon_thesis.pdf?sequence=1) (accessed, 15-Apr-2024)

Mąkosza, M. (2010). Nucleophilic substitution of hydrogen in electron-deficient arenes, a general process of great practical value. *Chemical Society Reviews*, 39, 2855.

Martyn, R. J. (1985). University of Canterbury, IPSO AROMATIC SUBSTITUTION. (https://core.ac.uk/download/pdf/3546 8597.pdf) (accessed, 15-Apr-2024)

Rosenblum, M. (1960). The mechanism of the von Richter reaction. *Journal of the American Chemical Society*, 82(14), 3796–3798.

Smith, M. B., & March, J. *March's Advanced Organic Chemistry Reactions, Mechanisms and Structure, Chapter 11, Aromatic Electrophilic Substitution, Chapter 13, Aromatic Nucleophilic Substitution*, 5th Ed. John Wiley & Sons, Inc., 2001, 675, 686, 850.

Snieckus, V. (1990). Directed ortho metalation. Tertiary amide and O-carbamate directors in synthetic strategies for polysubstituted aromatics. *Chemical Reviews*, 90(6), 879–933.

Sykes, P. *A Guidebook to Mechanism in Organic Chemistry*, 6th Ed. John Wiley & Sons, Inc., 1985, 130, 167.

Vasilyev, A. V., et al. (2002). Halogen bonding in supramolecular chemistry: minireviews. *New Journal of Chemistry*, 26, 582–592.

Wayne, F. K. Schnatter, et al. (2013). Electrophilic aromatic substitution: enthalpies of hydrogenation of the ring determine reactivities of C_6H_5X. The direction of the $C_6H_5–X$ bond dipole determines orientation of the substitution. *Journal of Physical Chemistry A*, 13079. https://doi.org/10.1021/jp409623j

10 Reaction Dynamics and Reaction Kinetics

In this chapter, we will focus on reaction dynamics. Let us begin our quest to decipher the million-dollar question "how to reach the destination?" (Figure 10.1)

When you travel, sometimes you are concerned about when will you reach the destination and some other times you will be thinking about how you will reach the destination may be by bus, auto or taxi. If you think about philosophy, for some people reaching a destination is important and for some other people it is the path they follow is crucial.

For an organic chemist, both the path followed and the destination are important. I mean, the product, which is the destination and the reaction mechanism, which is the path followed are both crucial. If you know the reaction mechanism, you can extend that to similar situations. If you know the product, you can attempt to predict the mechanism by using experimental techniques or computational approach. Parameters affecting the course of the reaction include (solvent, reagent, catalyst, temperature, concentration, substituents, etc.)

10.1 BASIC TERMINOLOGIES TO DESCRIBE A REACTION

10.1.1 REACTION COORDINATE DIAGRAM

Reaction coordinate is a graphical representation of snapshots that represent a **set of atomic motions** that are essential and form the basis of the change from reactants to products. The axis may be time, structural parameter, or group of parameters, for example, fraction of bond dissociation. When the reaction coordinate is plotted against the **potential energy or free energy of the system**, it is called the reaction coordinate diagram. They may have peaks (high energy, transition state) and troughs (low energy, intermediate/product).

10.1.2 FREE ENERGY AND REACTION

Here we have three different situations (Figure 10.2) A, B, and C. In A, the starting material is lower in energy compared with the product. The free energy change for this reaction is greater than zero. This reaction is neither favorable nor spontaneous. In B, the starting material is higher in energy compared with the product. The free energy change for this reaction is less than zero. This reaction is favorable and spontaneous. In C, the starting material is much higher in energy compared with the product. The free energy change for this reaction is largely negative. This reaction is highly favorable and spontaneous. Here we are looking at the feasibility of the reaction to occur. We do not know how long it will take for the reaction to complete.

10.1.3 ACTIVATION ENERGY AND REACTION

Let us look at the relationship between activation energy and the progress of a reaction. In A and C, the activation energy is high. So the reaction is slow. In B, the activation energy is low, so it is fast.

Comparison between free energy and activation energy

In Table 10.1, we see the comparison between Gibbs free energy and activation energy for a reaction.

When we talk about stability (free energy) what we mean is what products are formed and when we talk about the rate of reaction (activation energy) what we mean is how fast the products will be formed.

The free energy change can be positive, negative, or even zero. When ΔG is positive, the reaction is non-spontaneous, and when it is zero, it is an equilibrium reaction, and when it is –ve, it is a spontaneous process. On the other hand, activation energy is always needed for organic reactions. There are a few biochemical processes where the activation energy is zero or negative. Some of the examples include pressure-induced unfolding and the urea-induced unfolding of proteins at ambient pressure (negative activation energies), that is, these processes go faster at lower temperatures.

The following two statements look confusing or contradicting:

FIGURE 10.1 How to reach the destination?

DOI: 10.1201/9781032631165-10

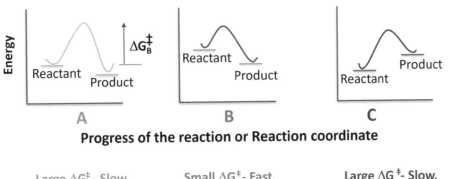

FIGURE 10.2   Relationship between activation energy and progress of a reaction.

TABLE 10.1
Comparison between Gibbs Free Energy and Activation Energy for a Reaction

Free Energy ($\Delta G°$)	Activation Energy (ΔG)
Related to enthalpy ($\Delta H°$) and entropy ($\Delta S°$). ($\Delta G° = \Delta H° - T\Delta S°$)	It is the energy difference between ground-state energy (starting material) and the transition state energy
Related to stability	Related to the rate of reaction
Related to thermodynamic control	Related to kinetic control
Change can be positive, negative, or zero for a reaction	Mostly positive, in rare cases it may be negative or zero for a reaction
Changes with temperature	Independent of temperature

When we increase the temperature generally many reactions become faster.

Activation energy is independent of temperature.

These two statements are correct. Activation energy is independent of temperature. When we increase the temp, we are not reducing the energy barrier. But we are increasing the number of molecules having a certain threshold energy to undergo reaction. An increase in temperature leads to an increase in kinetic energy, but it does not change the activation energy.

Let me explain this with an analogy. You are traveling from place A to place B. The distance between them is 10 km. If you walk (5 kmph), it will take 2 hours to reach place B. But if you travel by car (60 km/h), it will take 10 minutes. Reaching the destination faster does not reduce the distance between the places. The distance is the activation energy which remains same. You reach quickly when you have a faster mode of transportation (smaller activation energy).

10.1.4   KINETIC AND THERMODYNAMIC CONTROLLED REACTIONS

Let us look at a chemical reaction compound A giving compound B. Compound A undergoes the reaction through a high-energy transition state TS1 to form product B. The free energy difference between A and B is represented as $\Delta G_B°$. The activation energy is represented as $\Delta G^{\ddagger}_B$ and the energy of activation, E_a.

Note: According to IUPAC Gold book (https://doi.org/10.1351/goldbook.T06470) the symbol for Gibbs energy of activation is given as $\Delta^{\ddagger}G_{rtn}$ (where the double dagger ($\ddagger$) symbol is a prefix). In this book, we are using a more widely followed conventional representation, where the symbol double dagger ($\ddagger$) is used as suffix and not as prefix.

Let us look at another chemical reaction A giving C. Compound A undergoes the reaction through a high-energy transition state TS2 to form the product C. The free energy difference between A and C is represented as $\Delta G_C°$. The activation energy is represented as $\Delta G^{\ddagger}_C$ (Figure 10.3)

Let us compare these two reactions: A giving B and A giving C. Based on the reaction profile, we can make a few observations:

1. Let us compare the activation energies for products B and C. E_a for reaction giving product C is smaller than reaction giving product B. We can say, that product C will be formed faster than product B. We call this kinetic control.
2. Let us compare the free energies for products B and C. Product B has much lower energy than product C. We call this thermodynamic control. It is related to the stability of the product.

Thermo means heat. You can remember that in most cases, high temperature favors thermodynamically controlled products.

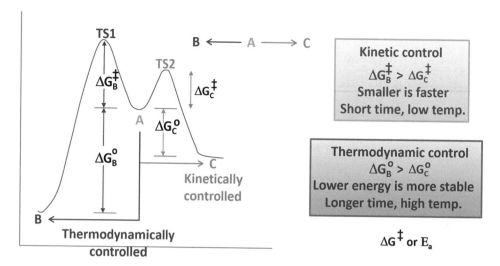

FIGURE 10.3 Thermodynamic and kinetic control of a reaction.

Equilibrium is the basic requirement for studying the thermodynamic control of the system.

10.2 TYPES OF MECHANISMS (BONDS BREAKING)

Bond breaking can be homolytic (free radical), heterolytic (ionic), concerted (bond breaking and bond making occur simultaneously), or pericyclic for reactions. In the concerted mechanism both the bond breaking and forming occurs simultaneously. Heterolytic bond breaking is also called ionic mechanism; homolytic bond breaking is otherwise free radical mechanism.

In an *exothermic* reaction (**A**), the transition state is closer in energy and structure to the **reactants** than to the products (e.g., chlorination). In this reaction, the transition state is reached early, so it resembles reactants more than the products. It is called an early transition state.

In an **endothermic** reaction (**B**), the transition state is closer in energy and structure to the **products** than to the reactants (bromination). In this case, the transition state is reached relatively late, so it resembles the products more than the reactants. It is called late transition state (Figure 10.4).

10.3 HAMMOND POSTULATE

The Hammond postulate states that the transition state of a reaction resembles the structure of the species (reactant or product) to which it is closer in energy.

The Hammond postulate is a useful, qualitative tool that interrelates structural similarities between reactants, transition structures, and products with the exo- or endothermicity of reactions.

We will take the simple example of halogenation of alkanes, specifically chlorination and bromination of *iso*-butane. Chlorination gives two products, and bromination gives only one product. This can be explained by Hammond's postulate.

Why chlorination gives two products and is non-selective? This is an exothermic or exergonic reaction. As per Hammond's postulate, in this reaction, the transition state will have structure and energy similar to the reactants. This being a radical reaction, the chlorine radical can abstract either the primary hydrogen or the tertiary hydrogen from isobutane. The difference in activation energy for these two transition states is not very high, but they are close. Since this reaction has an early transition state and the

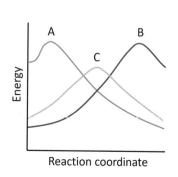

FIGURE 10.4 Exothermic and endothermic reactions.

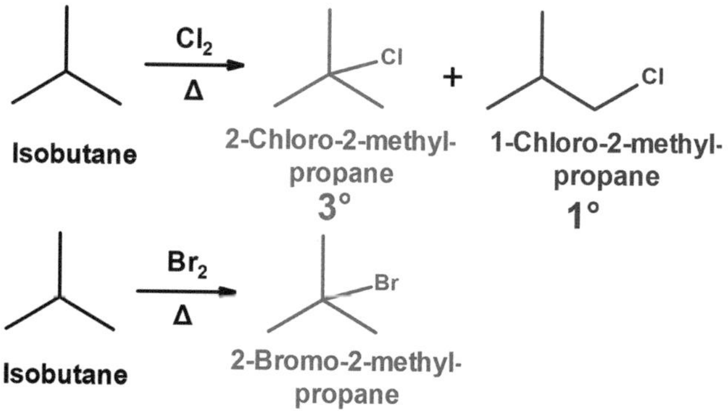

SCHEME 10.1 Chlorination and bromination of isobutene.

difference between the transition states is not high, both reactions occur.

On the other hand, in the bromination reaction, it is an endothermic or endergonic reaction. The transition state is the primary or tertiary radical which then reacts with the halogen to give the final product. Now there is a considerable difference in the energy between these two transition states due to their differing stability. This reaction has a late transition state, and the transition state will have structure and energy similar to the radical intermediate. So, there is excellent selectivity to get the more stable tertiary radical which in turn leads to the tertiary bromo derivative (Figure 10.5).

10.3.1 S_N1 Reaction (Transition States)

- The rate of S_N1 reaction increases as the number of alkyl groups on the carbon having the leaving group increases
- The stability of carbocation increases as the number of alkyl groups on the carbocation increases

The Hammond postulate relates reaction rate to stability. We already know that if the activation energy is low, reaction rate will be higher. Based on that, we can also say that for the S_N1 reaction, activation energy is inversely proportional to the stability of the carbocation. Hammond's postulate provides a quantitative estimation of the energy of a transition state (Figure 10.6).

We have a carbocation intermediate in the S_N1 reaction. This is formed by the loss of the leaving group, which is the first transition state. We also know the stabilities of the carbocations formed by this dissociation will be in the order tertiary > secondary > primary > methyl. According to Hammond's postulate, the tertiary carbocation intermediate will have a structure similar to the starting material because it has energy close to the R-X reactant. On the other hand, the energy of a methyl carbocation is very high. In other words, the structure of the transition state is more similar to the intermediate carbocation than to the starting material. So, the methyl transition state is called the late transition state and it lies far to the right.

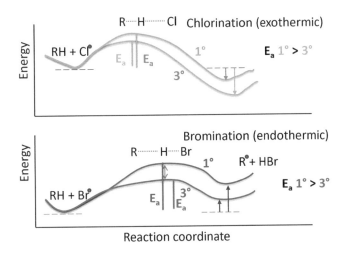

FIGURE 10.5 Potential energy diagram for chlorination and bromination.

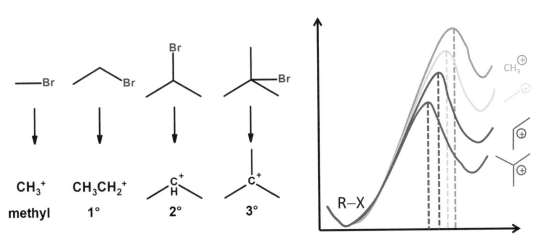

FIGURE 10.6 Relationship between stability of carbocation and energy of a transition state.

10.3.2 Electrophilic Substitution of Benzene

In the first step, there is a temporary loss of aromatic π conjugation. It is a slow step, that is, rate-determining step, because the activation energy is very high for this transition state. This transition state leads to the intermediate I which is shown in the middle. According to Hammond's postulate, the transition state is close in energy to the intermediate I, it will have a structure similar to the intermediate and not the benzene, which is the reactant. In this intermediate I lose a proton to give the product. Loss of proton regenerates the aromatic π conjugation, hence it is a fast step (Figure 10.7).

We need to understand how the activation energy, ΔG‡ changes when R is either an electron-donating substituent or an electron-withdrawing substituent. We only need to know the relative energy of the transition state. We can assume that the transition state looks more like the intermediate than like the starting material because it is close in energy to the unstable intermediate. With this preliminary information we can look at the different intermediates that could be formed by attack at the *ortho, meta,* and *para* positions, and from that transition states, we can try to identify which one will have higher energy.

Let us take the case of electrophilic attack when the benzene ring has an electron-donating group (EDG, here OMe), the following intermediates are possible, depending on whether the electrophile attacks *ortho, meta,* or *para* to the group already present.

Each intermediate is stabilized by delocalization of the positive charge to three carbon atoms in the ring. If the electrophile attacks *ortho* (or *para*) to the EDG, OMe, the positive charge is further delocalized directly onto OMe, but the intermediate in *meta* substitution does not enjoy this extra stabilization. We can assume that the extra stabilization in the intermediate in *ortho* (or *para*) substitution means that the transition state is similarly lower in energy than that in *meta* substitution. Due to the delocalization of a lone pair of electrons from oxygen to the aromatic ring, the *ortho* and *para* positions become moderately electron rich, so they can easily compensate for the loss of electron density due to the electrophilic attack in *ortho* or *para* positions. In addition, the transition states resulting from *ortho* and *para*-attack are lower in energy than the transition state for *meta*-attack. In other words, ΔG‡ is smaller for *ortho/para*-attack and that the reaction is faster than *meta*-attack (Figure 10.8).

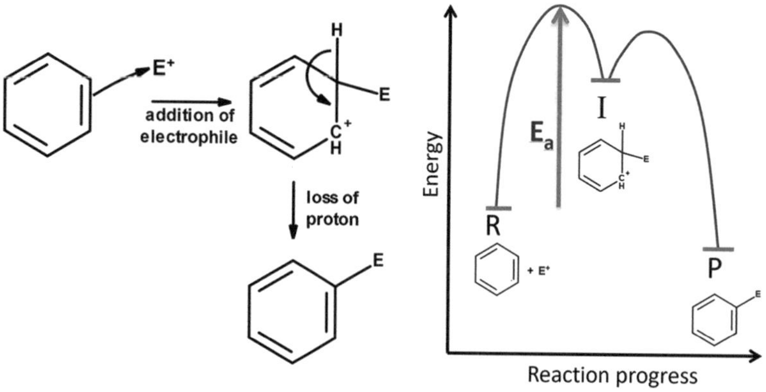

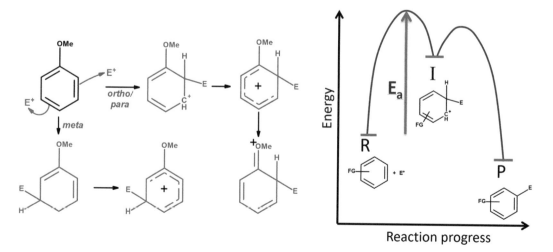

FIGURE 10.7   Electrophilic substitution of Benzene.

FIGURE 10.8   Electrophilic substitution of anisole.

10.3.3 LIMITATIONS OF HAMMOND POSTULATE

Although Hammond's postulate can explain various reaction mechanisms, it fails in certain cases. The Hammond postulate makes a connection between rate (kinetics) and equilibrium (thermodynamics) that has no theoretical basis. Transition states involve kinetics ($\Delta G^{\ddagger}$), and intermediate species involve thermodynamics (ΔG°). Use of the Hammond postulate is based primarily on enthalpy considerations and neglects the effects that activation entropies can have on reaction rates

10.4 LINEAR FREE ENERGY RELATIONSHIP

Here we will study the parameters, substituent constant, and reaction constant that affect the stability and reactivity of transition states or intermediates in a chemical reaction. This will help us in understanding the probable reaction mechanisms. We will be studying Hammett's equation.

When we want to learn about reaction mechanisms, some important questions come to our mind. What parameters are interrelated? By knowing structural effects, orientation, electronic effects, concentration, temperature, degree of dissociation, rate constant, equilibrium constants, and energy changes can we understand reaction mechanisms? In fact, the most famous observation and correlation was made by Hammett. What was Hammett's observation about general reactivity? What Hammett did and how he did it? Why to study LFER? Why Hammett chose a particular type of compound? We will try to find answers to some of these questions.

Earlier, we saw there is a relationship between the energy of the transition state and the structure of either the reactant or product. That is only a *qualitative* observation and not a quantitative one. Now we will approach the reaction mechanism puzzle from a different angle.

As a curious organic chemist, Hammett wanted to *quantify the electronic effects* that affect a chemical reaction. He found that there is a relationship between the nature of the substituent and the acidity constants of substituted benzoic acids. The relationship is based on the ability of the substituents to donate or withdraw electron density. The intriguing question is how Hammett arrived at this relationship. For Kekülé, the benzene structure came in his dream (snake biting its tail).

Hammett did a couple of different experiments: (1) hydrolysis of ethyl esters of aliphatic acids (mainly substituted acetic acids) and (2) hydrolysis of ethyl esters of aromatic acids. It may be due to the following reasons Hammett used hydrolysis of esters for his studies. (1) A large number of substituted benzoic acids are readily available. (2) It is not very difficult to determine the ionization constants for various substituted benzoic acids. (3) The distance from the reaction center and *ortho, meta,* or *para* substituents does not deviate much.

Hammett plotted the rate of hydrolysis of ethyl esters with respect to dissociation of the respective acids. He used

$$ArCO_2Et + {}^-OH \xrightarrow{k} ArCO_2{}^- + EtOH$$

$$ArCO_2H + {}^-OH \underset{}{\overset{K}{\rightleftharpoons}} ArCO_2{}^- + H_2O$$

SCHEME 10.2 Hydrolysis of ethyl esters and dissociation of acids.

logarithmic scale for both axes. In the X axis he used pK_a and in the Y axis, he used log k (rate of ethyl ester hydrolysis). The reason to use a log scale is simple. This will facilitate the coverage of hydrolysis rates to a greater extent. We can also use pK_a directly (because $pK_a = -\log [K_a]$).

When Hammett plotted the entire set of results, he did not find any unique feature. The points were all over the graph. However, he did observe a couple of patterns. (1) The points from the aliphatic esters occupied the top portion of the graph. (2) The *ortho*-substituted benzoic acids occupied the bottom portion of the graph. In other words, they hydrolyzed slowly than the *meta* and *para* isomers. (3) The *meta* and *para* isomers were falling in a narrow range. He finally did the curve fitting for those points and got a straight line. In other words, there is a correlation between the rate of hydrolysis of substituted benzoic acids and their acid dissociation constants at least for *meta* and *para* substituents (Figure 10.9).

On carefully looking at the data, Hammett arrived at the following relationship. There is a linear free energy relationship between the acidity constant and rate constant (Figure 10.10).

Hammett also studied the rate constants of hydrolysis of benzoates and substituted benzoates. Here again, he found out that there is a relationship between the rate of hydrolysis and the electronic properties of the substituents.

The acidity constant can be related to free energy, similarly, the rate constant can be related to free energy. The acidity constant and rate constant vary with respect to the nature of the substituent. The net effect is there may be a linear free energy relationship existing between the substituents and the reaction. Hammett suggested that the free energy change can be considered the sum of three terms: the entropy change, and changes in the kinetic and in the potential energies. When the reactions occur at the side chain of aromatic systems, the entropy term is not appreciably affected by substituents in the *meta-* or *para* position.

For the two reactions (1) hydrolysis of substituted benzoates and (2) equilibrium dissociation of substituted benzoic acids, we can plot, the rate constant for hydrolysis and equilibrium constant for dissociation. This gives a straight line which follows $y = mx + C$ (straight line equation). Here the slope is called the reaction constant, ρ (rho) (Figure 10.11).

10.4.1 DERIVATION OF HAMMETT EQUATION

We can derive the Hammett equation from the following relationship (Figure 10.12).

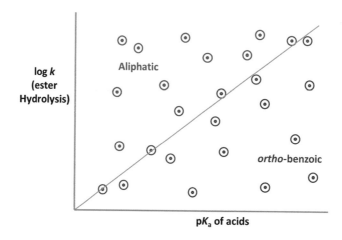

FIGURE 10.9   A plot of the rate of hydrolysis of ethyl esters with respect to dissociation of the respective acids.

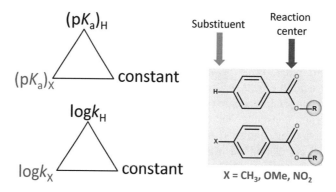

FIGURE 10.10   Relationship between acidity constant and rate constant.

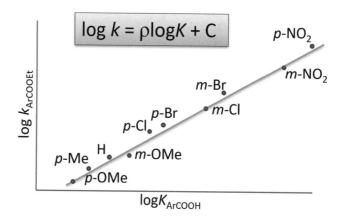

FIGURE 10.11   Plot, the rate constant for hydrolysis and equilibrium constant for dissociation.

$$\log k_X = \rho \log K_X + C \qquad \text{This is for the substituted derivatives}$$

If we subtract the rate and equilibrium constant for unsubstituted derivatives

$$\log k_X - \log k_H = \rho \, (\log K_X - \log K_H)$$

$$\log (k_X/k_H) = \rho \, (\log K_X - \log K_H) \implies pK_a = -\log [K_a]$$

$$\log (k_X/k_H) = \rho \, (pK_{a(H)} - pK_{a(X)})$$

$$\boxed{\log (k_X/k_H) = \rho \, \sigma} \qquad \text{where } \sigma = (pK_{a(H)} - pK_{a(X)})$$

FIGURE 10.12   Derivation of Hammett equation.

Rho (ρ) and sigma (σ) are called reaction constant and substituent constant, respectively.

The Hammett equation is applicable to both *meta* and *para* substituents. Rho is a constant for a given reaction under a given set of conditions, and sigma is the characteristic of a particular substituent. For his generalization, Hammett arbitrarily assigned the value of rho as 1 for ionization of XC_6H_4COOH in water at 25 °C. With this assumption, he calculated σ_m and σ_p for each substituent. If we know the σ for 10–15 substituents, we can then calculate rho for a particular reaction. Once rho is calculated for a particular reaction, we can predict what will be σ for an unknown substituent.

We can easily understand the Hammett equation if we can relate it to cause and effect. Change in the electronic nature of the substituent is the cause that leads to a change in the equilibrium or rate of the reaction which is the effect. We can say there is a linear free energy relationship with *respect to substituent.*

The electronic effect of the substituent leads to the substituent constant and the reaction constant is affected by the type and condition of the reaction. Hammett used the first letter for substitution "s" and reaction "r" in Greek to identify his constants as σ and ρ.

The substituent constant σ is a measure of the total polar effect exerted by a substituent X (relative to no substituent) on the reaction center. This tells about the ability of the substituent to influence or perturb the electron density of the molecule. This perturbation has some effect on the reaction and the reaction constantly talks about it.

We can also write the Hammett equation based on free energy changes. We already know $\Delta G = -RT \ln K$. We can write the Hammett equation as $\log K - \log K_0 = \rho \sigma$. When we replace natural logarithm [ln] with base 10 logarithm, we can arrive at a modified Hammett equation. For a given reaction under a given set of conditions, ρ, R, T, and G_0 are all constant, so that σ **is linear with G** (Figure 10.13).

We can plot $\log k/k_0$ or log k (rate) on the vertical axis vs σ value on the horizontal axis. If we get a straight line, then we have a linear free energy relationship. The slope is ρ for the reaction.

We will see one of the early examples where Hammett identified the linear free energy relationship. Earlier, we saw the activation energy or the free energy of activation is the deciding factor for the kinetics of the reaction. We also knew in an equilibrium reaction, the free energy change is related to stability of the species involved in the equilibrium. In the graph, we have a straight line. It implies that there is a linear relationship. We can infer that the change in free energy of activation for a series of substituent groups is directly proportional to the change in the free energy of dissociation for the same series of substituents on benzoic acid. If the dissociation of substituted benzoic acid is easy, then hydrolysis of benzoates with the same substituent will also be easy (Figure 10.14).

When there is a linear free energy relationship between two sets of reaction series, any one of the following three settings is possible: (1) ΔH is constant and the ΔS terms

$$\Delta G = -RT \ln K$$
$$\Delta G_0 = -RT \ln K_0$$
$$\log K - \log K_0 = \sigma \rho$$
$$\frac{-\Delta G}{2.3RT} + \frac{\Delta G_0}{2.3RT} = \sigma \rho$$
$$\Delta G = \Delta G_0 - \sigma \rho 2.3RT$$

FIGURE 10.13 Relationship between free energy and equilibrium constant.

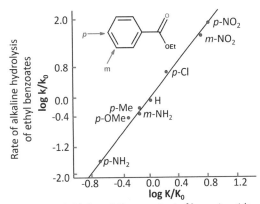

FIGURE 10.14 Plot of $\log k/k_0$ and $\log K/K_0$.

are proportional for the series, (2) ΔS is constant and the ΔH terms are proportional, or (3) ΔH and ΔS are linearly related. In LFER, the third situation prevails. At a constant temperature, entropy changes linearly with enthalpy. In fact, for the validity of the Hammett equation, the *entropy should remain constant.*

If a substituted benzoic acid dissociates faster than an unsubstituted benzoic acid, then the hydrolysis of substituted benzoate will also be faster than an unsubstituted benzoate.

10.4.2 Applicability of Hammett Equation

The Hammett equation had wide applicability. Here we have hydrolysis of methyl ester, halogen exchange, and nitration of aromatic system. The Hammett equation is very useful, as long as the mechanism whether it is electrophilic, nucleophilic, or free radical is the same within a given reaction series.

10.4.3 Substituent Constants (σ)

The σ values are numbers that sum up the total electrical effects (resonance plus field) of a substituent X when attached to a benzene ring. For hydrogen, the σ value is arbitrarily assigned as zero. A negative value of σ means an EDG, and a positive value indicates an electron-withdrawing group (EWG). If you look carefully, the σ value for the *para* substituent is higher than the *meta* (Figure 10.15). The exceptions are chloro and acetyl. This means the *para* substituents affect the π electron cloud of the aromatic system more than the *meta* substituents. Many books report slightly different values for σ. Do not worry about the accurate values. What we are concerned about is the **sign and magnitude**. Look at the σ values for the amino derivative.

X	σ_{para}	σ_{meta}
-O⁻	-0.81	-0.47
-NMe₂	-0.63	10.10
-NH₂	-0.57	-0.09
-OMe	-0.28	0.10
-CMe₃	-0.15	-0.09
-CH₃	-0.17	-0.07
-H	0.00	0.00

X	σ_{para}	σ_{meta}
-COOH	0.44	0.35
-Cl	0.23	0.37
-CF₃	0.53	0.46
-COMe	0.36	0.47
-CN	0.70	0.62
-NO₂	0.81	0.71

FIGURE 10.15 σ values for various substituents.

There is a huge difference between the *para* and *meta* substituents. Let us see what is the reason.

The value of σ includes contribution from both resonance and inductive effects. When the amino group is in the *para* position it can donate the electron to the reaction center through resonance. The same is the case with this. In *meta* position, these resonance effects are not predominant. But still, they are electron donating (σ negative). When oxygen substituents are in *para* position, they donate electrons through resonance. But when they are in *meta*, the inductive effect kicks in. It now becomes electron-withdrawing and has a positive σ value. When we look at halogens, they are inductively electron withdrawing so the effect is seen remarkably when they are in the *meta* position Figure 10.16.

When we compare the σ values for EDGs like NH₂, it is observed that the *para* substituents have higher values than *meta* substituents. This is due to the resonance effect. The lone pair on nitrogen can take part in the resonance contribution leading to more electron density at the reaction center.

The same effect is seen for oxygen substituents also. For example, the lone pair on oxygen in the methoxy substituent

also can participate in the resonance contribution. From these two examples, it becomes clear that when an EDG is present at the *para* position, it donates electrons to the aromatic system through resonance.

Let us take an example of the same electron-donating group, the methoxy group present in *meta* position. When it is in the *meta* position, it inductively withdraws electrons. So, the sign of σ changes from negative to positive. For all the halogens, the σ value is positive. This implies that they are electron withdrawing. Their inductive withdrawal is more pronounced when they are in the *meta* position (see Figure 10.17).

In the case of *m*- or *p*-substituted substrates, in the rate-limiting step, there is no steric repulsion. Because the substituents are far away from the reaction center. These compounds show a linear free energy relationship.

In the case of o-substituted substrates, in the rate-limiting step, there is **steric repulsion**. Due to this, there is no linear free energy relationship for *ortho* substituents (see Figure 10.18).

10.4.4 σ AND EWG/EDG

Let us see a comparison of the rates of base-catalyzed hydrolysis of *m*-NO₂, and *m*-Me, substituted ethyl benzoates with that of the unsubstituted ester. The initial attack on the ester by the OH is the rate-limiting step. The σ for the *m*-NO₂ derivative is =+0·71, and this ester is hydrolyzed 63·5 times faster than the unsubstituted ester. In this reaction, the transition state is a tetrahedral carbon with *negative charge*. The electron-withdrawing nature of the nitro group stabilizes the reaction center by pulling the electron away from the reaction center toward itself. So the rate of hydrolysis is faster in the *m*-nitro derivative.

When we look at the *meta* methyl benzoate, the methyl group is inductively electron donating. Hence, it destabilizes the reaction center where a negative charge is generated, hence the hydrolysis is slower in the *meta* methyl derivative than the unsubstituted benzoate.

In Figure 10.16, we saw the σ values for both *para* and *meta* methoxy benzoate changed the sign. Similar to the nitro derivative, the *meta* methoxy group also inductively withdraws electrons and stabilizes the tetrahedral intermediate for the hydrolysis reaction.

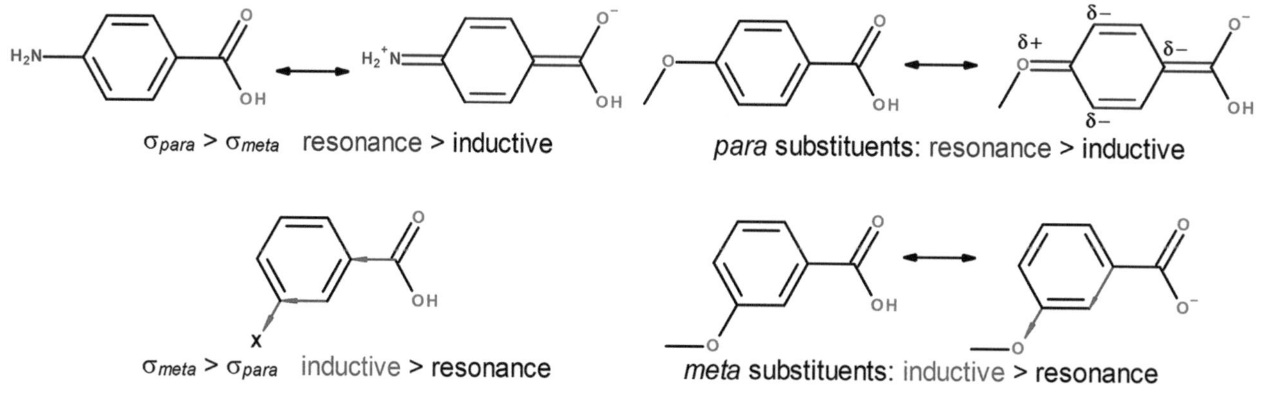

Group	σ_{para}	σ_{meta}	Group	σ_{para}	σ_{meta}
NH₂	-0.57	-0.09	F	0.06	0.34
S⁻	-1.21	-0.36	Cl	0.23	0.37
OMe	-0.28	0.10	Br	0.23	0.37
OH	-0.37	0.12	I	0.18	0.35

FIGURE 10.16 σ values for N, O, or X substituents.

$\sigma_{para} > \sigma_{meta}$ resonance > inductive

para substituents: resonance > inductive

$\sigma_{meta} > \sigma_{para}$ inductive > resonance

meta substituents: inductive > resonance

FIGURE 10.17 Comparison of σ values and electronic effects for *meta* and *para* substitutions.

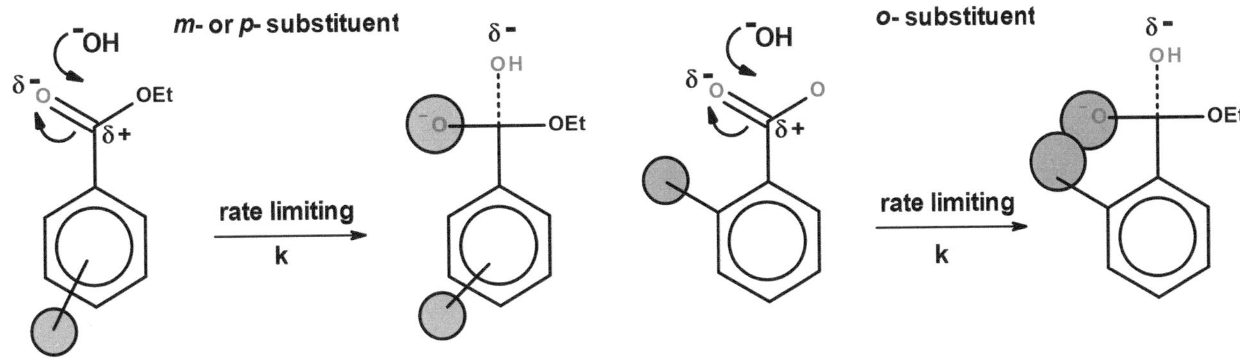

FIGURE 10.18   Steric effect of ortho substituents affecting σ values.

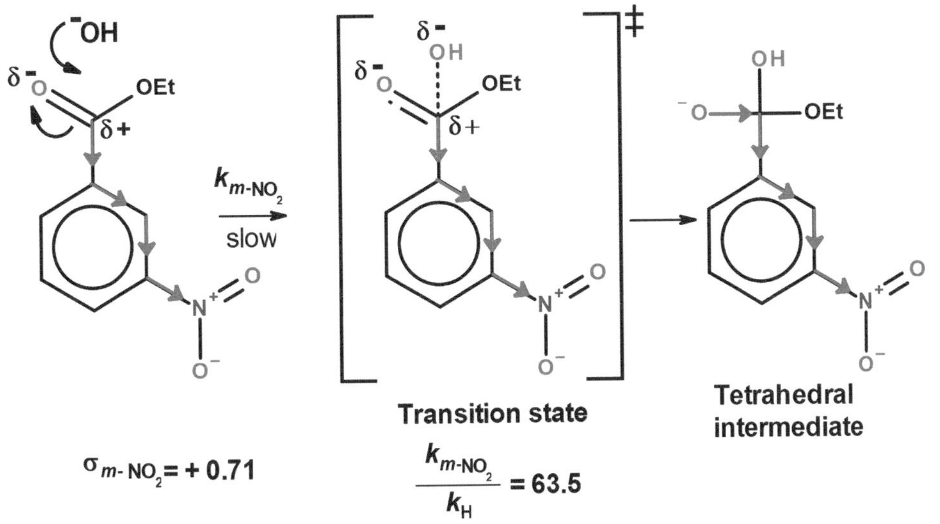

$$\sigma_{m\text{-}NO_2} = +0.71$$

$$\frac{k_{m\text{-}NO_2}}{k_H} = 63.5$$

SCHEME 10.3   Comparison of base-catalyzed hydrolysis of m-NO$_2$ ethyl benzoates with that of the unsubstituted ester.

$$\sigma_{m\text{-}Me} = -0.07$$

$$\frac{k_{m\text{-}Me}}{k_H} = 0.66$$

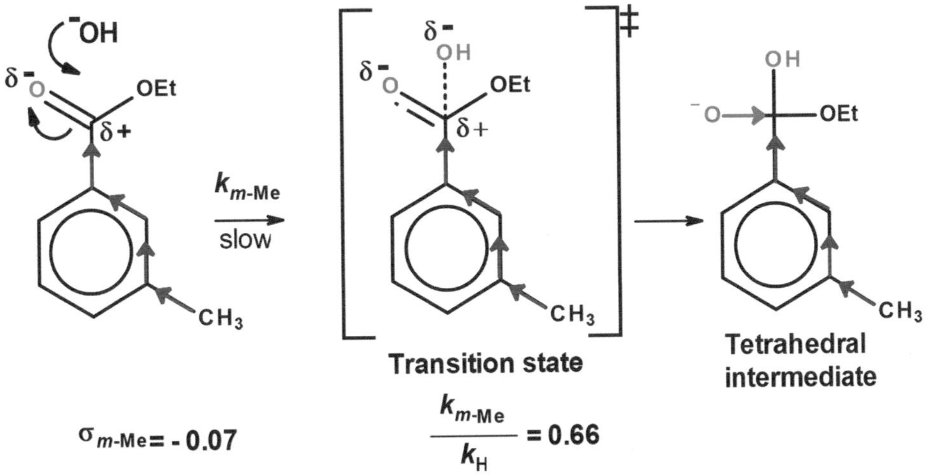

SCHEME 10.4   Comparison of base-catalyzed hydrolysis of m-Me ethyl benzoates with that of the unsubstituted ester.

But when the same methoxy is present in the *para* position, it donates electrons through resonance and withdraws through inductive effect. We already know the resonance effects are more pronounced than the inductive, and moreover, the number of intervening bonds is also more for the *para*-substitution, we can expect a very weak inductive electron withdrawal. Due to the strong electron donation which destabilizes the intermediate, the rate of *para* methoxy benzoate hydrolysis is lower than the benzoate.

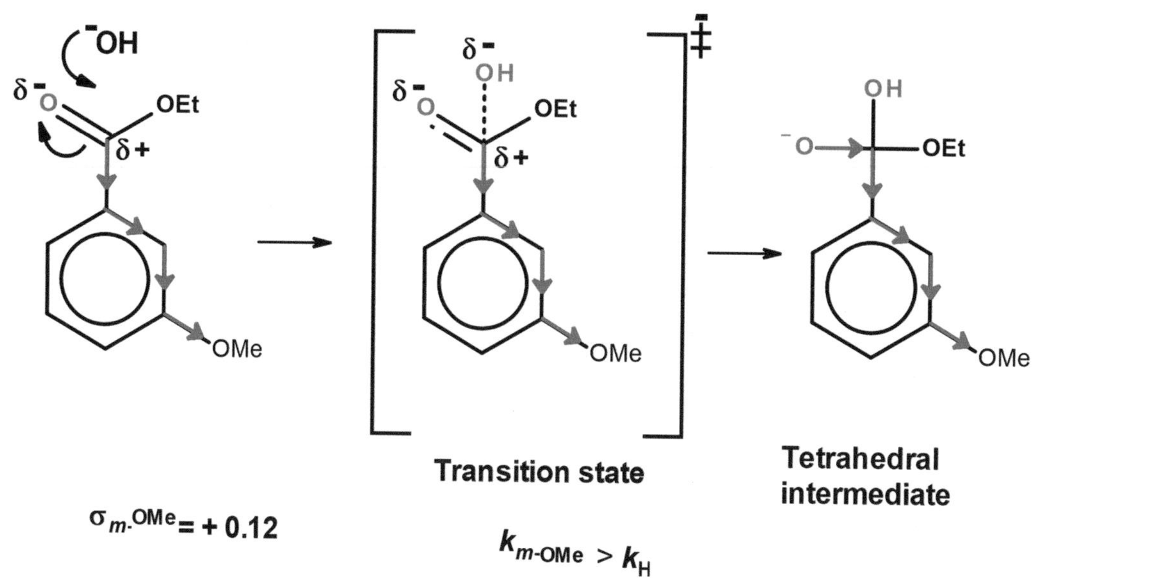

$\sigma_{m\text{-OMe}} = +0.12$

$k_{m\text{-OMe}} > k_H$

SCHEME 10.5 Comparison of base-catalyzed hydrolysis of *m*-OMe ethyl benzoates with that of the unsubstituted ester.

$\sigma_{p\text{-OMe}} = -0.27$

$k_H > k_{p\text{-OMe}}$

OMe - electron withdrawing (- inductive effect)
OMe - electron donating (mesomeric)

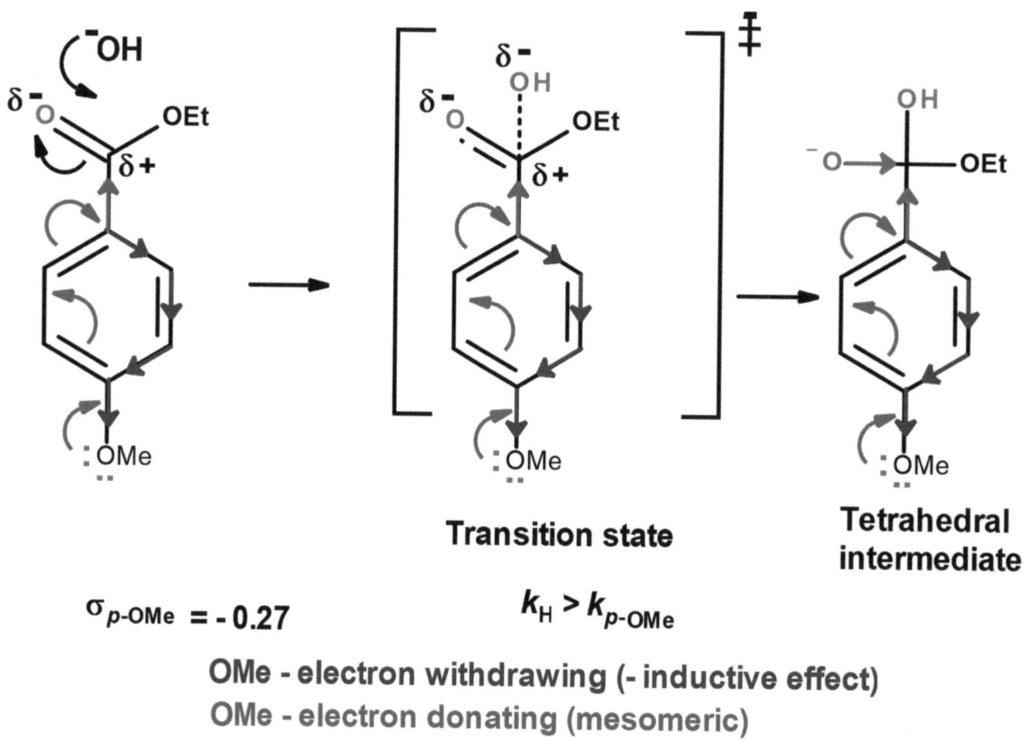

SCHEME 10.6 Comparison of base-catalyzed hydrolysis of *p*-OMe ethyl benzoates with that of the unsubstituted ester.

Let us take the example of benzoylation of substituted anilines. The reaction constant for this reaction is −2.69. The attack of the nucleophilic nitrogen on the electrophilic carbon is the rate-limiting step. When the attack takes place, the nitrogen develops a partial positive charge. It is obvious that any electron-releasing substituent on the aromatic ring will stabilize the intermediate. Similarly, any EWG on the aromatic ring will destabilize the intermediate. We can conclude, this behavior is found to hold in general for reactions with -ve rho values.

The reaction constant value for the base-promoted hydrolysis of substituted benzoates is +2·51. In this reaction, negative charge developed at the reaction center in the rate-limiting step. So any EWG will stabilize this intermediate and any electron-releasing group will destabilize the intermediate.

Let us recap, when rho is less than 0 or negative, there is a development of a positive charge at the reaction center during the transition state in the rate-limiting step. These reactions are favored by EDGs (they stabilize positive charge). Reactions with $\rho > 0$ are favored

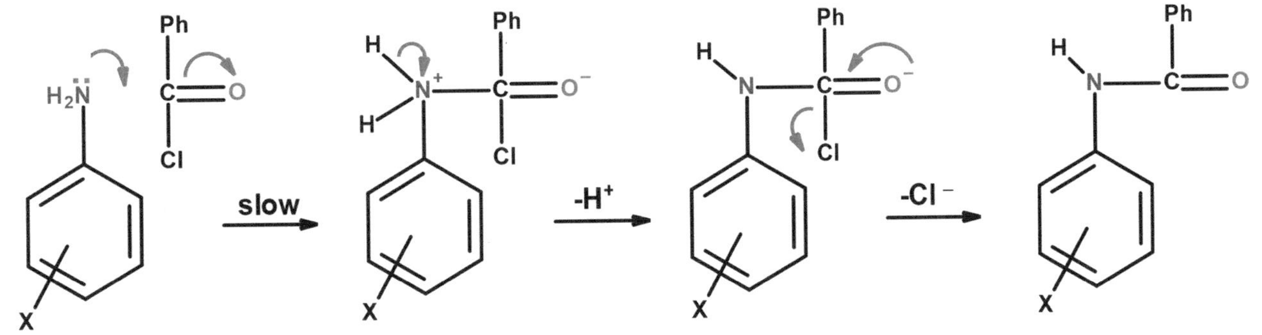

SCHEME 10.7 Benzoylation of substituted anilines.

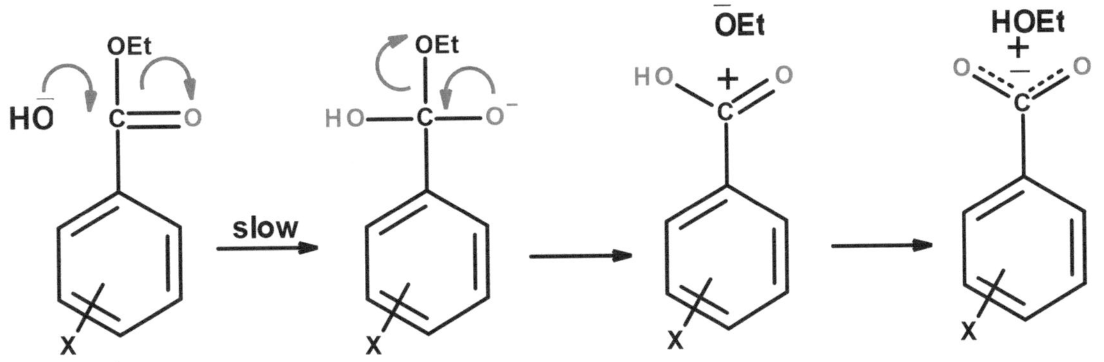

SCHEME 10.8 Base-promoted hydrolysis of substituted benzoates.

by electron-withdrawing groups (they stabilize negative charge). In addition, the magnitude of the value denotes the change in the charge distribution at the reaction center in the transition state. The reaction constant depends on the nature of the reaction and the experimental conditions (solvent, temperature, etc.).

Let us look at some of the deviations of the Hammett equation.

10.4.5 ρ AND EWG/EDG

When there is a carbocation intermediate involved in the solvolysis (S_N1) reaction of tertiary halides as in 2-aryl-2-chloropropane, the carbocation can undergo two types of reactions: (1) reaction with nucleophilic solvent (water) to give the alcohol and (2) elimination to give an alkene.

10.4.6 APPLICATION OF HAMMETT EQUATION

When σ>0 for EWG, those EWG favor the ionization of benzoic acid. When ρ>0, all reactions are favored by EDG. If there is LFER, the sign and magnitude of ρ give information about the transition state.

Let us take the example of saponification of substituted methyl benzoates. This reaction has ρ=+2.38. This indicates that EWG facilitates the reaction. This is in agreement with the accepted mechanism for saponification of ester. The tetrahedral intermediate is negatively charged. Its formation should therefore be favored by electron-withdrawing substituents that can stabilize the developing charge.

On the other hand, if we take the solvolysis of diarylmethyl chloride in ethanol, it has ρ=−5.0. This indicates that EDG greatly increases the rate of the reaction.

10.5 DISCREPANCIES IN CORRELATION OF LINEAR FREE ENERGY RELATIONSHIP

Many reactions do not fit into the linear free energy relationship. These are mostly reactions (1) where the attack is directly on the aromatic ring and (2) where the X group can enter into direct resonance interaction with the reaction site in the transition state. We are sure when the solvolysis of the tertiary halide occurs, it is not a reaction that occurs on the aromatic ring. For example, strongly electron-donating substituents don't fall on the line predicted by the Hammett correlation. The difference is due to the hyperconjugation. When we plot the $σ_x$ values against log (k/k$_0$) for the solvolysis of tertiary halide, there are some discrepancies observed. To compensate for the discrepancies, we need to account for hyperconjugation and is given by the term σ⁺. In addition to the $σ_m$ and $σ_p$, values used with the classical Hammett equation there are σ⁺ and σ−. These are substituent constant sets which reflect a recognition that the extent of resonance participation can vary for different reactions (Figure 10.19).

When the substituent constants are plotted against the rate of solvolysis of tertiary halides, most of the substituents as shown here follow LFER. But p-methyl and p-methoxy do not fall on a straight line. p-Methoxy undergoes solvolysis 800 times faster than the unsubstituted derivative.

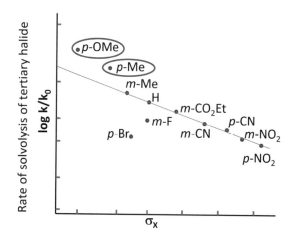

SCHEME 10.9 Solvolysis S_N1 reaction of tertiary halides.

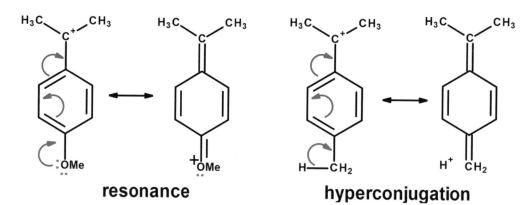

FIGURE 10.19 Plot the σ_x values against log (k/k_0) for the solvolysis of tertiary halide.

FIGURE 10.20 Stabilization of the carbocation intermediate.

This is due to stabilization of the carbocation intermediate. In the case of methyl group, the stabilization of the carbocation occurs through hyperconjugation. This additional stabilization is denoted by the sigma plus constant (Figure 10.20).

If we take the rate of formation of phenoxide ions from various substituted phenols, for many substituents, there is a linear free energy relationship. But for p-CN and p-NO$_2$ substituents, the values are above the line; this suggests that compounds with these substituents act as stronger acids than we would have predicted from log(K/K$_0$) values. Conjugation exists in these cases (Figure 10.21).

How can we find the σ^- values. First, we need to develop a line with ρ value based on m-substituents only, which

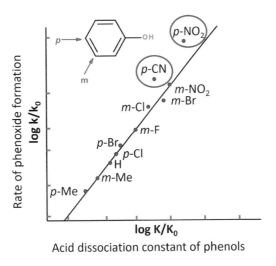

FIGURE 10.21   Rate of formation of phenoxide ion from various substituted phenols.

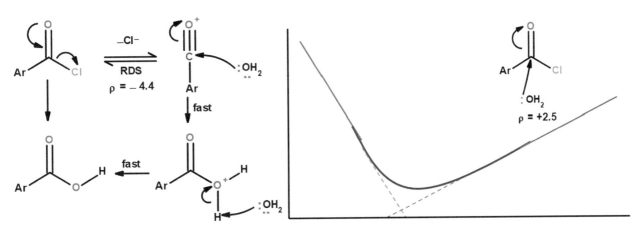

FIGURE 10.22   Hydrolysis of acid chlorides in solvent water.

cannot exhibit mesomeric effects. The amount by which certain substituents deviate from the line can be added to their σ values to produce a new scale of σ⁻ value.

10.6   HAMMETT EQUATION DISCREPANCIES

We have two other discrepancies for the Hammett equation. Both relate to transition states. In one case, we have a completely different mechanism when the substituents change. This is called concave upwards. Another case is when the substituent changes, the rate-limiting step also changes. This is called concave downwards.

10.6.1   CONCAVE UPWARDS

Let us take the hydrolysis of acid chlorides in solvent water. For one set of substituents, we have a reaction constant −4.4. In the rate-limiting step, the chloride ion leaves, and a triple-bonded oxonium ion is formed. For another set of substituents, the reaction constant is +2.5. In this case, the reaction mechanism is different. The nucleophile attacks the electrophilic carbon with concomitant removal of the leaving group. When electron-releasing

groups greatly increase the reaction rate, supporting a mechanism involving ionization as the rate-determining step. Electron-releasing groups can facilitate ionization by a stabilizing interaction with the electron-deficient carbocation that develops as ionization proceeds ($\rho = -4.4$). When EWG is present on carbonyl carbon, it will pull the electron toward itself. The net result will be poor electron density on the carbonyl carbon. So strong nucleophiles can easily attack ($\rho = +2.5$) (Figure 10.22).

Let us take another reaction, the hydrolysis of aryl esters. Compare the Hammett plots for the hydrolysis of $ArCO_2R$ (R= Me and Et) carried out in 99.9% H_2SO_4. Me esters show a well-behaved plot with $\rho = -3.25$, while Et esters showed ρ of +2.5. What do we understand from ρ changing from negative to positive? The reaction center is developing a considerable positive charge (ρ is negative) by the electron-donating groups in the rate-determining step changes to the poor positive charge (ρ is positive) due to the electron-withdrawing groups in the aromatic ring. This can also mean two different mechanisms operating when the leaving group changes from methyl to ethyl. The mechanism for hydrolysis is given below (Figure 10.23).

ρ = −3.25
ArCO$_2$Me
log k_x
ρ = −3.25
ArCO$_2$Et
ρ = +2.0
σ_x

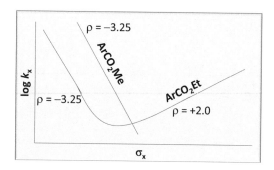

FIGURE 10.23 Hydrolysis of aryl esters.

X, Z = Me
log k_{obs}
ρ = +2.67
ρ = −2.51
X, Z = Cl
X, Z = OMe
Σσ+
0
Σσ

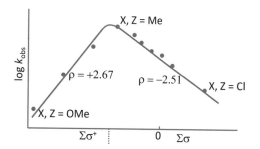

FIGURE 10.24 Cyclodehydration of 2-phenyl-triaryl methanol.

10.6.2 Concave Downwards

Here when the substituent changes, the rate-limiting step also changes. Cyclodehydration of 2-phenyl-triaryl methanol gives the corresponding tetraaryl methanes. In this reaction, two of the benzene rings in the reactants each carry a *para* substituent (X and Z, respectively). Since we have two substituents, the sigma value is the sum of the individual sigma values. Here we have two straight lines, one with a reaction constant of +2.67 and another with −2.51. We will see the detailed mechanism and reasoning for this trend (Figure 10.24).

There are four steps in the overall reaction mechanism. The first two steps constitute an E1 elimination of water to yield a carbocationic intermediate, which then, in the last two steps, effects internal electrophilic substitution on the 2-phenyl nucleus to yield the product. We know initial protonation (Step 1) and loss of proton (step 4) are very fast, so they cannot be rate determining. That leaves with either step 2 or 3 as the rate-determining step. In step (2), +ve charge increases at the reaction center, whereas in step (3), +ve charge decreases at the reaction center. Now what we have to do is simply find out which substituent will stabilize +ve charge at the reaction center. For those substituents step 2 is going to be fast. Because there is stabilization. That leaves step 3 to be rate-determining or slow step. This matches well with the obtained results. The methoxy substituent which is electron-releasing can stabilize the carbocation.

Apart from the methoxy group, the methyl group which also has electron-donating ability appears at the top of the straight line. These two groups come at different places due to the difference in their electron-donating capacity. When we move to the right-hand side of the graph, the carbocation formation is very difficult when there are electron-withdrawing groups in the aromatic ring. So, this step is very slow or rate determining. The plot confirms our reasoning. For halogens, step 2 is rate determining.

Let us recap, till now we have seen four types of σ values. σ_m, and σ_p, for *meta* and *para* substituents where inductive and resonance effects predominate, respectively. σ^+ for EDG for the positive reaction center where hyperconjugation is compensated and σ^- for EWG for the negative reaction center where mesomeric effect respectively. These values are substituent-specific and not reaction-specific.

10.7 TAFT EQUATION

One of the major areas addressed by the Taft equation is application toward aliphatic esters. In addition, this also deals with steric, inductive, and resonance effects. The steric contribution is given by the term δEs. Here Es is the substituents steric effects constant and d is the sensitivity factor for the reaction to steric effects. We already know what is σ and ρ (Figure 10.25).

$\log(k_s/k_{CH3})$ = ratio of the rate of the substituted vs reference reaction

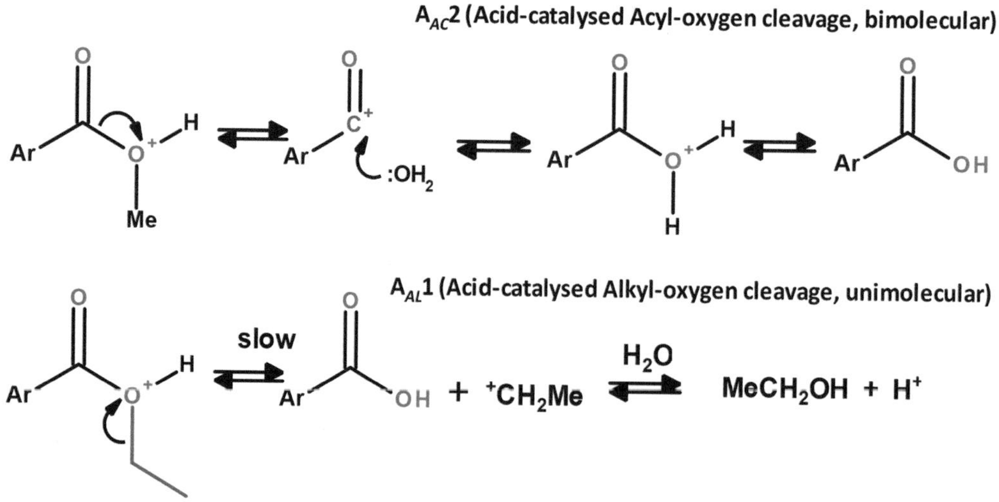

$A_{AC}2$ (Acid-catalysed Acyl-oxygen cleavage, bimolecular)
Ar
H
Me
:OH$_2$
$A_{AL}1$ (Acid-catalysed Alkyl-oxygen cleavage, unimolecular)
slow
OH + $^+CH_2Me$
H_2O
$MeCH_2OH$ + H^+

SCHEME 10.10 Mechanism of cyclodehydration of 2-phenyl-triaryl methanol.

$$log\left(\frac{k_s}{k_{CH_3}}\right) = \rho^*\sigma^* + \delta E_s$$

FIGURE 10.25 Taft equation.

σ^* = the polar substituent constant (describes the field and inductive effects of the substituent) or σ_I

E_s = the steric substituent constant

ρ^* = the sensitivity factor for the reaction to polar effects

δ = the sensitivity factor for the reaction to steric effects.

It is used in the study of reaction mechanisms and in the development of quantitative structure–activity relationships (QSAR) for organic compounds.

However, because of the difference in charge buildup in the rate-determining steps, it was proposed that polar effects would only influence the reaction rate of the base-catalyzed reaction since a new charge was formed. He defined the polar substituent constant σ^* (Figure 10.26).

The value of σ^* is calculated as per the equation given here (Figure 10.27):

- $log(k_s/k_{CH3})_B$ is the ratio of the rate of the base-catalyzed reaction compared with the reference reaction.
- $log(k_s/k_{CH3})_A$ is the ratio of the rate of the acid-catalyzed reaction compared with the reference reaction.
- ρ^* is a reaction constant that describes the sensitivity of the reaction series.

- The factor of 1/2.48 is included to make σ^* similar in magnitude to the Hammett σ values.

The Hammett equation had considerably contributed to QSAR. QSAR or quantitative structure–activity relationship is a regression or classification model used in the chemical and biological sciences and engineering. This is used in drug development.

10.8 CURTIN–HAMMETT PRINCIPLE

According to the IUPAC Gold Book, the Curtin–Hammett principle is given as follows:

In a chemical reaction that yields one product (A1) from one conformational isomer (A2) and a different product (A4) from another conformational isomer (A3) (and provided these two isomers are rapidly interconvertible relative to the rate of product formation, whereas the products do not undergo interconversion) the product composition is not in direct proportion to the relative concentrations of the conformational isomers in the substrate; it is controlled only by the difference in standard free energies ($\delta\Delta\ddagger G$) of the respective transition states.

A_2 and A_3 are interconverting isomers, and they may be conformational isomers. A_1 and A_4 are the products formed from the isomers A_2 and A_3, respectively (Figure 10.28).

The reaction profile is given below. Between A_2 and A_3, A_2 is more stable, and between A_1 and A_4, A_1 is more stable. The activation energy for the conversion of A_3 to A_4 is smaller than the activation energy for the conversion of A_2 to A_1.

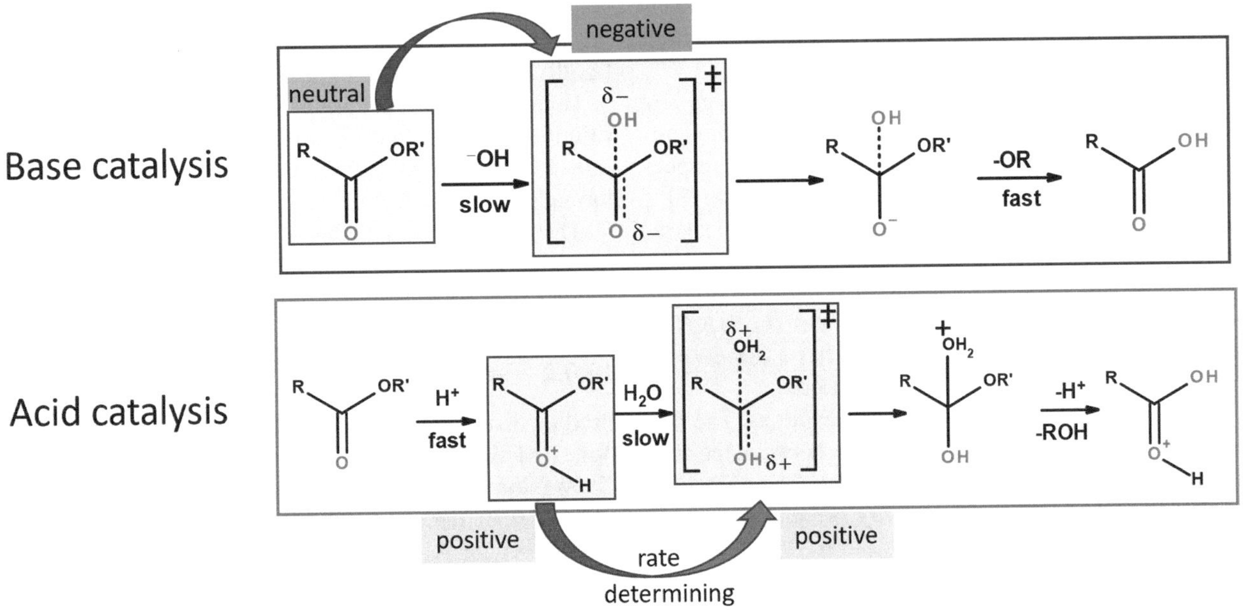

FIGURE 10.26 Acid- and base-catalyzed hydrolysis.

$$\sigma^* = \left(\frac{1}{2.48\rho^*}\right)\left[log\left(\frac{k_S}{k_{CH_3}}\right)_B - log\left(\frac{k_S}{k_{CH_3}}\right)_A\right]$$

FIGURE 10.27 Equation for the polar substituent constant.

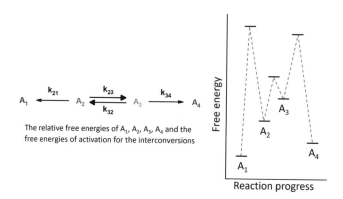

FIGURE 10.28   The relative energies of the transition states for conversion of A_2 and A_3 to A_1 and A_4, respectively.

$$\frac{d\,[A_1]}{dt} = k_{21}[A_2] \quad \cdots\cdots [1] \qquad \frac{d\,[A_4]/dt}{d\,[A_1]/dt} = \frac{d\,[A_4]}{d\,[A_1]} = \frac{k_{34}[A_3]}{k_{21}[A_2]} \quad \cdots\cdots [3]$$

$$\frac{d\,[A_4]}{dt} = k_{34}[A_3] \quad \cdots\cdots [2] \qquad \int d\,[A_4] = \frac{k_{34}}{k_{21}} \int \frac{[A_3]}{[A_2]} d\,[A_1] \quad \cdots\cdots [4]$$

$$\frac{[A_4]}{[A_1]_\infty} = K\frac{k_{34}}{k_{21}} \; when \; [A_4]_0 = [A_1]_0 = 0 \; \cdots\cdots [5]$$

FIGURE 10.29   Derivation of the Curtin–Hammett equation.

According to the C–H principle, A_4 will be the major product, because the conversion of A_3 to A_4 has the smaller activation energy.

10.9   CURTIN–HAMMETT EQUATION

The exact solution to the Curtin–Hammett equation allows the determination of the concentration of reactants and products (A_1 to A_4) (C–H principle) and permits the calculation of the rates of formation of A_4 (W–H equation), both as a function of time.

Rates of formation of A_1 and A_4 are given by Equations 10.1 and 10.2 and k_{21} and k_{34} are the rate constants for the conversion of A_2 to A_1 and A_3 to A_4, respectively. When we divide Equation 10.2 by Equation 10.1, we get Equation 10.3. Rearranging followed by integration from initial to final concentration gives Equation 10.4. This can be simplified to Equation 10.5. Here the product ratio depends on the ratio of reaction rates (k_{34}/k_{21}). The ground-state conformational preference has a direct (proportional) role in the value of the product ratio.

We have multiple Curtin–Hammett conditions. The first three are related to the *rate of conversion* between the conformers and the rate of product formation. The next three conditions are related to the *stability* of conformers and their rate of conversion to products.

1. $k_{23}, k_{32} >> k_{21}, k_{34}$
2. $k_{23}, k_{32} << k_{21}, k_{34}$
3. $k_{21}, k_{34} \approx k_{23}, k_{32}$
4. Stability $A_2 > A_3$, $k_{21} > k_{34}$
5. Stability $A_2 > A_3$, $k_{21} < k_{34}$
6. Stability $A_2 \approx A_3$, $k_{21} \approx k_{34}$

10.9.1   WHEN $k_{23}, k_{32} >> k_{21}, k_{34}$

The very first condition is commonly observed in most of the reactions. Because the barrier to interconversion is very small compared with product formation. Conversion of A_2 to A_3 has very small activation energy ($\Delta G^{\ddagger}_{23}$) so it occurs much faster than the conversion of A_2 to A_1 or A_3 to A_4. The activation energy barrier for conversion of A_3 to A_4 and A_2 to A_1 are similar (Figure 10.30).

According to Equation 10.5, we can say that the product ratio $[A_4]/[A_1]$ is equal to the product of the equilibrium constant K multiplied by the ratio of the two reaction rate constants k_{34} and k_{21}. The ground-state conformational preference has a direct (proportional) role in the value of $[A_4]/[A_1]$

Here we have a couple of examples for condition 1. Elimination of cyclohexyl tosylate. In both the reactions shown below, the interconversion is faster than the product formation (Figure 10.31).

The tosylates can undergo ring flip (conformational isomerism). One of the isomers gives the product, while another does not give the elimination product.

10.9.2   WHEN $k_{21}, k_{34} >> k_{23}, k_{32}$

Product distribution reflects initial conformer concentration. Reaction is faster than conformer interconversion. Conversion of A_2 to A_3 has very high activation energy ($\Delta G^{\ddagger}_{23}$) so it occurs much slower than the conversion of A_2 to A_1 or A_3 to A_4. The activation energy barriers for the conversion of A_3 to A_4 and A_2 to A_1 are similar. One of the boundary conditions within which the Curtin-Hammett Kinetics operate is **Kinetic quench.** Ex. Proton transfer of a tertiary amine with acid, or excess of highly reactive reagent produced suddenly in a flash photolysis. For a kinetic quenching system, the ratio of products is directly and solely related to the ground-state conformational distribution. It is commonly not observed (Figure 10.32).

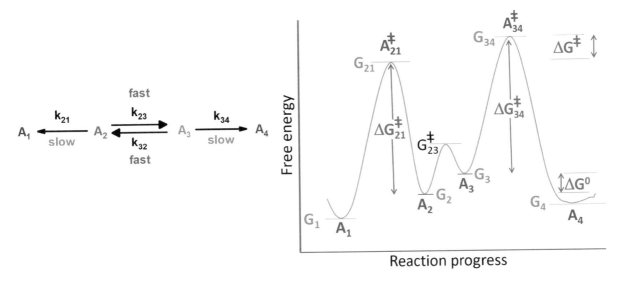

FIGURE 10.30   Rate of conversion between the conformers and rate of product formation.

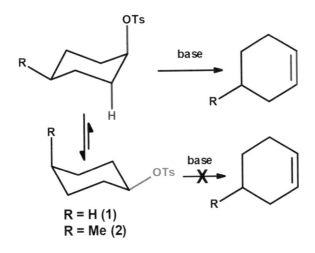

FIGURE 10.31   Elimination of cyclohexyl tosylate.

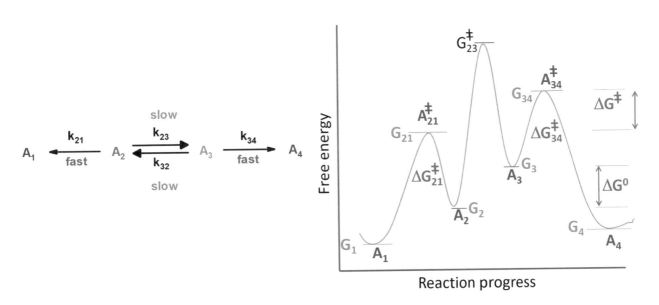

FIGURE 10.32   Reaction is faster than conformer interconversion.

For this condition, we have the following example, protonation of a tertiary amine with trifluoroacetic acid. In the case of 1,3,5-trimethylpiperidine as shown here, the inversion at nitrogen between diastereoisomeric conformers is much slower than the rate of protonation on nitrogen. The conformer with the axial methyl group suffers from 1,3-diaxial interaction. Due to that, the protonation of that isomer is very slow hence that is formed as a minor isomer only (Figure 10.33).

10.9.3 Transition State Energy vs Stability

We have three cases and we will see one by one in more detail: Case 1 More stable isomer/conformer reacts faster, Case 2 Less stable isomer/conformer reacts faster, and Case 3 Both isomer/conformer reacts at a similar rate.

10.9.4 Case 1 More Stable Isomer/ Conformer Reacts Faster

Here we have a reaction where the more stable conformer reacts faster. H_2O_2-mediated oxidation of 4-tert-butyl-1-methyl piperidine gives the corresponding N-oxide. Due to steric reasons, the bulky tert-butyl group prefers to be in the equatorial position. The N-methyl group can occupy either axial or

equatorial orientation. In the case of axial orientation, there is 1,3-diaxial interaction so that is less stable compared with the equatorial methyl which is more stable. The relative activation energy of the equatorial methyl oxidation is lesser than the activation energy of axial methyl oxidation. Hence, the major product obtained in the reaction is from the more stable conformer (Figure 10.34).

10.9.5 Case 2 Less Stable Isomer/ Conformer Reacts Faster

Here we have a reaction where the less stable conformer reacts faster. An example is N-methylation of 8-methyl-8-azabicylco[3.2.1]octane giving the corresponding N,N-dimethyl compound (Figure 10.35).

In this reaction, the stability of the transition state decides the product formation and not the ground-state stability. Out of the axial and equatorial alkylation (relative to the six-membered ring), the isomer with a methyl group in the proximity to the six-membered ring suffers from steric repulsion from the ring hydrogen in the ground state, but in the transition state, the methyl group can approach the nitrogen from the less hindered site and hence the reaction takes place more readily. If we take the more stable isomer in the ground state, for the alkylation to occur, the methyl group now has to

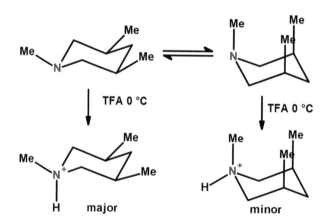

FIGURE 10.33 Protonation of 1,3,5-trimethylpiperidine.

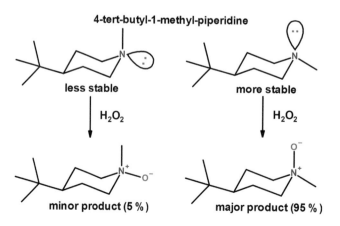

FIGURE 10.34 Oxidation of 4-tert-butyl-1-methyl piperidine.

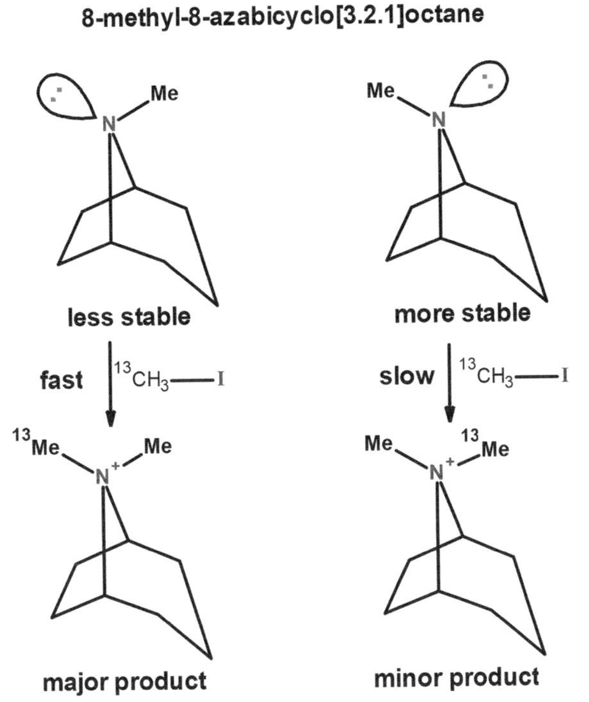

8-methyl-8-azabicyclo[3.2.1]octane

less stable

fast | $^{13}CH_3$——I

^{13}Me $_{N^+}$ Me

major product

more stable

slow | $^{13}CH_3$——I

Me $_{N^+}$ ^{13}Me

minor product

FIGURE 10.35   N-methylation of 8-methyl-8-azabicylco[3.2.1]octane.

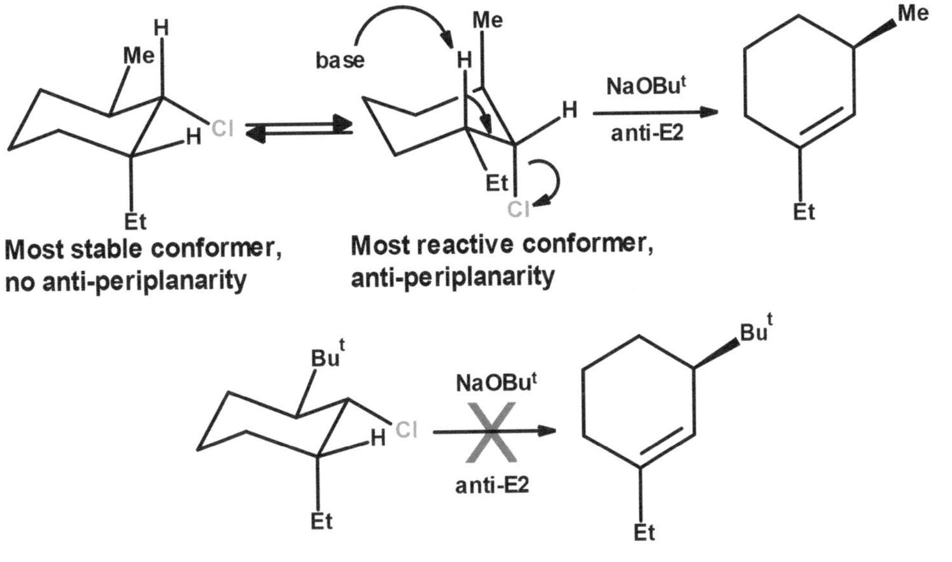

Most stable conformer,
no anti-periplanarity

base

Most reactive conformer,
anti-periplanarity

NaOBu^t
anti-E2

Me

Et

NaOBu^t
anti-E2

Bu^t

Et

FIGURE 10.36   Example of less stable isomer reacting faster.

approach from the hindered side which is quite unfavorable. In the ground state the more stable conformer will be present in large excess but as the reaction proceeds the product mixture will be consisting of the product formed from the less stable conformer. This is an important example to disprove the common misconception that a ground-state stable conformer reacts faster than a less stable conformer. In other words, it disproves the idea that ground-state conformational distribution (K) is equal to product distribution ([A₄]/[A₁]).

Here we have an example of elimination. The most stable conformer is the one where the chloro group is in the equatorial position. In this case, the leaving group chloride and the hydrogen are not in the same plane. There is no anti-periplanarity for elimination in the most stable conformer. When it undergoes ring flip, the chloro group occupies the less stable axial position. Now it can undergo elimination due to the presence of anti-periplanarity with the hydrogen (Figure 10.36).

10.9.6   CASE 3 BOTH ISOMER/CONFORMER REACTS AT A SIMILAR RATE

In this radical methylation reaction, two conformers are possible. Due to 1,3-non-bonding interaction, one of the conformers is present in a very small amount. It is expected that the radical methylation will give one of the products as the exclusive one (Figure 10.37).

But due to the presence of steric hindrance associated with incoming R radical, the product ratio is 60:40 (Figure 10.38).

A simple way to understand/visualize the reaction using kinetic studies. It is possible to relate the final product ratio (for a pair of products) obtained from a pair of equilibrating starting materials. It talks about the effect of conformational change on chemical reactivity. The rate constant and equilibrium constant are physically measurable. From this, we can understand how the reaction proceeds. Application of Curtin–Hammett principle to the following dynamic kinetic resolution of stereoisomers (enantiomers and diastereomers).

Not every reaction follows Curtin–Hammett's principle. But for those reactions that follow, we can predict how the products are formed from the equilibrating starting material which may be a conformer, tautomer, or stereoisomer.

The product composition is not solely dependent on the relative proportions of the conformational isomers of the substrate; however, it is controlled by the difference in standard Gibbs energies of the respective transition states. According to Curtin–Hammett, we can conclude that there is no relationship between stability and reactivity.

10.10 QUESTIONS

1. Using Hammond's postulate write the reaction energy diagram for nitration of toluene.
2. Explain the product formation in the following two reactions from the same starting material using Hammond's postulate. (hint Jerry March P 974) (Figure 10.39)
3. Using Hammond postulate write the reaction energy diagram for bromination of phenol.
4. The substitution constant (s_p) for $-NH_2$ and $-S^-$ groups are −0.66 and −1.21. Based on this data predict which one of these substituents is a strong electron-donating group.
5. With a suitable example explain why the inductive effect is more pronounced for meta substituents.
6. Although methoxy substituent is electron donating, this effect is completely absent in meta methoxy benzoic acid ester when this undergoes hydrolysis. Explain with a suitable electron flow diagram.
7. From the reaction constant data can we predict the effect of substituents on a particular reaction? Explain with examples.
8. For the following reaction can you predict (Figure 10.40):

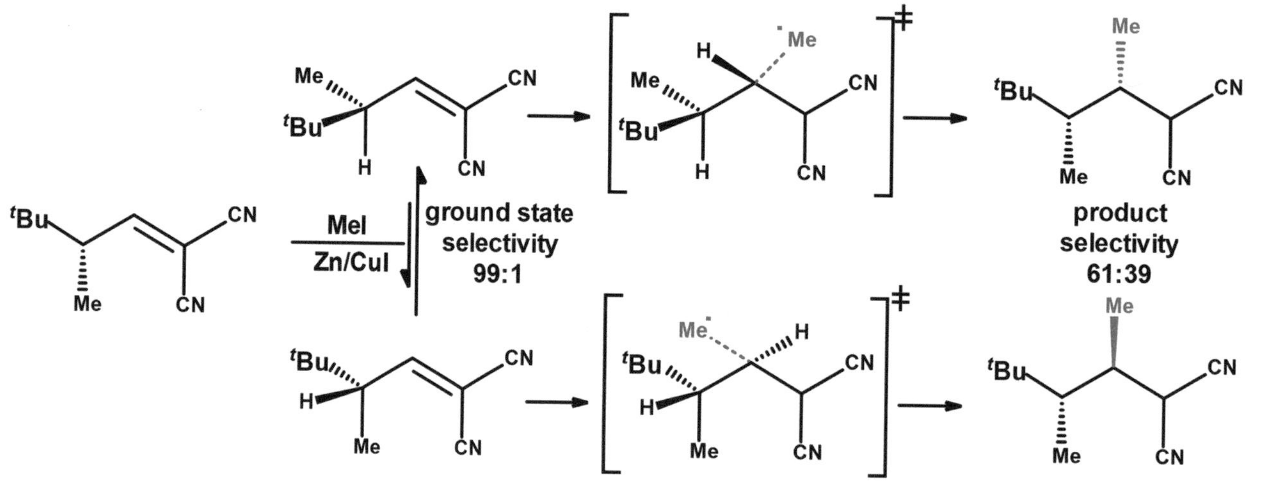

FIGURE 10.37 Example of each conformer reacts at a similar rate.

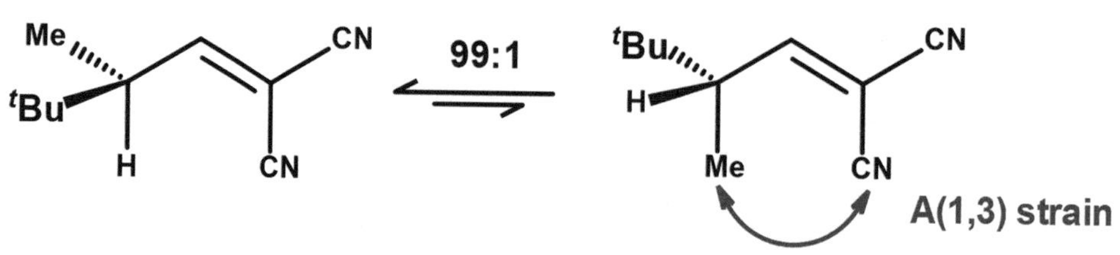

FIGURE 10.38 Example of steric hindrance affecting product distribution.

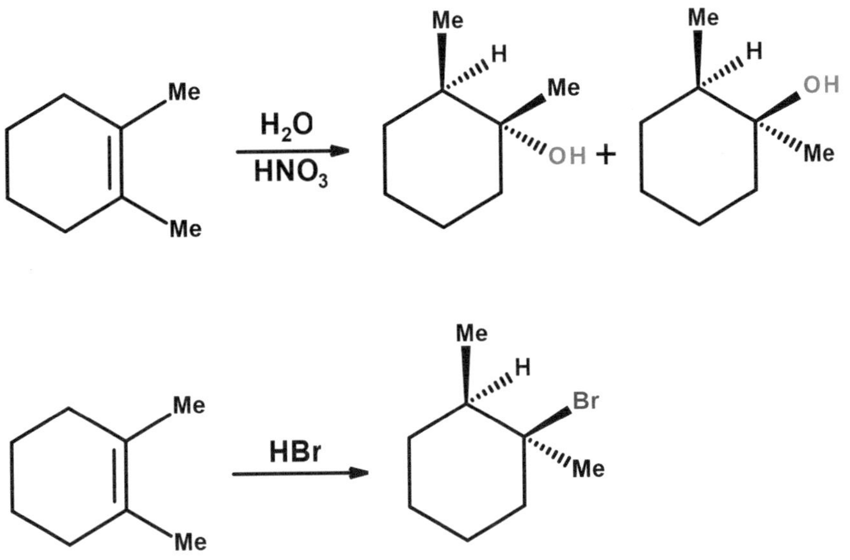

FIGURE 10.39   Question 2.

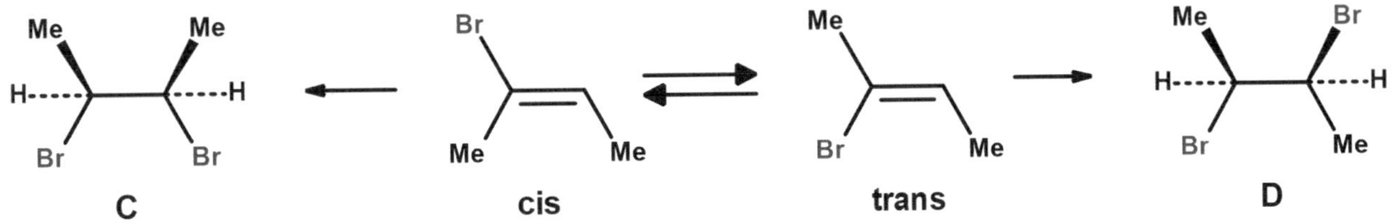

FIGURE 10.40   Question 8.

a. What are the rates of product formation for each conformation in these reactions?

b. What are the factors that influence these rates?

c. What is the ground-state distribution of the reactants?

d. Does the reactant ratio change as a function of time?

e. What are their rates of interconversion?

9. With suitable examples explain the deviation of Hammett Linear free energy relationship.

10. In what way Taft equation differ from Hammett's equation?

BIBLIOGRAPHY

Abraham, D. J. *Burger's Medicinal Chemistry and Drug Discovery, Volume 1*, 6th Ed. John Wiley&Sons, Inc., 2003. (https://media.wiley.com/product_data/excerpt/03/04712709/0471270903.pdf) (accessed, 15-Apr-2024)

Andraos, J. (2008). Quantification and optimization of dynamic kinetic resolution. *Canadian Journal of Chemistry*, 86(4), 342–357. (https://www.nrcresearchpress.com/doi/10.1139/v08-020) (accessed, 15-Apr-2024).

Ashenhurst, J. (2021). A Primer on Organic Reactions, Hammond's Postulate (https://www.masterorganicchemistry.com/2011/09/28/hammonds-postulate/) (accessed, 15-Apr-2024).

Blackmond, D. (Imperial College, London), Reactivity of Organic Compounds, 2004, (https://www.ch.ic.ac.uk/local/organic/tutorial/db2.pdf) (accessed, 15-Apr-2024).

Blackmond, D. (Imperial College, London), Linear Free Energy Relationships, 2004, (https://www.ch.ic.ac.uk/local/organic/tutorial/db3.pdf) (accessed, 15-Apr-2024).

Carey, F. A., & Sundberg, R. J. *Advanced Organic Chemistry, Part A-Structure and Mechanisms*, 4th Ed. Kluwer Academic/Plenum Publishers, 2000, 204, 210.

Chakraborty, S., & Saha, S. (2016). The Curtin–Hammett principle: A qualitative understanding. *Resonance Journal of Science Education*, 21, 151–171. https://www.ias.ac.in/article/fulltext/reso/021/02/0151-0171

Chemical Reactions-How Far and How Fast (2016) (https://www.cstf.kyushu-u.ac.jp/~furutalab/kougi/2016-AOR/AOR_No3.pdf) (accessed, 15-Apr-2024).

Clayden, J., Greeves, N., & Warren, S. *Organic Chemistry, Chapter 22, Electrophilic Aromatic Substitution*, 2nd Ed. Oxford University Press, 2014, 1090.

Eliel, E. L. (1975). Conformational analysis-The last 25 years. *Journal of Chemical Education*, 52(12), 762.

Evans, D. A. A Brief Introduction to the Curtin-Hammett Principle, (https://www.haldiagovtcollege.org.in/wp-content/uploads/2017/12/PG-Sem-IV-402O-Unit-1.pdf) (accessed, 15-Apr-2024).

Farmer, S. Sonoma State University, 7.11: The Hammond Postulate and Transition States, in Organic Chemistry (OpenStax) (https://chem.libretexts.org/Bookshelves/Organic_Chemistry/Map%3A_Organic_Chemistry_(Wade)/07%3A_

Alkyl_Halides%3A_Nucleophilic_Substitution_and_ Elimination/7.11%3A_The_Hammond_Postulate) (accessed, 15-Apr-2024).

Francis A. Carey, Richard J. Sundberg, Study and description of organic reaction mechanisms. In: *Advanced Organic Chemistry. Advanced Organic Chemistry.* Springer, 2002, 187–261. https://doi.org/10.1007/0-306-46856-5_4

Hammett, L. P. (1937). The effect of structure upon the reactions of organic compounds. Benzene derivatives. *Journal of the American Chemical Society*, 59(1), 96–103.

Hammond, G. S. (1955). A correlation of reaction rates. *Journal of the American Chemical Society*, 77(2), 334–338.

Heremans, K. *Chemistry at Extreme Conditions, Chapter 1 - Pressure - Temperature Effects on Protein Conformational States*, Elsevier Science, 2005, 1–27.

Jaffe, H. H. (1953). A reexamination of the Hammett equation. *Chemical Reviews*, 53(2), 191–261. https://www.ch.ic. ac.uk/local/organic/tutorial/db3.pdf

Kennepohl, D. Athabasca University, 5.6: Reaction Energy Diagrams and Transition States in Organic Chemistry (OpenStax) (https://chem.libretexts.org/Bookshelves/Organic_Chemistry/ Map%3A_Organic_Chemistry_(Wade)/04%3A_The_Study_ of_Chemical_Reactions/5.06%3A_Reaction_-_Energy_ Diagrams_and_Transition_States) (accessed, 15-Apr-2024).

Kinetic analyses, University of Ottawa (https://mysite.science.uottawa.ca/jkeillor/English/Teaching_files/Kinetic-analyses.pdf).

Lowry, T. H., & Richardson, K. S. *Mechanism and Theory in Organic Chemistry*. Harper& Row, Publishers, 1976, 60.

More O'Ferrall–Jencks diagram (https://goldbook.iupac.org/ html/M/M04030.html) (accessed, 15-Apr-2024).

Oki, M. (1984). Reactivity of conformational isomers. *Accounts of Chemical Research*, 17(5), 154–159.

Seeman, J. I. (1983). Effect of conformational change on reactivity in organic chemistry. Evaluations, applications, and extensions of Curtin–Hammett Winstein-Holness kinetics. *Chemical Reviews*, 83(2), 83–134.

Seeman, J. I. (1986). The Curtin–Hammett principle and the Winstein-Holness equation: New definition and recent extensions to classical concepts. *Journal of Chemical Education*, 63(1), 42.

Seeman, J. I., & Farone, W. A. (1978). Analytical solution to the Curtin–Hammett/Winstein-Holness kinetic system. *Journal of Organic Chemistry*, 43(10), 1854–1864.

Smith, M. B., & March, J. *March's Advanced Organic Chemistry Reactions, Mechanisms and Structure*, 5th Ed. John Wiley & Sons, Inc., 2001, 284, 368.

Sykes, P. *A Guidebook to Mechanism in Organic Chemistry*, 6th Ed. John Wiley & Sons, Inc., 1985, 358, 372.

Tropsha, A., et al. (2014). QSAR modeling: Where have you been? Where are you going to? *Journal of Medicinal Chemistry*, 57, 4977–5010.

Winstein, S., & Holness, N. J. (1955). Neighboring carbon and hydrogen. XIX. t-Butylcyclohexyl derivatives. Quantitative conformational analysis. *Journal of the American Chemical Society*, 77(21), 5562–5578.

11 Reactive Intermediates
Carbocations

11.1 CARBOCATIONS: STRUCTURE AND GEOMETRY

The study of reaction intermediates is divided into four major categories, namely, carbocation, carbanions, free radicals, and carbenes. In this chapter, we will be focusing on carbocations and we will be studying their structure, geometry, stability, reactions, and rearrangements.

11.2 NOTATIONS TO REPRESENT SUBSTITUENTS

11.2.1 C Substituents

Let us look at some of the basic notations C substituents. (C=conjugating). Conjugated systems contain alternating single and double bonds (mainly carbon atoms) ex. vinyl, phenyl. If it is π donor they stabilize electron demand and if it is π acceptor they supply electron excess. Their effect on the σ framework is small, and they are σ neutral (Figure 11.1).

11.2.2 Z Substituents

Z substituents are conjugated systems that are also electron withdrawing, like acetyl, cyano, nitro, sulfonyl, and carboxy. Most of the groups have electronegative heteroatoms, when they are in conjugation they withdraw electrons from the double bonds. They are also weakly electron withdrawing by an inductive effect within the σ framework. Such substituents are, therefore, strong π electron acceptors and usually weak σ acceptors. There is another group of π electron-withdrawing substituents. Metals, and metalloids like the silyl group, are π acceptors but, because metals are more electropositive than carbon, they are σ donors (Figure 11.2).

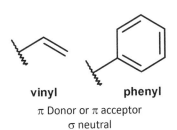

vinyl   **phenyl**
π Donor or π acceptor
σ neutral

FIGURE 11.1 Examples of C substituents.

FIGURE 11.2 Examples of Z substituents.

11.2.3 X Substituents

X substituents are typically electronegative heteroatoms like nitrogen, oxygen, or sulfur, and they have a lone pair of electrons. They donate their lone pairs to a π system, and those based on electronegative heteroatoms withdraw electrons from the σ framework. They are therefore π donors and σ acceptors, exactly the opposite of the metals and metalloids. We usually include simple alkyl groups in the category of X substituents, because they are able, by overlap of the C—H (or C—C) bonds with the π system (σ conjugation or hyperconjugation) to supply electrons to a conjugated system. Alkyl groups are therefore π donors, but they are largely neutral with respect to the σ framework. The electronegative halogen atoms are anomalous; technically they are X substituents, but their effect in the π system is weak because the lone pair of electrons are held so tightly that they are strong σ acceptors (Figure 11.3).

11.3 TYPES OF CARBOCATIONS

There are two types of carbocations: classical and non-classical. They are also called the carbenium ion in which the positive carbon has coordination number 3, as in *t*-butyl cation; and the carbonium ion in which the positive carbon or carbons have coordination number 4 or 5, as in the bridged structure. The former structures, for many years, are designated as classical ions and have ordinary two-electron bonds. The latter, known as non-classical ions, have a three-center, two-electron bond (Figure 11.4).

π donor but σ acceptor
π Donor σ neutral

FIGURE 11.3 Examples of X substituents.

DOI: 10.1201/9781032631165-11

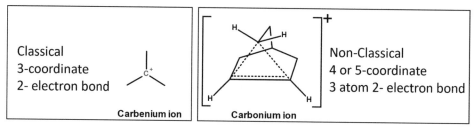

FIGURE 11.4 Types of carbocations.

11.4 GENERATION OF CARBOCATIONS

There are several methods for the generation of carbocation out of which only three of the following are considered:

1. Heterolytic fission of neutral species
2. Addition of cation to neutral species
3. From other cations

Reactions in which the carbocation is an intermediate (S_N1 or E1) are not considered here. But those cases where the reactive species can be generated albeit momentarily or can be formed at lower temperatures. In fact, when carbocations are formed in solvolysis reactions, the solvents will have sufficient nucleophilicity due to which the carbocation readily undergoes further reactions. If we want to synthesize carbocations, we need to do it either under the gas phase or in solvents of low nucleophilicity.

11.4.1 HETEROLYTIC FISSION OF NEUTRAL SPECIES

Ionization of halo derivatives (R^+X^-) in high polar-ion solvating medium, reactions of alkyl halide in the presence of Ag^+ (change of mechanism from S_N2 to S_N1) are some examples of heterolytic fission of neutral species. *Tert*-butyl bromide or benzyl chloride can undergo heterolytic fission to give the corresponding carbocation.

We can also use Lewis acid like BF_3 to carry out the ionization. When acetyl fluoride is treated with BF_3, an acyl carbocation is formed. When pivaloyl chloride is treated with $AlCl_3$, an acyl cation is initially formed, and this unstable acyl cation loses carbon monoxide to give a stable *tert*-butyl carbocation (Figure 11.5).

Other cases include the pioneering work by the Nobel laureate George Olah. He used super acid SbF_5 with either liquid SO_2 or excess of SbF_5 as a Lewis acid to generate carbocations from alkyl halides or alkenes (Figure 11.6).

Ex. 1 different starting materials lead to the same carbocation. Both 1-fluropropane and 2-fluropropane give *iso*-propyl cation.

Ex. 2 Primary, secondary, or tertiary-flurobutanes give *tert*-butyl cation. This is due to the stability of the *tert*-butyl carbocation.

For generating methylcyclopentyl carbocation, we need extremely strong acids. When we have a good leaving group

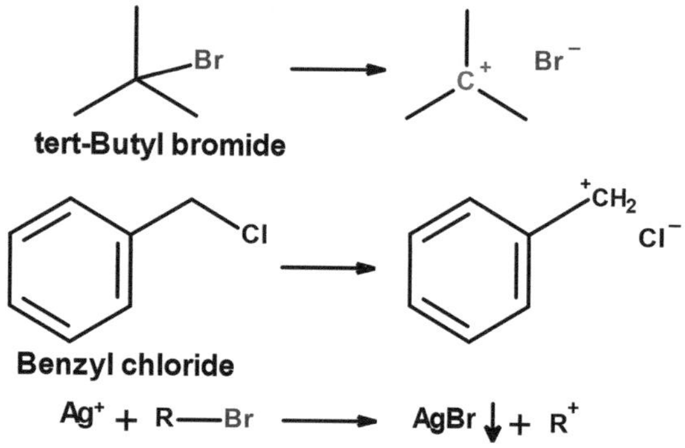

tert-Butyl bromide

Benzyl chloride

$$Ag^+ + R\!-\!Br \longrightarrow AgBr\downarrow + R^+$$

FIGURE 11.5 Heterolytic fission of neutral species.

$$MeCOF + BF_3 \rightleftharpoons MeCO^+BF_4^-$$

$$Me_3CCOCl + AlCl_3 \rightleftharpoons Me_3CCO^+AlCl_4^- \rightleftharpoons Me_3C^+AlCl_4^- + CO\uparrow$$

$$RF + SbF_5 \longrightarrow R^+ SbF_6^-$$

FIGURE 11.6 Generation of carbocations by Lewis acid.

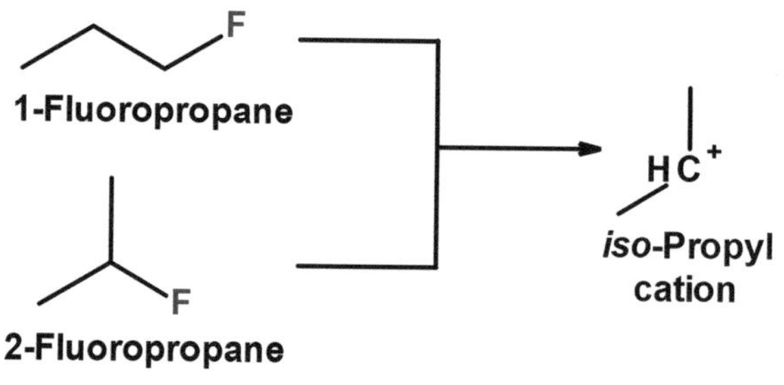

1-Fluoropropane

2-Fluoropropane

iso-Propyl cation

SCHEME 11.1 Formation of *iso*-propyl cation.

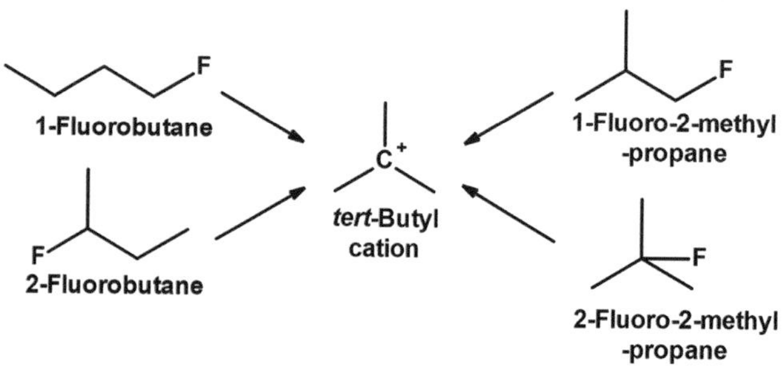

1-Fluorobutane

2-Fluorobutane

tert-Butyl cation

1-Fluoro-2-methyl-propane

2-Fluoro-2-methyl-propane

SCHEME 11.2 Formation of *tert*-butyl cation.

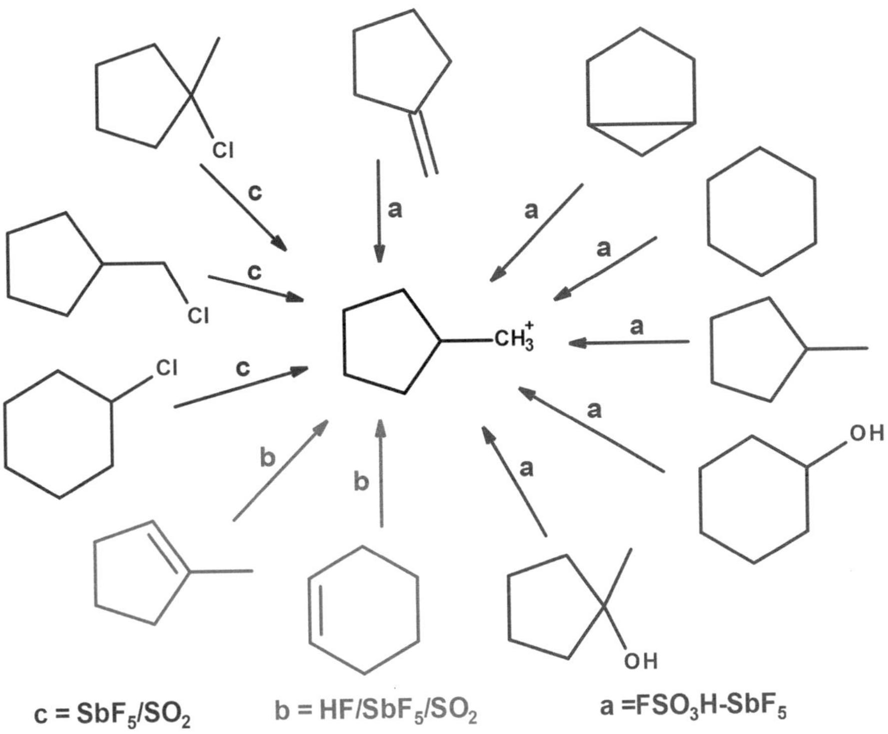

c = SbF₅/SO₂   b = HF/SbF₅/SO₂   a =FSO₃H-SbF₅

SCHEME 11.3   Generation of methylcyclopentyl carbocation.

like chloride, endocyclic double bond, cycloalkanes, or alcohols, the cation can be generated using a strong acid.

Unlike what we saw earlier, here we have 11 different starting materials (Scheme 11.3) giving the same methylcyclopentyl carbocation.

11.4.2 Addition of Cations to Neutral Species

The simplest cation is a proton. We can add this cation to unsaturated bonds like alkene (C=C) or carbonyl compounds (C=O), acid-catalyzed hydration of alkenes, or carbonyl compounds. When a carbonyl compound is treated with concentrated H_2SO_4 in the absence of water, there is two-fold depression of freezing point, indicating the formation of two ionic species. One is the cation, and the other is the HSO_4^- ion.

Other instances include aromatic electrophilic substitution reactions like nitration ($^+NO_2$), sulfonation (SO_3), and so on, where cationic intermediates are produced.

11.4.3 From Other Cations

Examples include hydrogen transfer, decomposition of other cations, and rearrangements (Figure 11.7).

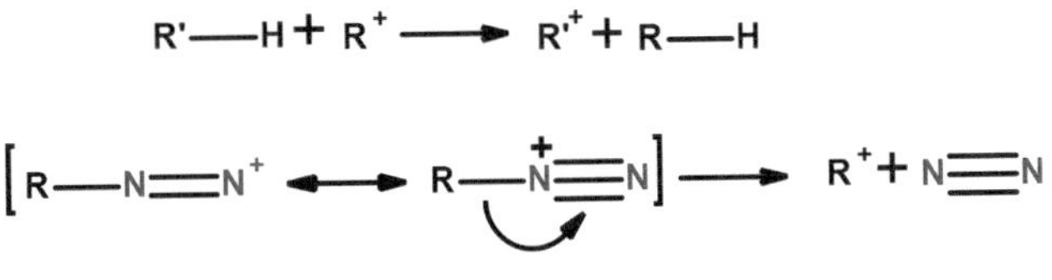

$$R' \!\!-\!\!H + R^+ \longrightarrow R^+ + R \!\!-\!\!H$$

$$\left[R \!\!-\!\! N \!\!\equiv\!\! N^+ \longleftrightarrow R \!\!-\!\! \overset{+}{N} \!\!=\!\! N \right] \longrightarrow R^+ + N \!\!\equiv\!\! N$$

FIGURE 11.7   Carbocation from other cations.

11.5 STRUCTURE AND GEOMETRY OF CARBOCATIONS

We already know carbocation is sp^2 hybridized with an empty p orbital. It has a planar structure with a 120° H-C–H bond angle (NMR and X-ray crystallography). Another evidence is the racemization of chiral alkyl halides under solvolysis conditions.

Due to the inductive effect, more electron cloud is present near the electrophilic center. The carbocation is stabilized through delocalization of the positive charge. When we talk about the ease of formation of carbocations, it is 3°>2°>1°>methyl.

We have some additional evidence for the planar structure of the carbocations. Trimethylboron (Me_3B) and *tert*-butyl carbocation (Me_3C^+) are isoelectronic. According to VSEPR theory, trimethylboron is planar so does *tert*-butyl carbocation.

There is indirect evidence that comes from the difficulty in forming the carbocation at bridgehead position. Similarly, stereomutation (isomerization) of 1,3-disubstituted allyl cation system (^{1}H NMR) is another example. The isomerization of 1,3-*cis*-*cis*-dimethylallyl cation to 1,3-*trans*-*trans*-dimethylallyl cation occurs due to its planar nature.

Most of the evidences we have seen so far are for quite unstable species. Triphenylmethyl ions are sufficiently stable (enantiomeric) but leads to racemic modification when reacted with nucleophiles. This is due to the fact that planar carbocation can be attacked from either side with equal probability leading to racemic modification (Figure 11.8).

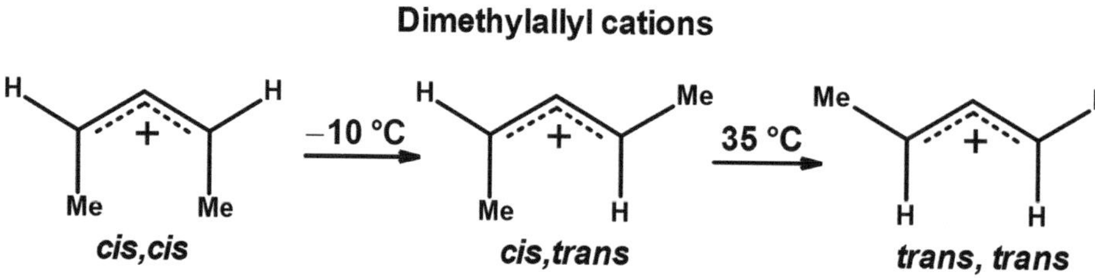

Dimethylallyl cations

SCHEME 11.4   Isomerization of dimethylallyl cation.

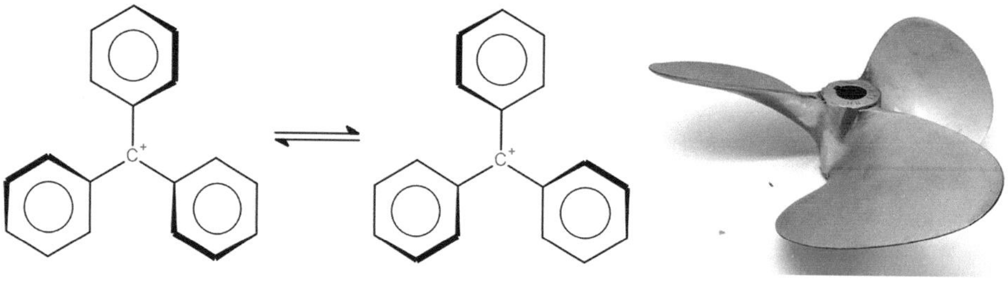

FIGURE 11.8   Triphenylmethyl cation.

Triphenylmethyl chloride (trityl chloride) gave conducting solutions when dissolved in liquid SO_2, a polar non-nucleophilic solvent. Trityl chloride also reacts with Lewis acids, such as $AlCl_3$, to give colored salt-like solids. In contrast to triphenylmethyl chloride, which has the properties of a covalent compound, triphenylmethyl perchlorate behaves like an ionic compound. Triphenylmethyl cation has a propeller shape as shown above.

Why the triphenylmethyl cation has a propeller shape instead of being planar?

There are two things we are discussing here. One is the structure of the whole species, and the second one is the structure of the point of interest. The central carbocation is planar, but the three phenyl rings are at an angle of 54° to the plane of the trigonal carbon so that the overall species has a propeller-like shape. In other words, the central carbocation and the three bonds attached to it are planar, but the substituents at the end of each bond are out of plane due to steric interactions.

11.6 STABILITY OF CARBOCATIONS

The stability of carbocation depends on the neighboring atoms, groups, or bonds. It can be stabilized by resonance, delocalization, or hyperconjugation. Resonance stabilization includes C=C bonds (allyl, benzyl), C=Q bonds (heteroatom with lone pair), delocalization (N, O, S, aromatization), hyperconjugation (alkyl), number of substitution (3°, 2°, and 1°), and so on. Many researchers have used the triphenylmethyl carbocation to study the stability with respect to various substituents.

11.6.1 Hydride Affinity

There is another way to determine the stability. Here we are looking at the ease with which the carbocation combines with hydride ion to form neutral species. Smaller value means more stable carbocation.

$$R^+ + H^- \rightarrow R\text{-}H - \Delta G° = \text{Hydride affinity}$$

Let us look at the hydride affinity of methyl, ethyl, iso-propyl, and *tert*-butyl cations. They are 314, 274, 247, and 230 kcal/mol, respectively. Since a lower value means more stable, *tert*-butyl cation is more stable than secondary which in turn is more stable than primary. In fact, this is the same order in stability as established on the basis of the solvolysis rate of various alkyl halides in solution. Within each structural class (primary, secondary, and tertiary), larger ions are more stable than smaller ones. Any structural effect that alters the electron density at the tricoordinate carbon will also alter the stability of the carbocation.

The π-electron delocalization requires proper orbital alignment. As a result, there is a significant barrier to the rotation of the carbon–carbon bonds in the allyl cation. When we compare benzyl and phenyl carbocation, the phenyl cation has hybridization in between sp and sp^2. It is having partial linear structure due to which there is a deviation from trigonal planar structure. This puts lots of strain when it is in the cyclic system. So, it is highly unstable than benzyl carbocation by a large margin. On the other hand, tropylium cation has as many as seven resonance structures so it is highly stable. The most controversial case is between the stability of *tert*-butyl and benzyl cation. We know *resonance* is more powerful than *hyper conjugation*. Based on that, we may tend to say benzyl cation is more powerful than *tert*-butyl. But the data show otherwise. Benzyl is actually a primary cation with a benzene substituent. *Tert*-butyl has nine hyperconjugative structures, whereas benzyl cation has four resonance forms only.

11.7   NON-BENZENOID AROMATIC CARBOCATIONS

Here we have a few examples from the aromatic system. Cyclopropenyl cation or the cyclobutadienyl dication or the tropylium cation all are stable species (Figure 11.9).

11.8   CARBOCATION STABILIZING GROUPS

C and X-substituents stabilize carbocation. C substituent (double bond) stabilizes through π conjugation. In the case of X substituent, it is lone pair donation by oxygen or nitrogen, and hyperconjugation by alkyl groups.

11.8.1   BENZYL CATION STABILIZATION

Stabilization of benzyl cation is strongly affected by the substituents on the benzene ring. Oxygen and nitrogen which are present in the *para* position to the carbocation can donate their lone pair through resonance (Figure 11.10).

11.8.2   HETEROATOM STABILIZATION

The next one is stabilization through heteroatom in aliphatic systems. Although these structures have a positive charge on a more electronegative atom, they benefit from an additional bond that satisfies the octet requirement of

the tricoordinate carbon. These "carbocations" are well represented by the doubly bonded resonance structures. In the first one, we have examples of ether oxygen and amine nitrogen stabilizing the carbocation. One indication of the participation of adjacent oxygen substituents is the existence of a barrier to rotation about the C–O bonds in these types of carbocation. In the last one, we have halogens stabilizing carbocations (Figure 11.11).

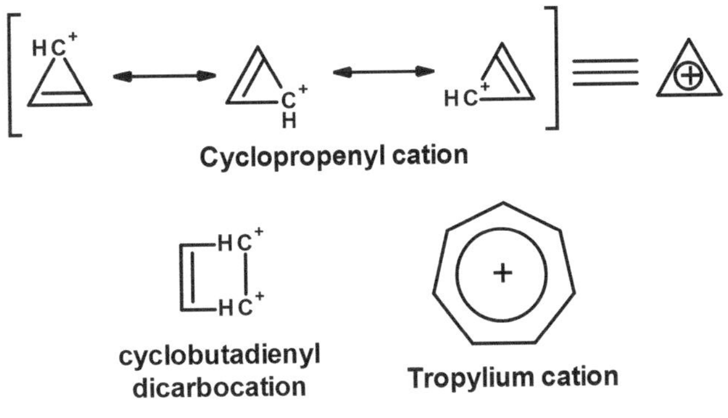

11.9   CARBOCATION DESTABILIZING GROUPS

The Z (EWG) substituents destabilize carbocations. When they are present adjacent to a carbocation which itself is electron deficient, there is destabilization. Even though Z substituents have π electrons, the donation of π electron by double bond does not reduce the energy of HOMO. Due to polarization of the carbonyl bond, there is a slight positive charge on carbonyl carbon, which leads to coulombic repulsion (Figure 11.12).

11.9.1   BENZYL CATION DESTABILIZATION

In the aromatic system, the presence of EWG (cyano or nitro group) in the *para* position to the carbocation creates a positive charge on the *ipso* carbon. Due to this, there is destabilization (Figure 11.13).

11.9.2   WHY TERT-BUTYL CATION IS HIGHLY STABLE?

It is observed that the alkyl carbocations are unusually more stable. In a carbocation, the electron-deficient carbon is more electronegative than any substituent carbons. The carbocation will attract electron density from the nearby alkyl substituents *through the σ bonds. This is the inductive effect.* Based on the increased stability due to alkyl substituents on the carbocation, we can say stability will increase from 1° to 2° to 3°. (Et<*iso*-Pr<*tert*-Bu). However, the higher stability of 3° cannot be explained by the relative increase in magnitude of inductive effect compared with 2° or 1° carbocation. Hyperconjugation is invoked to explain this phenomenon (Figure 11.14).

FIGURE 11.9   Non-benzenoid aromatic carbocations.

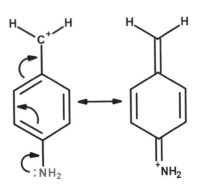

FIGURE 11.10   Benzyl cation stabilization.

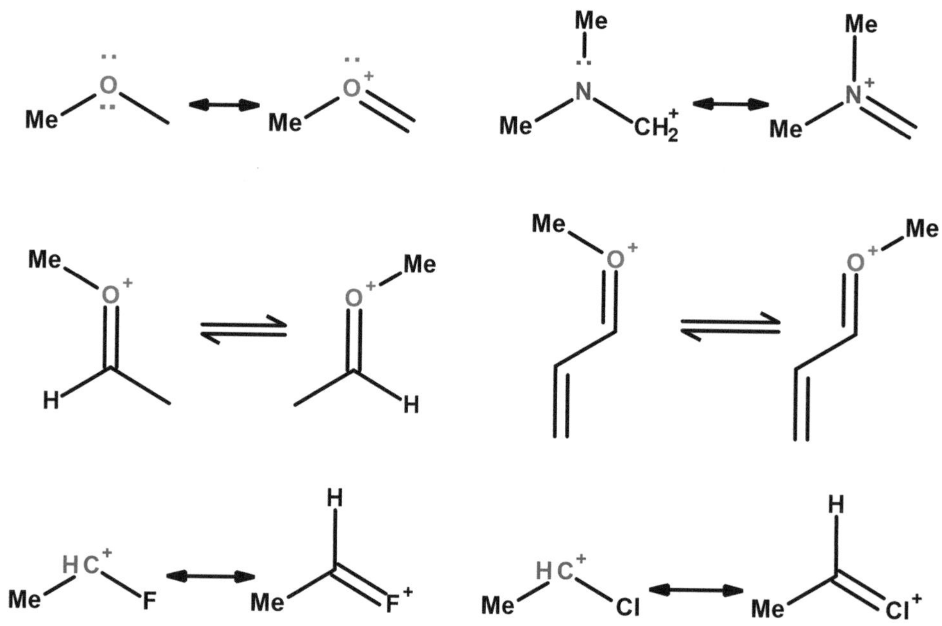

FIGURE 11.11   Heteroatom stabilization.

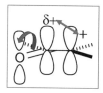

FIGURE 11.12   Carbocation destabilizing.

11.10   HYPERCONJUGATION

Mulliken in 1939 described hyperconjugation as "a mild sort of conjugation". In the example, (Figure 11.15) it is basically the overlap of σ electrons with the empty *p* orbital on the carbocation. This leads to bond- no bond resonance. Generally, for hyperconjugation, the presence of protons on the alpha carbon is required.

Hyperconjugation is the overlap of σ bonds with σ bonds or p orbitals that are close in energy (Ex. σ with σ overlap or σ* with σ*), that is electron donation of a filled bonding orbital to a nearby low-lying, unfilled orbital. This overlap requires anti-periplanarity of the groups that are involved in hyperconjugation. For σ bond category, hyperconjugation can involve C–H, C–C, C-hetero atoms (Si, Ge, Sn, Pb, and Hg) (Figure 11.16)

Let us recall the differences between inductive effect and hyperconjugation (Table 11.1).

11.10.1   CRITERIA FOR HYPERCONJUGATION

Here we will see some of the criteria for hyperconjugation:

- Low electronegativity of A.
- Inductive donation by R.
- No competing R–A conjugation.
- A strong B=C double bond.
- Planarity of the ABCX unit.

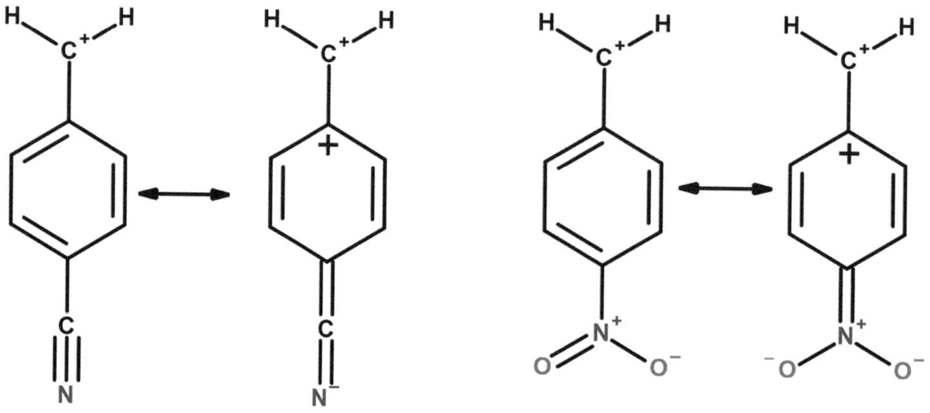

FIGURE 11.13   Benzyl cation destabilization.

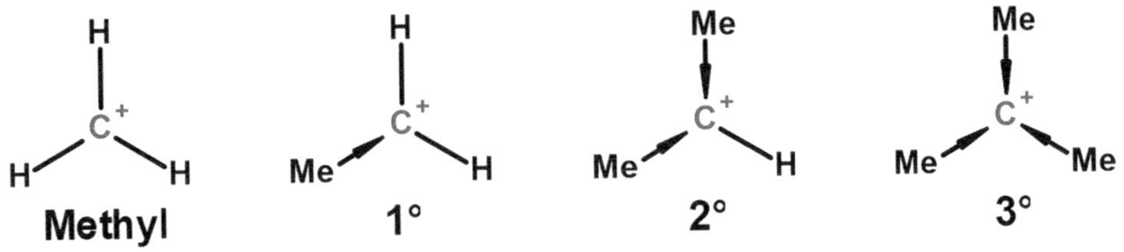

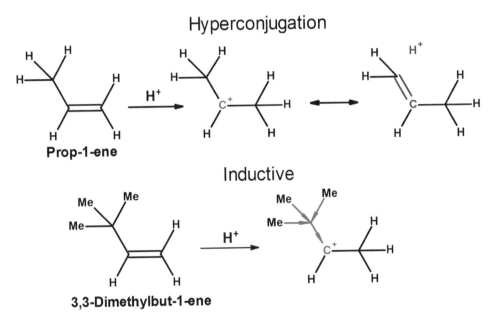

FIGURE 11.14 Inductive effect of 1°, 2° and 3° carbocations.

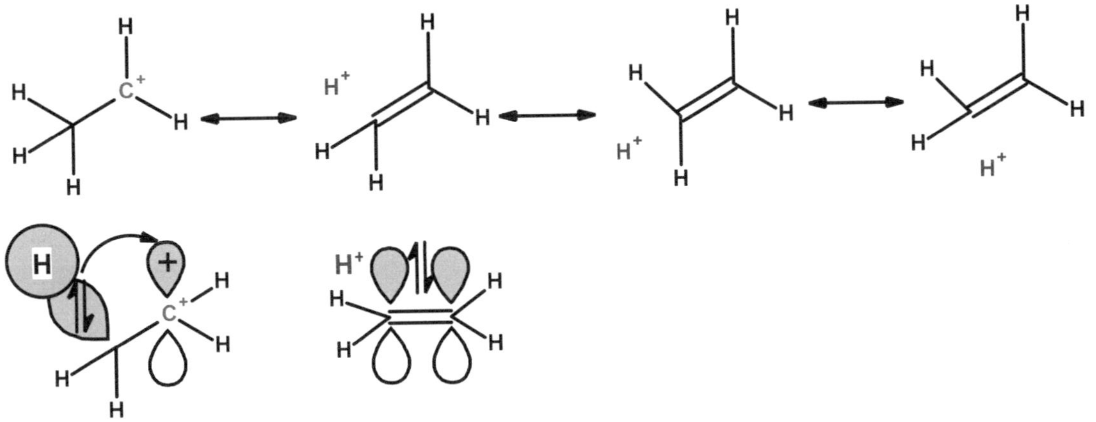

FIGURE 11.15 Inductive effect of carbocations (bond-no bond effect).

Hyperconjugation

Prop-1-ene

Inductive

3,3-Dimethylbut-1-ene

FIGURE 11.16 Inductive effect and hyperconjugative effect in carbocations.

TABLE 11.1
Comparison of Inductive Effect and Hyperconjugation

Inductive Effect	Hyperconjugation
Arise due to electronegativity difference between the two atoms	Arise due to overlap of an empty p orbital with an adjacent σ bond
Permanent state of polarization	Anti-periplanarity is required
Can be donation (+I) or withdrawal (−I) of electrons density along the sigma bonds	Partial shifting of sigma bond electrons
	Leads to dispersion of partial charge

11.11   NON-CLASSICAL CARBOCATIONS

The contribution of a neighboring group to the stability of a carbocation is called the neighboring group effect (non-classical). Instead of normal carbocation on the left, the bridged carbocation on the right is formed. Solvolysis of p-hydroxy phenylethyl bromide is 10^6 times faster than the corresponding methoxy derivative (Figure 11.17).

11.11.1   ACETOLYSIS OF 2-NORBORNYL BROSYLATE

The acetolysis of both exo-2-norbornyl brosylate and endo-2-norbornyl brosylate produces exclusively exo-2-norbornyl acetate. The exo-brosylate is more reactive than the endo isomer by a factor of 350. This involves a 3-atom 2-electron bond. It is sometimes called carbonium ion (Figure 11.18).

An open trivalent classical carbocation has less stability compared with the symmetrical bridging of cyclic carbocations. In fact, they are all transient species, and they are invoked to explain the unusual stability that is seen in some systems (Figure 11.19).

11.12   REACTIONS OF CARBOCATIONS

We have four major types of reactions of carbocations:

- Substitution with nucleophile
- Elimination of proton (deprotonation)
- Addition to multiple bonds
- Rearrangements

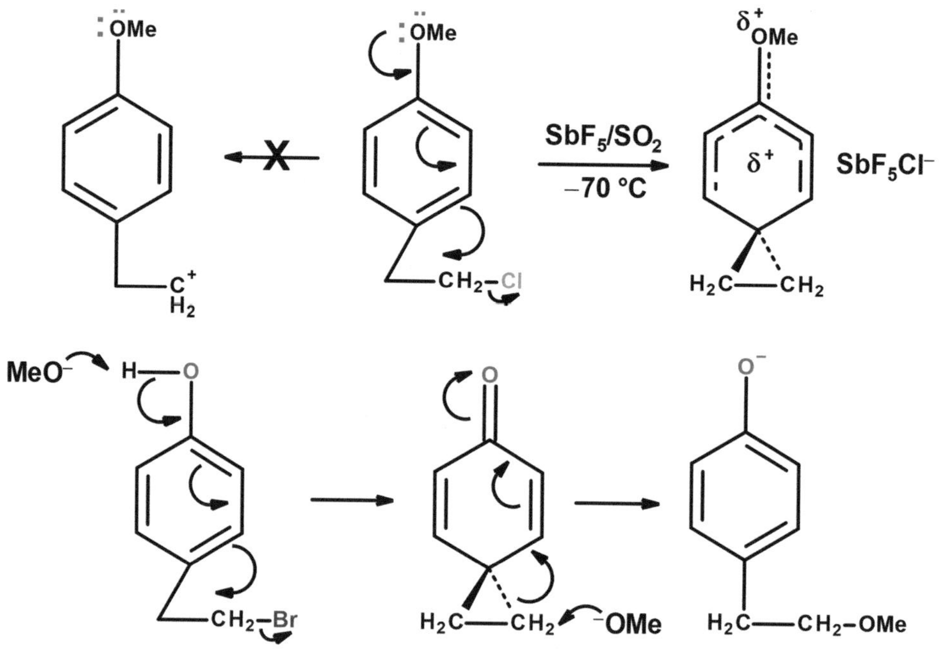

FIGURE 11.17   Examples of neighboring group effect, non-classical.

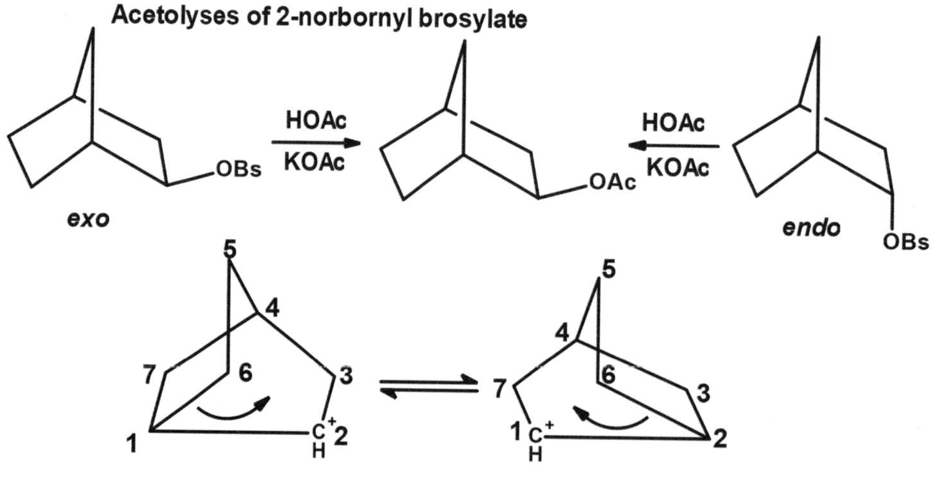

FIGURE 11.18   Acetolysis of 2-norbornyl brosylate.

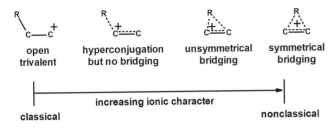

open trivalent | hyperconjugation but no bridging | unsymmetrical bridging | symmetrical bridging

increasing ionic character

classical   nonclassical

FIGURE 11.19 Classical to non-classical carbocation.

X =

X = halide   X = Alkyl halides
X = water   X = Alcohols
X = alcohol/phenol X = Ethers
X = acid   X = Esters
X = amine   X = Ammonium salt
X = thiol   X = Thioether

FIGURE 11.20 Examples of substitution with nucleophile.

In S_N1 reactions, based on the nucleophile, we get different types of products like alkyl halide, alcohol, ammonium salt, ethers, thioethers, and so on.

Decalin 2-ol undergoes carbonylation to give *cis*- and *trans*-decalin-9-carboxylic acid. The bridgehead *tertiary* carbocation is formed by the loss of water molecules followed by protonation. The addition of carbon monoxide leads to the acylium ion. Models show that the *trans*-acylium ion with the C–C=O$^+$ atoms in a straight line has very little axial–axial interaction, so can be easily formed without strain (Figure 11.21).

But the *trans* acid or its protonation product carboxyl appears more crowded than the methyl group, so is less stable than the *cis* isomer. In a strong acid (oleum) all the reactions are reversible, hence the more stable *cis* product is formed

Elimination of a proton from a carbocation gives olefin. It is facilitated by the abstraction of a proton by the base (Figure 11.22).

The exchange of proton leads to a new carbocation as shown below (Figure 11.23).

Carbocations can undergo additional reactions with alkenes. In many cases, it is possible that a cyclic carbocation intermediate may be formed (Figure 11.24, first set of equations).

11.12.1 Carbocations in Terpene Synthesis

The pinenes are among the most common monoterpenes produced by plants and are principal components of turpentine.

CO$_2$H   CO$_2$H

HCO$_2$H   O H   HCO$_2$H

98 % H$_2$SO$_4$ fuming H$_2$SO$_4$

H   H

trans-9-Decalincarboxylic acid   *cis*-9-Decalincarboxylic acid

decalin-2-ol

Kinetic   **Thermodynamic**

stable

less stable
kinetic

stable (Thermodynamic)

FIGURE 11.21 Carbonylation and its mechanism of Decalin 2-ol.

Terpene synthases utilize electrophilic reaction mechanisms involving carbocationic intermediates, a feature of terpenoid biochemistry. Formation of α-and β-pinene involves cyclization of the neryl cation to a monocyclic α-terpinyl intermediate which undergoes internal Markownikoff addition to a pinyl cation that is subsequently stabilized by two different reactions of proton loss. The isomerization step is essential because geranyl pyrophosphate (GPP) cannot cyclize directly, given the presence of the *trans*-double bond.

Let us look at how different monoterpene intermediates are formed. The initial intramolecular electrophilic attack of C1 of the neryl cation on the distal double bond gives a monocyclic (α-terpinyl) intermediate. This undergoes various reactions to give different skeletal products. Some of the reactions include internal electrophilic additions, hydride ion shifts, and Wagner–Meerwein rearrangements. All these reactions produce cationic equivalents of most known skeletal types like carene, fenchane, pinane, and bornane (Figure 11.25.

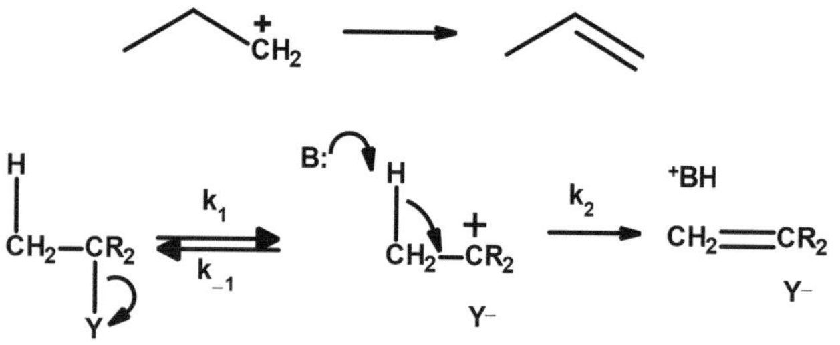

FIGURE 11.22 Elimination of a proton from a carbocation.

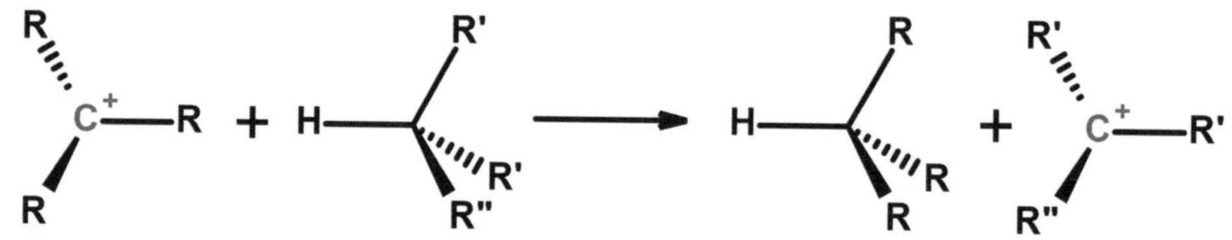

FIGURE 11.23 Exchange of proton giving a new carbocation.

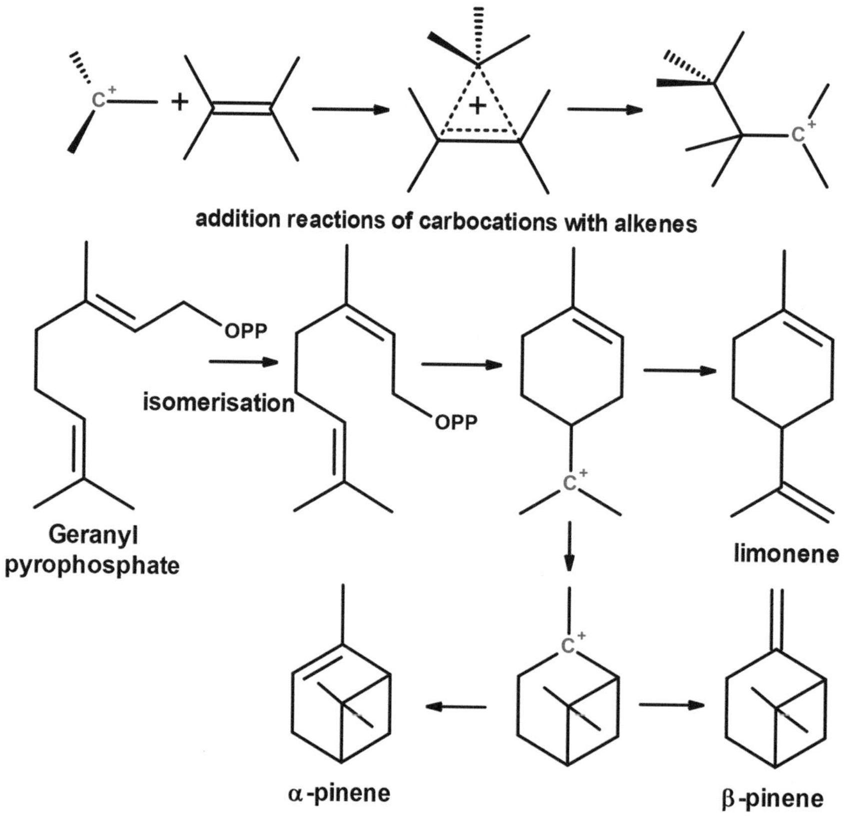

addition reactions of carbocations with alkenes

Geranyl pyrophosphate

isomerisation

limonene

α-pinene

β-pinene

FIGURE 11.24 Addition reactions of carbocations with alkenes and isomerization followed by cyclization of geranyl pyrophosphate.

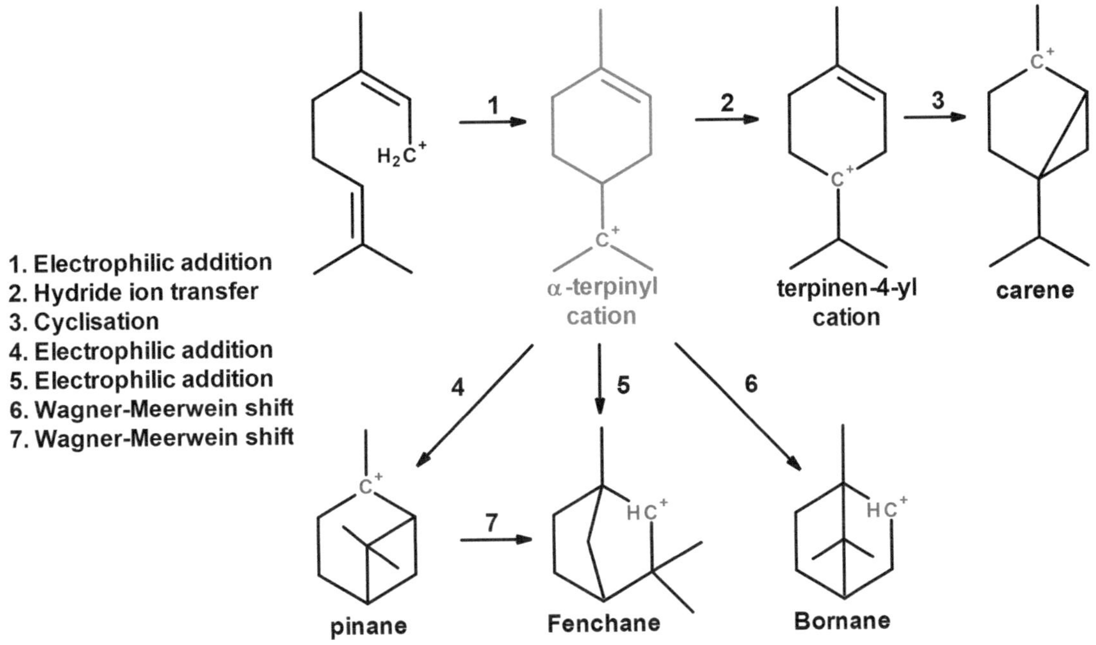

FIGURE 11.25 Various reactions of α-terpinyl cation.

1. **Electrophilic addition**
2. **Hydride ion transfer**
3. **Cyclisation**
4. **Electrophilic addition**
5. **Electrophilic addition**
6. **Wagner-Meerwein shift**
7. **Wagner-Meerwein shift**

Terpenoid synthases that produce cyclic products are also referred to as "cyclases", and *Limonene Synthase* is an example. Ionization of the enzyme-bound intermediate promotes cyclization to a six-membered ring carbocation (the α-terpinyl cation), which may undergo additional electrophilic cyclizations, hydride ion shifts, or other rearrangements before the reaction is terminated by deprotonation of the carbocation or capture by a nucleophile (e.g., water). Isomerization of the exocyclic to the endocyclic isomer raises the possibility of a single cyclization of an acyclic precursor to β-pinene, followed by enzymic (or nonenzymic) conversion to α-pinene (Figure 11.26).

11.12.2 CARBOCATIONS AND HAMMOND POSTULATE

For the rate of reaction of alcohols with hydrogen halides, the rate increases, in the order methyl < primary < secondary < tertiary. This reactivity order is similar to the stability of the carbocation. For an S_N1 reaction, the rate-determining step involves the formation of the carbocation intermediate. In this case, the rate-determining step in the S_N1 mechanism is the dissociation of the alkyloxonium ion to the carbocation.

Rate = k [alkyloxonuim ion].

The value of k is related to the activation energy for the dissociation of alkyloxonium ions.

The methyl substrate has the maximum activation energy and the *tert*-alkanol has the least activation energy for the carbocation formation. According to Hammnod's postulate, in this reaction, the transition state is closer in energy to the carbocation and more closely resembles it than the alkyloxonium ion. Thus, structural features that stabilize

carbocation, stabilize transition states leading to them. By looking at the potential energy curve, we can say, that alkyloxonium ions derived from tertiary alcohols have a lower energy of activation for dissociation and are converted to their corresponding carbocation faster than those derived from secondary and primary alcohols. In other words, more stable carbocations are formed faster than the less stable ones (Figure 11.27).

The Hammond postulate predicts that the transition state should resemble the ion pair. Hence any structural change that lowers the carbocation energy should lower transition state energy and increase the rate of the reaction. The evidence provides unequivocal confirmation of this prediction, for limiting S_N1 reactions, with no assistance to ionization by nucleophiles. The substitution of H by CH_3 on the reacting carbon accelerates the rate by a factor of 10^8, a difference in activation energy of about 11 kcal/mol (Figure 11.28).

Based on the Hammond postulate, for the nucleophilic substitution reaction of alkyl halide, the *tert*-alkyl carbocation (intermediate) resembles the energy of the starting material more than the product. We have an early transition state.

11.13 REARRANGEMENT REACTIONS OF CARBOCATIONS

Let us look at four major categories of rearrangement reactions:

1. No change in the carbon skeleton
2. Change in the carbon skeleton
3. Loss of groups
4. Rearrangement involving an electron-deficient center that is not carbon.

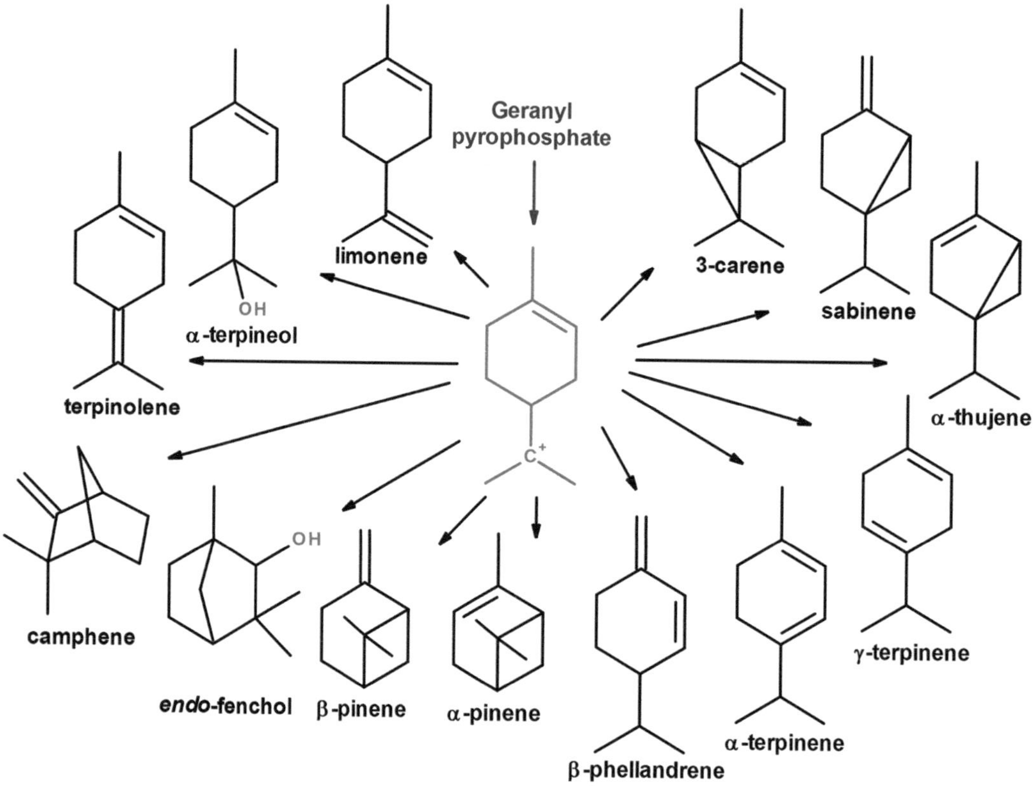

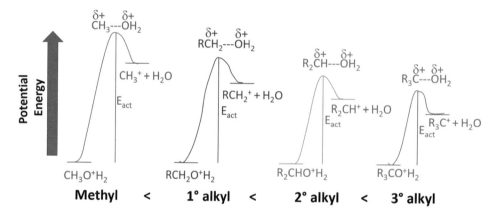

FIGURE 11.26   Various reactions of terpene biosynthesis.

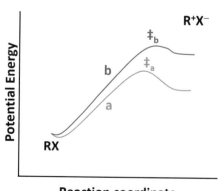

FIGURE 11.27   Activation energies of various cations.

Path a:
- Stable ion gives rapid reaction
- Early transition state
- Small charge separation

Path b:
- High energy ion does not give rapid reaction
- Late transition state
- Large charge separation

FIGURE 11.28   Activation energies of methyl and primary carbocations.

We will be studying these four types of reactions. Figure 11.29 gives a brief overview of all the reactions.

11.13.1 No Change in Carbon Skeleton

1,2-Hydride ion shift belongs to this category. We have two types. One is conversion of 1°–2° or 2°–3° carbocation. In either case, the most stable carbocation will be formed. Another one is allylic rearrangement.

11.13.2 1,2-Hydride Ion Shift

This is also called intramolecular reactions of carbocations. Since it is the hydrogen that is being shifted, there is no change in the carbon skeleton. Here is an example of *iso*-propyl cation rearrangement. A primary carbocation rearranges to a more stable secondary carbocation. If you

recollect, this is a major problem in Friedal–Crafts alkylation. When a primary or secondary alkyl halide whose carbocation can undergo rearrangement is used in FCR, we will end up with a mixture of rearranged products, in addition to *ortho*, *para* isomeric products.

Occasionally rearrangements in the reverse direction, that is, tertiary to secondary, may occur, if (1) the energy difference between them is not too large or (2) the carbocation that rearranges has no other possible rapid reactions available to it or (3) there is greater delocalization possibility exists (Figure 11.30).

Let us look at this transformation as an intramolecular reaction. Ethyl carbocation can form a bridged structure and is more stable than the open chain transition state. In the reaction of halogens with alkenes, we know a bridged halonium ion is possible as an intermediate. The loss of proton to form the alkene is a competing reaction (Figure 11.31).

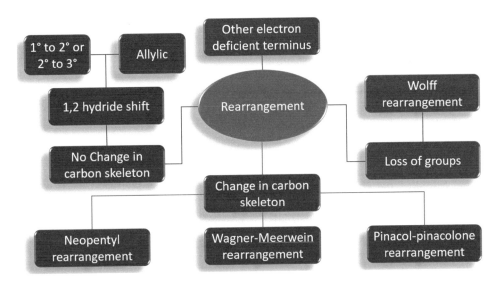

FIGURE 11.29 Overview of rearrangements of carbocations.

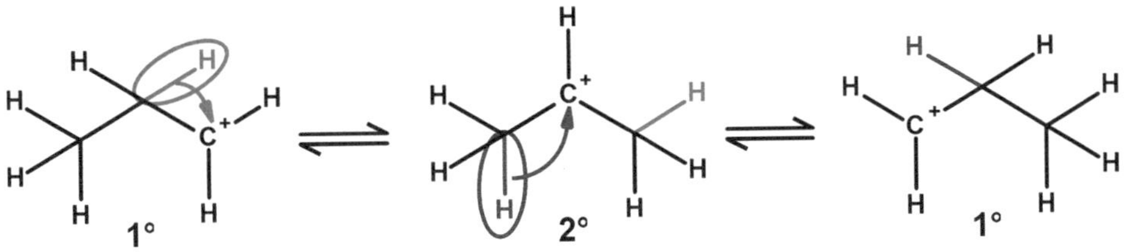

FIGURE 11.30 1,2-Hydride ion shift.

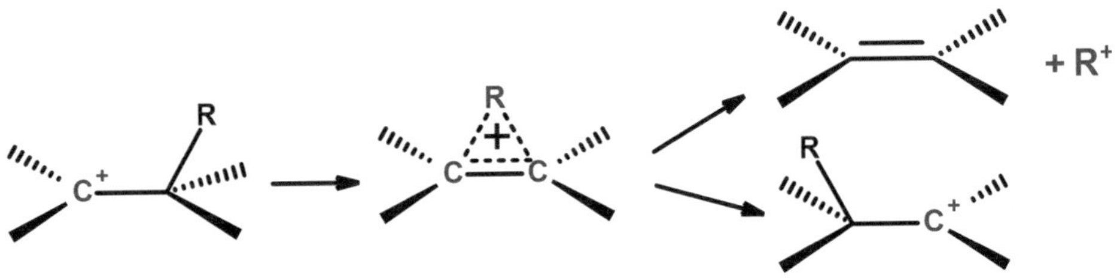

FIGURE 11.31 Intramolecular transformation of carbocation.

11.13.3 Criteria for 1,2-Shift

Let us look at the criteria for 1,2-shift: (1) the migration origin and migrating group must lie in the same plane (planarity) and (2) the dihedral angle must be zero, vacant p orbital and C-α–C-β bond must lie in the same plane (Figure 11.32).

11.13.4 Allyl Rearrangement

The solvolysis of 3-chloro-1-butene and 1-chloro-2-butene gave the same mixture of two isomeric ethers in a similar ratio. Unless these two alkenes have the same intermediate in the same ratio, we cannot explain the product formation (Figure 11.33).

We need to have two carbocations formed from the two alkenes and they also need to be in resonance. The primary carbocation formed from 1-chloro-2-butene undergoes rearrangement through stabilization from the neighboring π bond, this is possible because we know alkene carbons are planar and the carbocation is sp^2 hybridized. All three carbons are planar, and the movement of π electrons stabilizes the primary carbocation which in turn produces the secondary carbocation. It is fine if the less stable primary cation is converted to the more stable secondary carbocation. But the problem is how can we explain the primary carbocation formed from the secondary cation. There are exceptions to the stability rule. The movement of π electron from the adjacent double bond to the secondary carbocation produces primary carbocation. Why does this happen? Because the positive charge is now spread over three carbons instead of on a single secondary carbon. There is additional stabilization due to the delocalization of the positive charge. Moreover, the energy barrier for this transformation is very small. This is nothing but the *allyl rearrangement* (Figure 11.34).

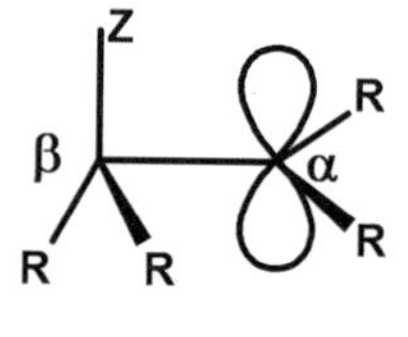

FIGURE 11.32　Criteria for 1,2-shift.

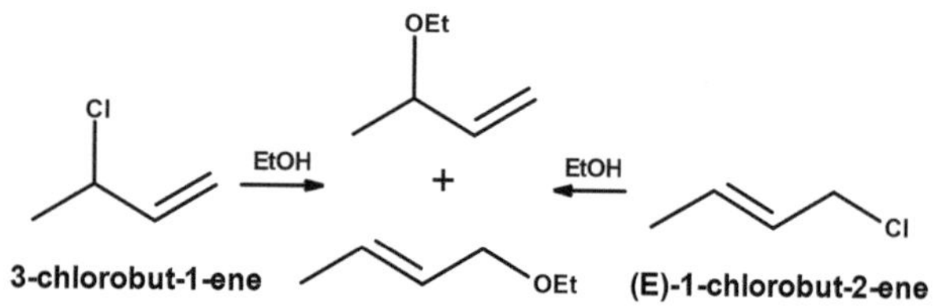

FIGURE 11.33　Allyl rearrangement.

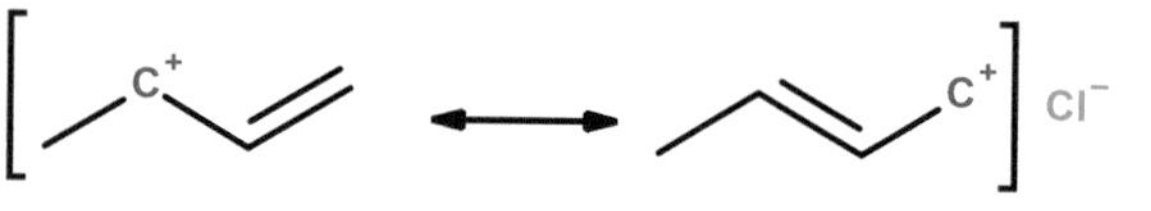

FIGURE 11.34　Formation of primary carbocation from secondary carbocation.

11.14 CHANGE IN CARBON SKELETON

11.14.1 Neopentyl Rearrangement

When neopentyl iodide is treated with silver nitrate, we get *tert*-amyl alcohol and 2-methylbut-2-ene. There are a few interesting things that happened, which are difficult to explain. (1) Neopentyl iodide is a *primary* halide and the alcohol obtained is a *tertiary* alcohol. (2) Neo-pentyl iodide has the leaving group at the terminal carbon, and there are no β-hydrogens. It cannot undergo elimination, when it cannot give a terminal olefin, and we cannot even think of an internal olefin.

Normal S_N2 is not possible because of molecular or steric crowding. S_N1 is not possible as it is the primary halide. (Primary halides generally undergo solvolysis through the S_N2 mechanism). When Frank Whitmore reported in 1939, he reported the positive silver ion removes the halogen atom, leaving the neopentyl system with 30 electrons where rearrangement is possible. One of the methyls from the β carbon move with its bonding pair of electrons to the primary cation. This results in the rearranged *tertiary* carbocation which can either be attacked by the nucleophilic solvent or can undergo elimination. For the migration of alkyl groups, a bridged cyclic intermediate can be anticipated (Figure 11.35).

11.14.2 Wagner–Meerwein Shift

The intramolecular migration, as shown below, of a hydrogen, an alkyl, or an aryl group with its bonding pair of electrons from a β-carbon (migration origin) to the adjacent carbocationic center (migration terminus) is called a 1,2-shift (or in the case of migration of an alkyl or an aryl group, a Wagner–Meerwein shift). In Wagner–Meerwein shift, preference or ease of migration will be aryl > alkyl.

The migration of an alkyl group to a cationic center is known as a Wagner–Meerwein rearrangement or Wagner–Meerwein shift. This rearrangement was first discovered in the bicyclic terpenes, mainly the conversion of isoborneol to camphene (Figure 11.36).

Let us move on to another example we have 1,2-migration, 1,3-migration, and Wagner–Meerwein shift happening in a bicyclic system. Although it looks complex, it is not. The carbocation is marked as 1. There are two adjacent carbons marked as 2. There is a distal carbon 3 with 2 hydrogens. In this example, we have C–H σ bond and C–C σ bond migration taking place. The first one which is a WM shift is a C–C σ shift. The 1,3- and 1,2-shifts are C–H σ shifts. The rearranged carbocations can further undergo 1,2- or WM shift (Figure 11.37).

Another interesting case is conversion of tricyclodecane to adamantane. The reaction starts with the formation of the carbocation. This undergoes a WM shift followed by various 1,3-shifts and WM shifts to finally yield the adamantane. You can clearly see two cyclohexane chair conformation in this molecule (Figure 11.38).

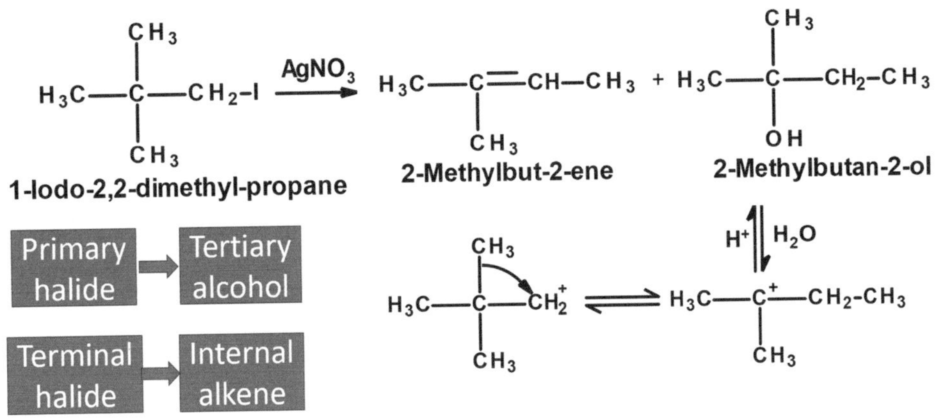

FIGURE 11.35 Neopentyl rearrangement.

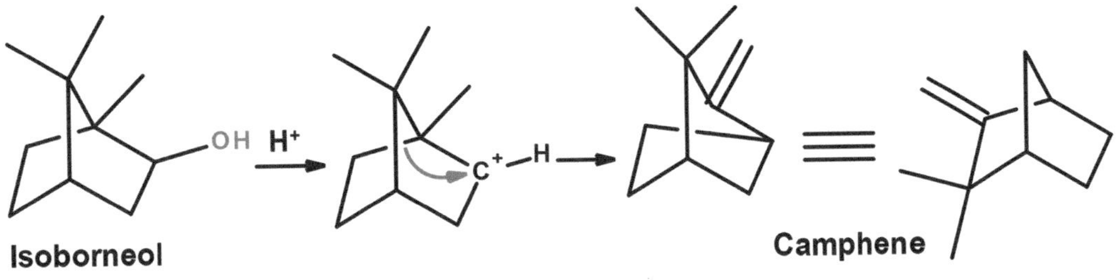

FIGURE 11.36 Conversion of isoborneol to camphene.

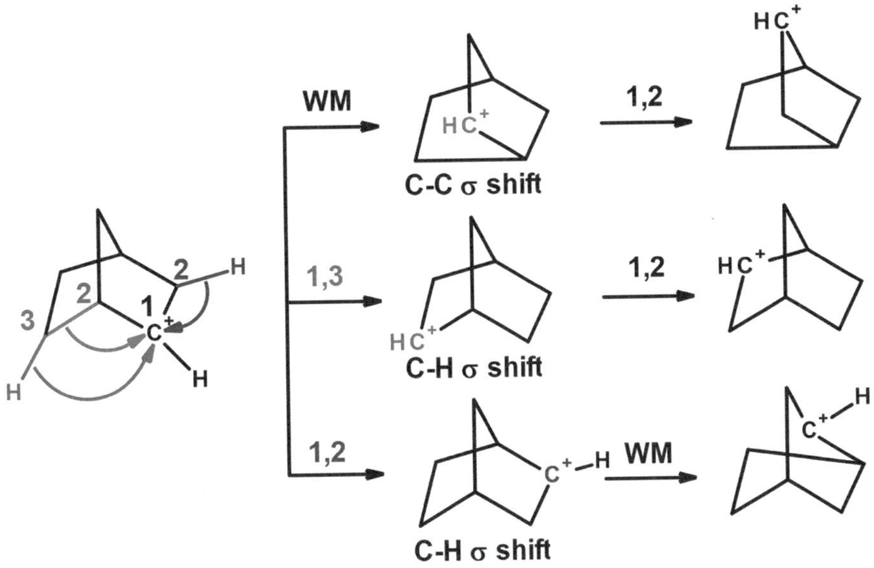

FIGURE 11.37 Rearrangements in bicyclic system.

The expected 1,2-hydride ion shift in the 2-adamantyl cation does not take place in a highly dilute solution. Likewise, the anticipated 1,2-methyl shift in the 2-methyladamantyl cation has been shown by isotope labeling to occur by a complicated skeletal rearrangement. In both these cases the C–Z bond and the vacant p orbital, in both the cases is perpendicular to the plane, they also form a dihedral angle of 90° (Figure 11.39).

11.14.3 PINACOL–PINACOLONE REARRANGEMENT

When vicinal-diols, that is, glycols are treated with acids, they can be rearranged to give aldehydes or ketones. Sometimes elimination without rearrangement is also possible. This reaction is called the pinacol–pinacolone rearrangement. The first step is the protonation of the hydroxyl. This leaves a water molecule and in the process generates carbocation. Now one of the alkyl groups from the

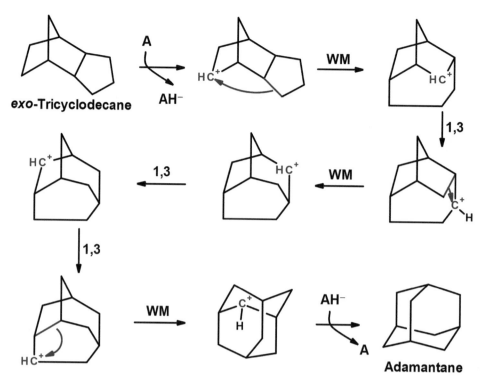

FIGURE 11.38   Conversion of tricyclodecane to adamantane.

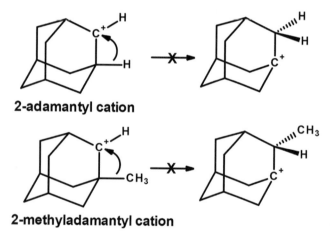

FIGURE 11.39   Examples of failure of 1,2-shift in adamantyl system.

neighboring carbon atom shifts with its bonding pair of electrons to the carbocation. The newly formed carbocation has the hydroxyl, which readily loses a proton to form the carbonyl compound. Since a H⁺ is lost, the molecule is now neutral or chargeless (Figure 11.40).

In the pinacol–pinacolone rearrangement, the replacement of methyl group in 1,2-diol by alkyl, aryl, or hydrogen, follows the order aryl>alkyl>hydrogen.

What is the driving force for this rearrangement? The most important one is the thermodynamic stability of the resulting carbocation. The hydride shift or the methyl shift happens when a primary or secondary carbocation is converted to a secondary or tertiary carbocation. In many

cases, the energy barrier to the shift is relatively low, so it readily happens.

Now you will be wondering, if the initial carbocation itself is *tertiary*, which is a stable carbocation, then, why the 1,2-shift takes place?

It is a valid question. In the pinacol rearrangement, the driving force for migration is the formation of a carbonyl group. According to the microscopic reversibility of organic reactions, the initially formed *tertiary* carbocation can be attacked by the nucleophile (⁻OH or H₂O) or it can undergo alkyl migration. If it is the former, then we will get back our starting material, if it is the latter, we will end up with the carbonyl compound. When we talk about the relative migratory aptitude, phenyl moves faster than>*tert*-butyl>*iso*-Pr>Me. In fact, there are more possibilities and we will see some of them here.

Pinacol rearrangement is not very simple. In an unsymmetrical diol, various possibilities exist. They are (1) Which hydroxyl is protonated first? (2) If all the four substituents are different, can we predict which group will migrate? (3) Which is more important, the stability of the intermediate or the stability of the final product? (Figure 11.41)

Let us look at a specific example as given below. The migratory aptitude is based on the stability of the resulting carbocation. Stability is a thermodynamic property and the ease with which the migration takes place is a kinetic property.

If the barrier to migration, that is, activation energy is not very high then, migration takes place easily. Depending on the reaction conditions like solvent, temperature, concentration, how long we allow the reaction to proceed,

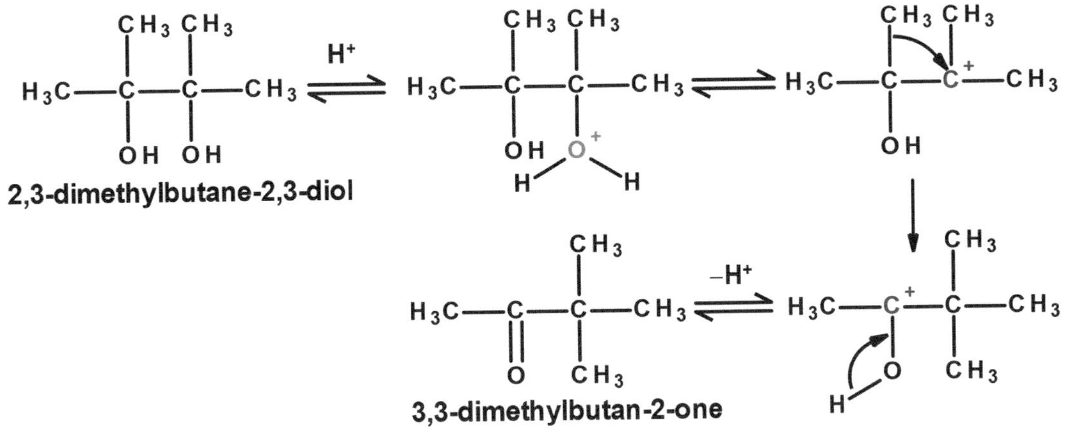

FIGURE 11.40   Pinacol–pinacolone rearrangement.

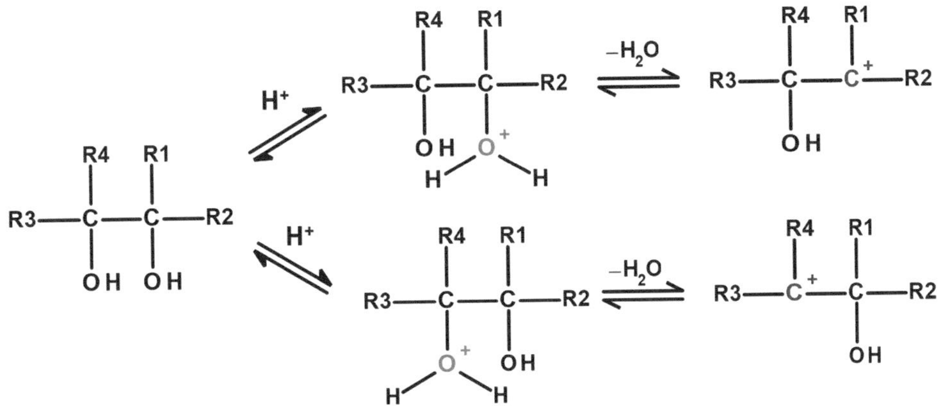

FIGURE 11.41   Relative migratory aptitude of different groups.

that is, allow it to equilibrate, and so on are all-important parameters that govern the product distribution.

To understand the preferential protonation, let us substitute R1 and R2 with methyl group and R3 and R4 with phenyl group in Figure 11.41. Here we have either 1st OH (R1-C-OH) protonates first or the 2nd OH (R4-C-OH) protonates first. If the 1st OH is protonates first, then it is lost as a water molecule and we get a carbocation that has both phenyl groups attached to it. In the other case, the resulting carbocation has both methyl groups attached to it. Based on the stability of the carbocation formed we can clearly say the reaction where the Ph group can stabilize the carbocation is the one that is preferred over the other path.

To answer the question of which group preferentially migrates, we can take the following two examples. Although the starting materials are not pinacols, the intermediate carbocation formed is similar to the one obtained in the pinacol–pinacolone rearrangement. Using ^{14}C labeling, we can identify the product and can find out the migratory aptitude of groups (Figure 11.42).

In the decomposition of tosylate, only the phenyl group migrates, while in acid treatment of the corresponding alkene, there is competitive migration of both methyl and phenyl. Although we can think of the same carbocation,

there is a subtle difference due to which both the carbocations give different products. The difference is due to the neighboring group participation. In the tosylate, the phenyl group can assist the leaving group, while no such process is possible for the alkene. This example clearly illustrates the difference between migration to a relatively free terminus and one that proceeds with the migrating group lending anchimeric assistance.

11.14.4   Stereochemistry of Rearrangements

When we talk about the stereochemistry of rearrangements, we have three places where the configuration at the reaction center can change:

1. At the migration origin
2. At the migration terminus
3. Occurring in (within) the migrating group itself

In these three places, we can have two different types of stereochemical outcomes, retention or inversion of configuration after the reaction. This is completely in contrast to our general understanding of carbocation reactions. We know that whenever there is a carbocation formation, they

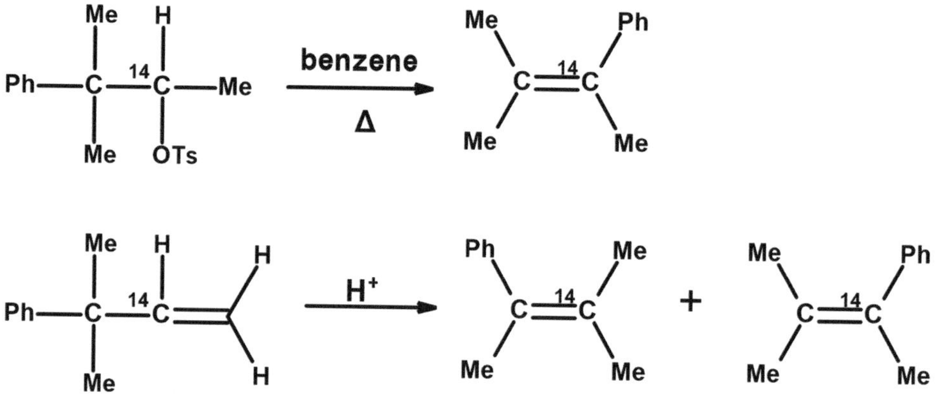

FIGURE 11.42 Preferential migration of different groups.

are planar and the incoming group (nucleophile) can attack from both sides to give a racemic product.

Although all three changes may not happen in a single reaction, it is still difficult to study more than one stereochemical change in a reaction.

One of the observations is that the migrating group does not become detached during the rearrangement. How do we know? Two pinacols having very similar structure, but having different migrating groups, are subjected to rearrangement in the same reaction flask (*crossover experiment*). When the products were analyzed, no cross-products were found. This is a clear indication that the migrating group does not detach itself during migration. In other words, the shift is an intramolecular one and not an intermolecular one. When this happens, there is no possibility of the migrating group undergoing any stereochemical modification, that is, configuration of the chiral group is retained (Figure 11.43).

What happens during the migration? In the given example, we can see predominant inversion of configuration occurs at both migration origin and terminus. This could be explained on the basis of a bridged intermediate. Once the carbocation is formed after the leaving group leaves, the R group migrates. From the crossover experiment, we already

knew the migrating group does not detach itself. When the migrating group is an aryl unit, there is a possibility that π electrons may assist in the stabilization of a bridged carbocation through delocalization. This leads to an inversion of configuration at both the migration origin and the terminus. In this reaction, the migrating group is present at the top, and the nucleophile is present at the bottom side. The leaving group is present at the bottom side and the migrating group is present at the top (Figure 11.44).

Although it is correct to say, bridged carbocation is formed in many cases, there are some exceptions to this. If you take the example of pinacolinic deamination of an optically active 1, 1-diphenyl-2-amino-1-propanol giving α-phenylpropiophenone, we have some interesting results. These deamination reactions proceed from a conformation (antiperiplanar) in which the migrating (Ph) and leaving (NH_2: as N_2) groups are *trans* to each other. If bridged carbocation is the only intermediate, then we would get 100% inversion at the migration terminus in the product ketone, irrespective of the initial conformation, of the amino-alcohol (Figure 11.45).

It was actually found, using ^{14}C labeling studies, that though inversion was predominant (88%), the product

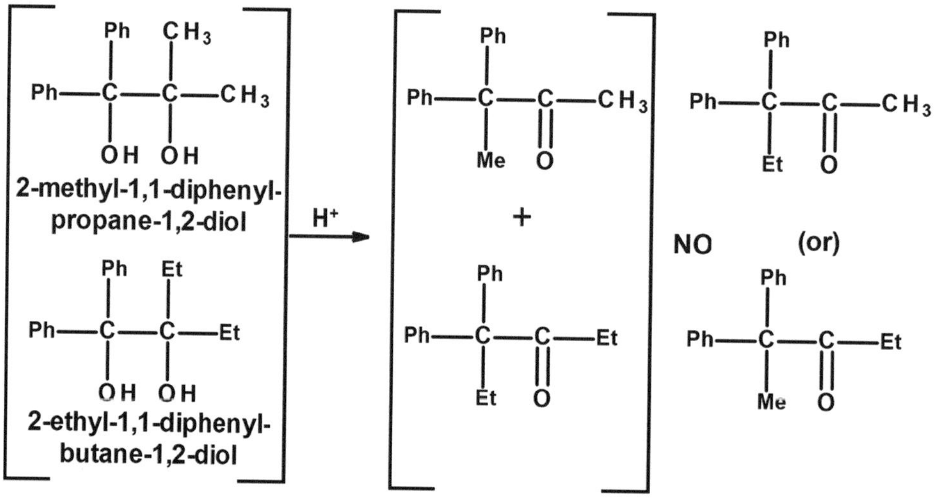

FIGURE 11.43 Example of crossover experiment.

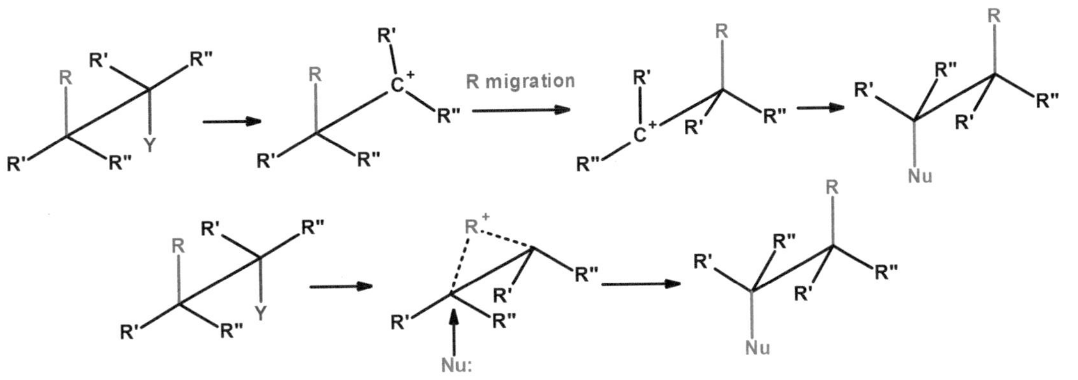

FIGURE 11.44   Example of inversion of configuration at both migration origin and terminus.

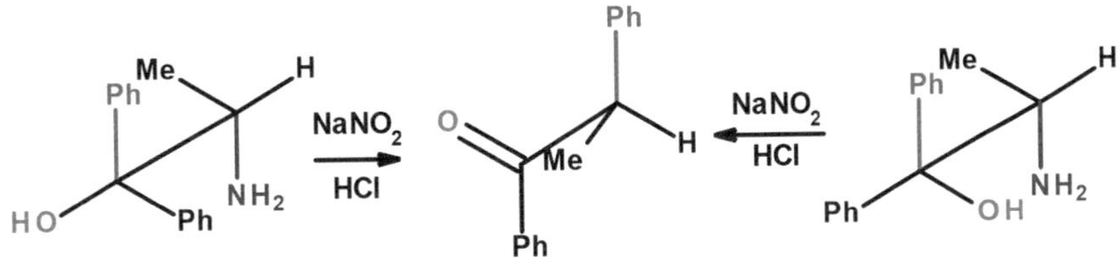

FIGURE 11.45   Example of rotation of the C–C bond between the migrating origin and the migrating terminus.

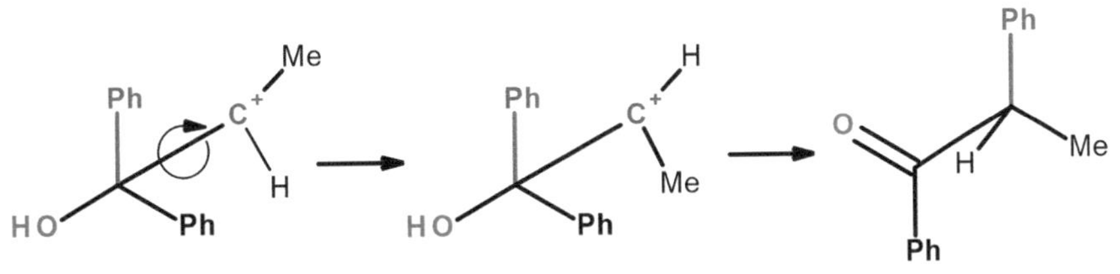

FIGURE 11.46   Example of how the rotation of the C–C bond between the migrating origin and the migrating terminus occurring.

ketone contained a significant amount of the mirror image (12%). Now how can we explain the 12% formation? It is obvious that there should be some other pathway other than the bridged carbocation. Since the carbocation is planar and also there is C–C single bond connecting the migration origin and terminus, we can expect rotation of the C–C bond between the migrating origin and the migrating terminus before the group actually migrates as shown below (Figure 11.46).

11.14.5   MIGRATORY PREFERENCE

If we need to visualize the placement of groups during rearrangement, as mentioned earlier the group that is *trans* to the amino group migrates. Since there is no restriction for the C–C bond rotations, the molecule assumes any one of the orientations say A, B, and C are possible orientations for the molecule. However, from the ^{14}C labeling experiments it was found

that the labeled phenyl ring migrates (conformation A, where the labeled ring is *anti* to NH$_2$) (Figure 11.47)

11.15   LOSS OF GROUPS

11.15.1   WOLFF REARRANGEMENT

Although strictly speaking this does not belong to carbocation rearrangement. But it does involve electron-deficient carbene-like carbon. The actual precursor to the intermediate is a diazoketone which can be prepared from acid chloride. The diazoketone when treated with water and silver oxide produces highly reactive ketene. This then reacts with water, alcohol, ammonia, or amine to give carboxylic acid, ester, or amides, respectively.

When we start with an acid and when we end with the homologous acid, this reaction is called Arndt–Eistert synthesis (Figure 11.48).

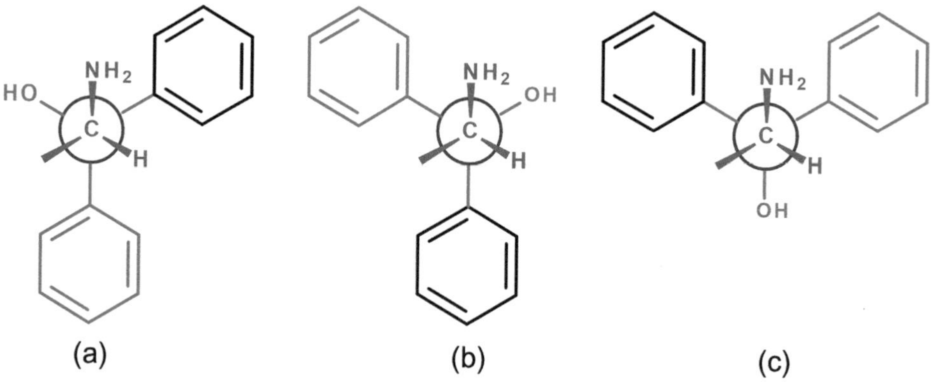

FIGURE 11.47 Example to visualize the migratory preference.

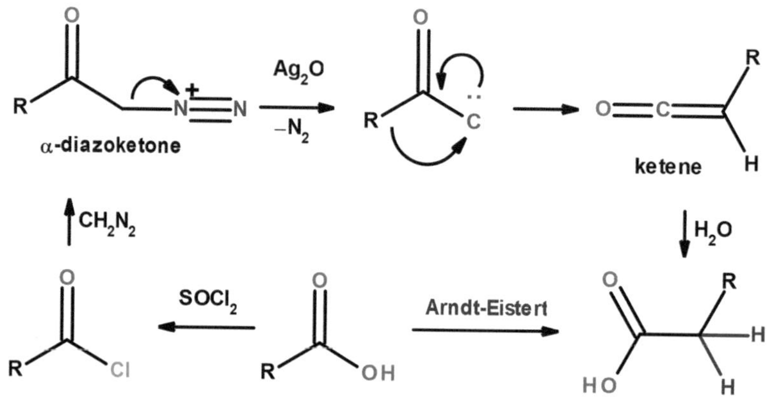

FIGURE 11.48 Arndt–Eistert synthesis.

11.16 CARBOCATION REARRANGEMENT INVOLVING HETERO ATOM

We will be studying Hofmann degradation, Lossen reaction, Curtius reaction, Schmidt reaction, Beckmann rearrangement, Baeyer–Villiger oxidation, neighboring group participation (anchimeric assistance).

11.16.1 CARBON-TO-NITROGEN MIGRATIONS OF R AND AR

Nucleophilic group migrations from carbon to an electron-deficient nitrogen atom, which has either six electrons in its outer shell or else loses a nucleofuge concurrently with the migration. It is very difficult to say whether it is a discrete process

or a concerted process. Some of the available evidence points to one of the two mechanisms shown here.

1. **Configuration** is **retained** in alkyl group
2. Reaction follows **first-order** kinetics
3. **Intramolecular** rearrangement is shown by labeling
4. No rearrangement occurs within the migrating group (Figure 11.49)

11.16.2 HOFMANN REARRANGEMENT

In the Hofmann rearrangement, an unsubstituted amide is treated with sodium hypobromite to give a primary amine that has one carbon less than the starting amide. The first step is the formation of an intermediate N-haloamide.

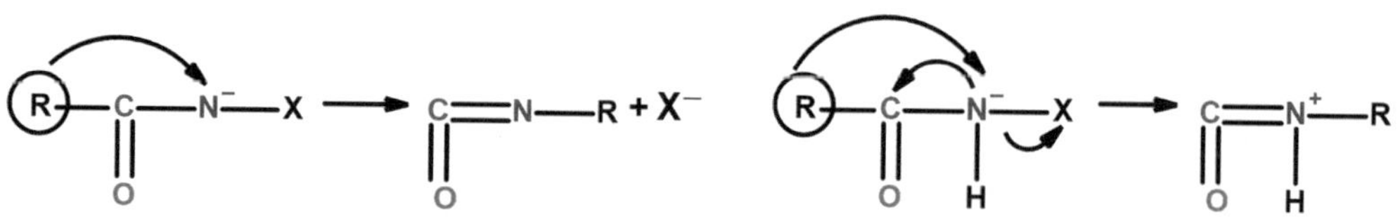

FIGURE 11.49 Example of carbon-to-nitrogen migrations of R and Ar.

In the second step, the proton attached to the nitrogen is lost. The base abstracts this proton, which results in the formation of a nitrogen-centered anion. This compound is acidic because of the presence of two electron-withdrawing groups (acyl and halo) on the nitrogen. The alkyl group moves with the bonding pair of electrons to the nitrogen with concomitant removal of the halide. This results in an unstable isocyanate. The addition of water leads to the carbamic acid which undergoes rapid decarboxylation to give the amine with one carbon less from the starting amide. Through crossover experiments, the non-formation of hydroxamic acid and rate enhancement of migration when the aryl ring has electron-donating substituents points to a concerted mechanism for the loss of bromide and the migration of the alkyl group (Figure 11.50).

11.16.3 LOSSEN, CURTIUS, AND SCHMIDT REACTIONS

The O-acyl derivatives of hydroxamic acids give isocyanates when treated with bases or sometimes even just on

heating, in a reaction known as the Lossen rearrangement. The Curtius rearrangement involves the pyrolysis of acyl azides to yield isocyanates. The reaction gives good yields of isocyanates, since no water is present to hydrolyze them to the amine. Of course, they can be subsequently hydrolyzed, and indeed the reaction can be carried out in water or alcohol, in which case the products are amines, carbamates, or acylureas. Addition of hydrazoic acid to carboxylic acids, followed by loss of nitrogen, produces isocyanate, and this process is called the Schmidt reaction (Figure 11.51).

11.16.4 BECKMANN REARRANGEMENT

When oximes are treated with PCl_5 or other acidic reagents, they rearrange to substituted amides in a reaction called the Beckmann rearrangement (Figure 11.52).

The group that migrates is generally the one *anti* to the hydroxyl, and this is often used as a method of determining the configuration of the oxime. In most cases, the oxime undergoes isomerization under the reaction conditions

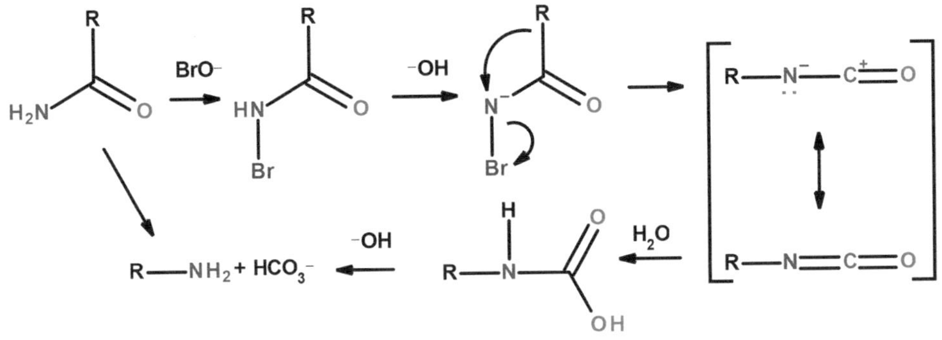

FIGURE 11.50 Example of Hofmann rearrangement.

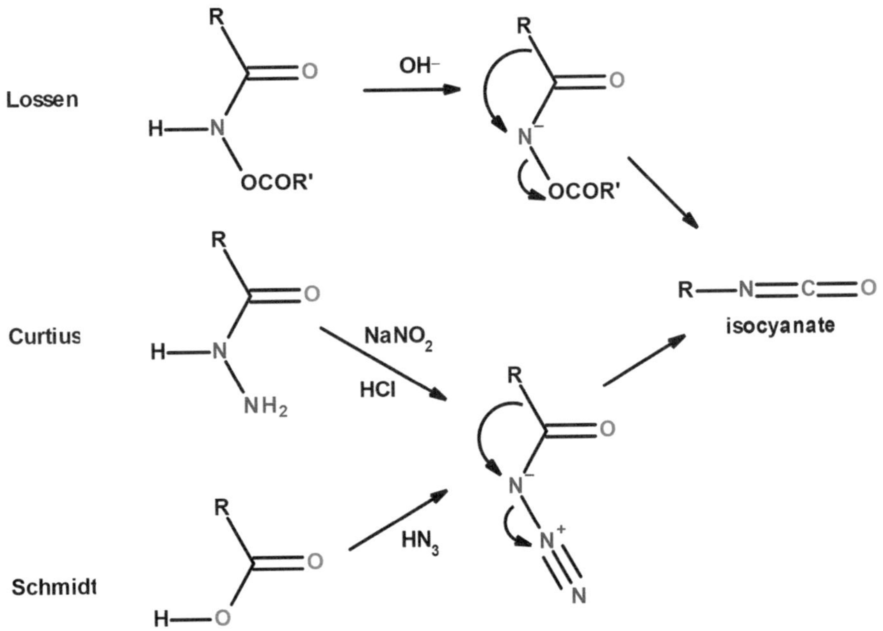

FIGURE 11.51 Example of Lossen, Curtius, and Schmidt's reactions.

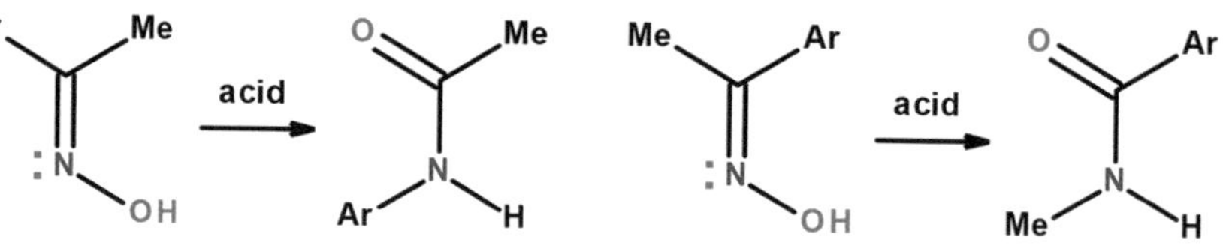

FIGURE 11.52 Example of Beckmann rearrangement.

before migration takes place. The scope of the reaction is quite broad. Both R and R′ may be alkyl, aryl, or hydrogen. However, hydrogen very seldom migrates, so the reaction is not generally a means of converting aldoximes to unsubstituted amides $RCONH_2$.

In the first step of the mechanism, the −OH group is converted by the reagent to a better leaving group, if it is acid then OH is converted to H_2O^+ and if it is an acid chloride or PCl_5, then the −OH is converted to the corresponding ester. Now migration of the alkyl or aryl group and the loss of the leaving group occurs simultaneously. This leads to the formation of a carbocationic intermediate, which is further attacked by the nucleophilic solvent to give the addition product. The loss of proton to form the enol, which then rearranges to the more stable amide (Figure 11.53).

One of the industrially important applications of Beckmann rearrangement is the synthesis of caprolactum. Global demand for this compound is approximately 5 million tons per year, and it is mainly used to make Nylon 6 filament, fiber, and plastics.

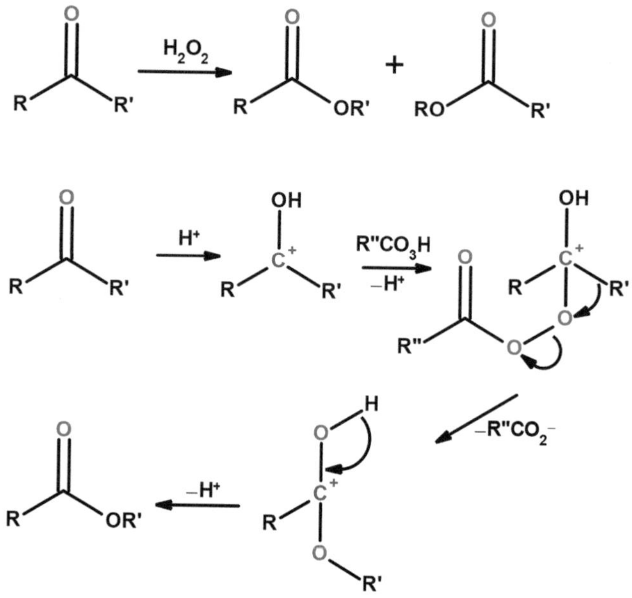

11.17 MIGRATION TERMINUS IS ELECTRON-DEFICIENT OXYGEN

11.17.1 BAEYER–VILLIGER OXIDATION

We will now see an example where the migration terminus is electron-deficient oxygen. Ketones, when treated with peroxide, give esters by oxygen insertion (Figure 11.54).

The first step is the protonation of carbonyl oxygen which leads to the formation of a carbocation. Peracid

FIGURE 11.54 Baeyer–Villiger oxidation and its mechanism.

oxygen forms an adduct with this carbocation. One of the alkyl groups from the carbonyl carbon now moves with its bonding pair of electrons to the electron-deficient oxygen, and there is a simultaneous loss of RCO_2^-. Then, the final loss of proton produces the ester. When there is competition between migration, the alkyl/aryl group which can stabilize the carbocation migrates.

11.18 ANCHIMERIC ASSISTANCE (NGP)

When there is a heteroatom with lone pair, it can stabilize the carbocation. A group adjacent to the leaving group can now act as an intramolecular nucleophile. What are the consequences of this arrangement? First, there is a lone pair of electrons available in close proximity to the electron-deficient reaction site. Two, with a very small degree of reorganization we can reach the transition state. Due to these favorable conditions, reactions of this type occur faster than intermolecular or unimolecular substitutions. We call this kind of support anchimeric assistance or neighboring group participation (Figure 11.55).

Why do we propose such a mechanism? This is to explain why there is a mixture of products in the substitution reaction. If we do not propose a cyclic three-membered transition state, we cannot explain the product formation.

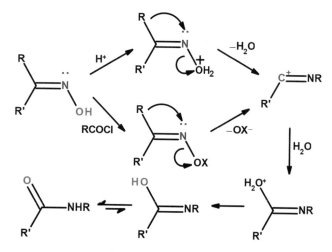

FIGURE 11.53 Mechanism for Beckmann rearrangement.

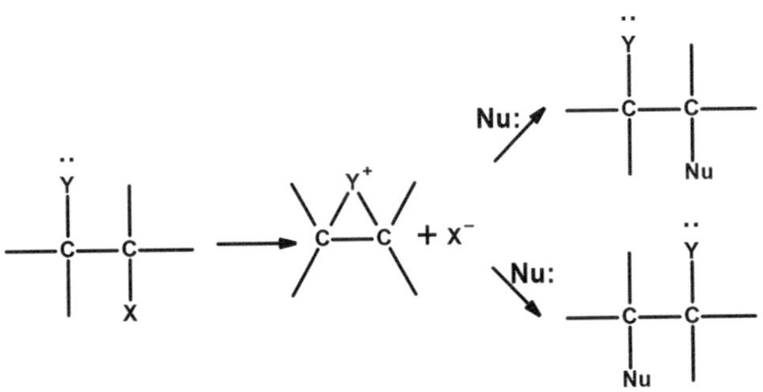

FIGURE 11.55   Example of anchimeric assistance.

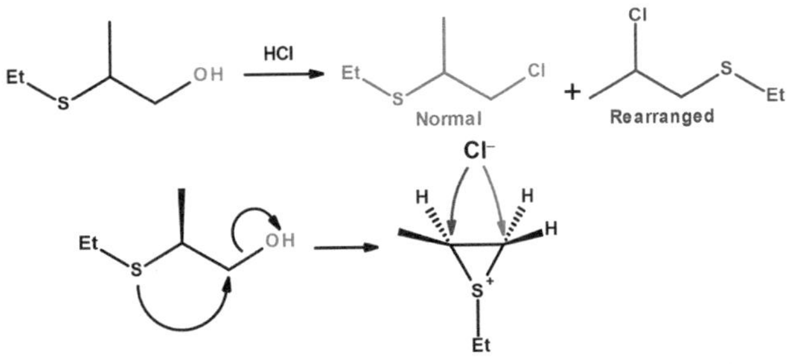

FIGURE 11.56   Example of anchimeric assistance by a heteroatom.

Remember, you need to watch out for NGP when you deal with these types of reactions.

The next difficult question we need to find answers is the time of electron donation:

1. When does the electron donation happen?
2. Does it occur at the time of departure of the leaving group or after the departure of the leaving group?
3. What happens to the stereochemistry of the migration terminus?

Let us find the answer to the question one by one. For the first one, electron donation cannot happen before the carbocation is formed. Before we find the answer to the second question, let us find the answer to the third one. As we had already seen in the case of stereochemistry of rearrangements, due to the formation of cyclic intermediate, there is inversion of configuration at the migration terminus.

For the second question, evidences point to both the possibility. In certain cases, before the heteroatom can donate the electron, if there is rotation of bond in the

carbocation ($C\alpha$–C_β) bond, we can see retention of configuration in the product. In those cases, we can say that donation happens after the leaving group leaves. But in all other cases where we have inversion, we can say that the donation of electrons and the departure of leaving group occurs simultaneously.

NGP can happen in both S_N2 and S_N1 mechanisms. In S_N2, NGP is invoked to explain in cases where there is retention of configuration. In S_N1 it is invoked to explain in cases where there is inversion of configuration. In simple terms, the stereochemistry of the product will tell whether NGP is involved or not. Various studies have shown that hydride, alkyl groups, open chains systems, and unstrained cyclic systems do not provide anchimeric assistance.

Here is an example of NGP by a heteroatom. The sulfur assists in the loss of leaving group. A cyclic three-membered intermediates is formed which can be attacked by the nucleophile from either of the carbons leading to the normal and the rearranged products (Figure 11.56).

Anchimeric assistance can also be given by unsaturated systems rather than lone pair. When we studied non-classical carbocations, we have already seen the anchimeric assistance by the phenyl ring (Figure 11.17).

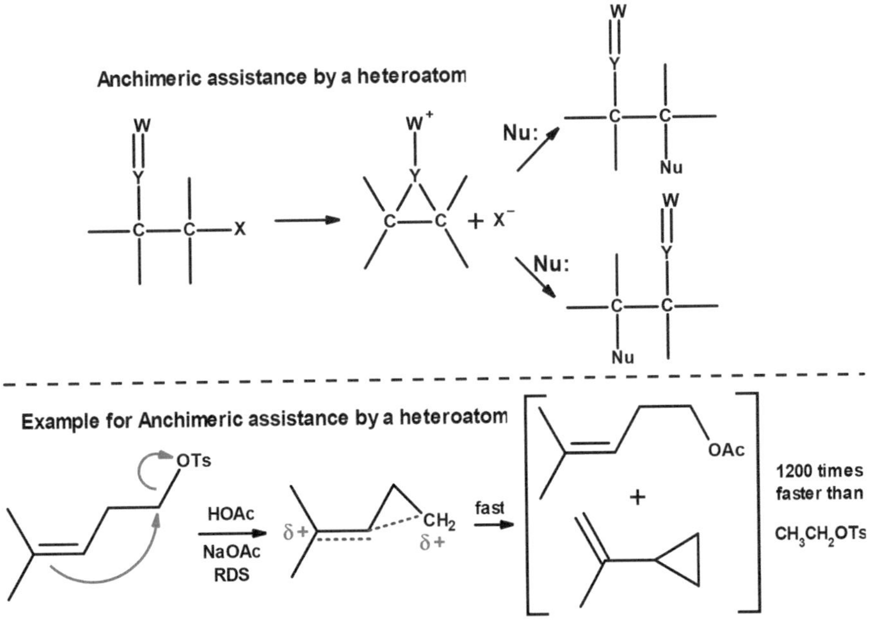

FIGURE 11.57   Example of Anchimeric assistance by unsaturated systems.

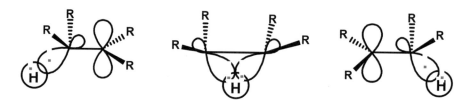

FIGURE 11.58 Example of formation of bridged structure.

In the given example, the solvolysis of tosylate by acetic acid is 1,200 times faster in the assisted case versus the non-assisted case (Figure 11.57).

In this figure, we can see how the bridged structure is formed and the smooth transition of protons from one carbon to the next one (Figure 11.58).

There are other examples of C–C bond formation involving carbocations. Friedel–Crafts alkylation, Friedel–Crafts acylation, and reaction of alkenyl and allyl stannanes and silanes with carbocations are some of the examples of it.

The reaction of carbocation with alkene is highly favorable. Because a weak π bond is converted into a strong σ bond. The only issue we have is an uncontrolled reaction that will lead to polymerization. It is very important to control this reaction by carefully manipulating the reaction conditions like the ratio of the reactants, solvent, temperate, rate of addition, etc.

11.19 QUESTIONS

1. Write briefly about the generation of carbocations, mainly based on heterolytic fission of bonds.
2. Write a short note on the stability of triphenylmethyl carbocation.
3. Based on hydride affinity values of benzyl (233 kcal/mol) and allyl (256 kcal/mol), can you compare the relative stability of these two carbocations?
4. How heteroatom can stabilize the carbocation?
5. Explain with suitable examples, the stabilization as well as the destabilization of benzyl cation by the substitutes present in the aromatic ring.
6. Why tert-butyl carbocation is more stable than iso-propyl carbocation?
7. What is hyperconjugation? How does it give stability to carbocations?
8. What are non-classical carbocations? Why they are called so? With suitable examples explain the stability of non-classical carbocations.
9. Write a short note on substitution reactions involving carbocations.
10. Using proton and carbocation as examples, explain the electrophilic addition of alkenes.
11. What is allyl rearrangement? Explain with a suitable example.
12. When vicinal-diols, that is, glycols are treated with acids they give aldehydes or ketones. Explain the mechanism by which this reaction takes place.
13. What is Wagner–Meerwein shift?
14. How adamantane is synthesized from *exo*-tricyclodecane.
15. Explain with a suitable example Wolff rearrangement. How is it related to Arndt–Eistert synthesis?
16. Write a short note on stereochemical changes that can occur at the migration origin.
17. What is Hofmann degradation?
18. Write a short note on Lossen, Curtius, and Schmidt reactions.
19. What is Beckmann rearrangement? Explain with a suitable example.
20. What is anchimeric assistance? Explain this with suitable examples.

BIBLIOGRAPHY

30.4: Anchimeric Assistance in Organic Chemistry (OpenStax) (https://chem.libretexts.org/Bookshelves/Organic_Chemistry/Book%3A_Virtual_Textbook_of_OChem_(Reusch)/30%3A_Cationic_Rearrangements/30.4%3A_Anchimeric_Assistance) (accessed, 15-Apr-2024).

Anchimeric Assistance (Neighboring Group Participation) (https://research.cm.utexas.edu/nbauld/Anchimeric.htm) (accessed, 15-Apr-2024).

Carey, F. A., & Sundberg, R. J. *Advanced Organic Chemistry, Part A-Structure and Mechanisms*, 4th Ed. Kluwer Academic/Plenum Publishers, 2000, 315, 335.

Chia Wu, J. I., & Schleyer, P. V. R. (2013). Strain energies of silicon rings and clusters. *Pure and Applied Chemistry*, 85, 921–940.

Clayden, J., & Greeves, N. *Organic Chemistry*, 2nd Ed. Oxford University Press, 2012, 156, 413, 868.

Croteau, R. (1987). Biosynthesis and catabolism of monoterpenoids. *Chemical Reviews*, 87(5), 929–954.

Fernandez, I., & Frenking, G. (2007). Correlation between Hammett substituent constants and directly calculated π-conjugation strength. *Journal of Physical Chemistry A*, 111, 8028–8035

Ian Hunt, University of Calgary, Chapter 4: Alcohols and Alkyl Halides (https://www.chem.ucalgary.ca/courses/351/Carey5th/Ch04/ch4-3-2-1.html) (accessed, 15-Apr-2024).

John A. Robinson, Biosynthesis of Natural Products - Terpene Biosynthesis (https://www.chem.uzh.ch/robinson/lectures/OC_V/OC-V.p2.2013.pdf) (biosynthesis) (accessed, 15-Apr-2024).

Lambert, J. B. (1990). Tetrahedron report number 273: The interaction of silicon with positively charged carbon. *Tetrahedron*, 46, 2677–2689.

Lambert, J. B., & Ciro, S. M. (1996). The interaction of π orbitals with a carbocation over three σ bonds. *Journal of Organic Chemistry*, 61, 1940–1945.

Reactive Intermediates

Reactive Intermediates

Lowry, T. H., & Richardson, K. S. *Mechanism and Theory in Organic Chemistry*. Harper& Row, Publishers, 1976, 214, 270, 271.

Olah, G. A. *A life of magic chemistry: Autobiographical reflections including post-nobel prize years and the methanol economy.* John Wiley & Sons, 2015, 166.

Pincock, R. E., Grigat, E., & Bartlett, P. D. (1959) Controlled formation of cis- and trans- decalin-9-carboxylic acids by carbonylation. *Journal of the American Chemical Society*, 81, 6332.

Smith, M. B., & March, J. *March's Advanced Organic Chemistry Reactions, Mechanisms and Structure, Chapter 11, Aromatic Electrophilic Substitution*, 5th Ed. John Wiley & Sons, Inc., 2001, 218, 421, 1384, 1411.

Sykes, P. *A Guidebook to Mechanism in Organic Chemistry*, 6th Ed. John Wiley & Sons, Inc., 1985, 101, 116.

Vincent, J. A. J. M. (1977). Neighbouring group participation in bicyclic systems [Phd Thesis 1 (Research TU/e /Graduation TU/e), Chemical Engineering and Chemistry]. Technische Hogeschool Eindhoven. https://doi.org/10.6100/IR104057 (https://pure.tue.nl/ws/files/1934641/104057.pdf) (accessed, 15-Apr-2024)

William Reusch, Michigan State University, Rearrangements Induced by Cationic or Electron Deficient Sites in Virtual Text of Organic Chemistry, 1999. (https://www2.chemistry.msu.edu/faculty/reusch/virttxtjml/rearrang.htm) (accessed, 15-Apr-2024)

Wolfe, S., Pinto, B. M., Varma, V., & Leung, R. Y. (1990). The Perlin Effect: bond lengths, bond strengths, and the origins of stereoelectronic effects upon one-bond C–H coupling constants. *Canadian Journal of Chemistry*, 68(7), 1051–1062.

Wu, J. I. C. & Schleyer, P. V. (2013). Hyperconjugation in hydrocarbons: Not just a mild sort of conjugation. *Pure and Applied Chemistry*, 85, 921–940. (https://publications.iupac.org/pac/pdf/2013/pdf/8505x0921.pdf) (accessed, 15-Apr-2024).

12 Reactive Intermediates
Carbanions

You will be wondering why a separate section for pK_a why not we study this wherever it is necessary. Since organic reactions involve movement of electrons, most of the reactions can be classified as redox reactions or acid–base reactions. We can use acid dissociation constant and hence we can associate pK_a with all the functional groups. Based on the value of pK_a, we can even predict the properties of the compounds, their chemical reactivity, the pH of the solution, their electrical conductivity, hybridization, resonance, induction, aromaticity, and so on. Because knowledge of pK_a is very essential for understanding organic chemistry, particular emphasis is given here and this will help in developing a sound chemical information skill.

pH talks about what is the H^+ ion concentration or the number of H^+ ion in a given solution. This depends on two factors. One, how much the acid is dissociated, that is, complete dissociation in the case of strong acids and incomplete or partial dissociation in the case of weak acids. The second one is the amount of solvent or the concentration.

Word of caution: when we say pH meter can be used to measure the pH of the solution, you should be aware that you are not actually measuring the H^+ ion concentration directly. What we measure is the potential difference that exists between the solution whose pH we are measuring and the std. KCl solution in the glass electrode. This potential measurement is based on the Nernst equation.

In fact, at pH above the pK_a of an acid, the acid will be more soluble in water. Any pH higher than the pK_a can be used to remove protons even from a strong base.

Many common drugs like aspirin (pK_a 3.5) can be made water-soluble above pH 3.5. What actually happens is at higher pH aspirin is converted to the corresponding conjugate bases. Water being amphoteric in nature stabilizes the carboxylate anion which in turn increases its solubility of aspirin in water.

When we measure pH it is for the solution as a whole. But we can talk about pK_a, we can have different pK_a for different hydrogens in a molecule. Knowing the relationship between pH and pK_a helps in understanding biological processes, drug development, organic synthesis, and so on.

Many organic reactions start with the addition or abstraction of proton. We need to know the strength of the acid or base to employ in that reaction. Using a reagent with a different strength than required may give unwanted side reactions.

Let me give a simple analogy. We have four types of hammers (plastic, metal, rubber, and wood) and four types of sticks (plastic, metal, glass, and wood). Which is the right one to do the job depends on which stick you want to break. Not only the tool but also the amount of force you apply also crucial. Application of excessive force will lead to complete destruction and not the intended simple breaking.

When a Brønsted acid HA is treated with a base B^-, it gives HB and conjugate base of HA. The equilibrium reaction will lie toward which side depends on the stability of the base. In this case A^- being more stable, it will be formed. When K_a is large, this equilibrium reaction will shift toward the right; in other words, HA will be easily dissociated and hence it will be the strong acid. We just saw the relationship between K_a and pK_a. pK_a is $-\log [K_a]$. A higher value of K_a means a lower value for pK_a.

When the HA bond is weak, it breaks easily so the compound is more acidic.

When we talk about bases, we talk about proton abstraction. We talk about thermodynamic control of the reaction. We talk about equilibrium-driven reactions (Figure 12.1).

Charged species is more soluble in water (polar solvent) than organic solvents (non-polar solvents). For example, anionic conjugate base (A^-) and cationic acid (HB^+) are more soluble in water. Neutral acid (HA) and neutral conjugate base (B) are more soluble in organic solvents.

Lawrence Joseph Henderson was a medical doctor working on blood and its respiratory function (1). Back in 1908, it was known that blood resists changes in acidity and basicity, but the relationship between the composition of a buffer, its buffering capacity, and the hydrogen ion concentration was not known. He came up with a simple relationship. He showed that this equation worked extremely well when K_a was near 10^{-7}. In 1916, Hasselbalch converted the Henderson equation into logarithmic form. This equation was applicable for weak acid/base and for dilute solutions. Another limitation was using initial molarities of H^+ and A^- that is before dissociation or hydrolysis (Figure 12.2).

$$K_{eq} = \frac{[H_3O^+][A^-]}{[HA][H_2O]}$$

$$K_{eq} = 10^{(pK_a{}^{pdt} - pK_a{}^{reat})}$$

$$K_a = \frac{[H_3O^+][A^-]}{[HA]}$$

FIGURE 12.1 Relationship between K_a and pK_a.

$$[H^+] = K_a \frac{[acid]}{[salt]} \xrightarrow{\log} pH = pK_a + \log \frac{[base]}{[acid]}$$

FIGURE 12.2 Relationship between Henderson equation and Hasselbalch equation.

DOI: 10.1201/9781032631165-12

In the case of pK_a measurement in a particular solvent, we have two limitations. We cannot measure the pK_a of weak acids that have pK_a more than the solvent or strong acids that have pK_a lower than the solvent. In other words, the strongest acid that can exist in water is H_3O^+ only. The conjugate base or the counter ion (Cl^-, SO_4^{2-}, and NO_3^-) has no effect. Similarly, the strongest base that can exist in water is ^-OH only. It is given in the equation.

Imagine you have four cups with sizes 100, 150, 200, and 250 mL. How much you can measure is based on the maximum capacity of the cup. Similarly, the solvent determines both the maximum and minimum pH that can be measured in its solution. In ammonia, we can measure pH up to 35, and in HF we can measure pH down to –20.

12.1 PARAMETERS THAT AFFECT pK_A

1. Inductive effect,
2. Electronegativity,
3. Resonance stabilization,
4. Electron-withdrawing groups,
5. Hybridization (sp vs sp^2 vs sp^3),
6. Attaining aromaticity,
7. Solvent.

12.1.1 RELATIONSHIP BETWEEN pK_A AND INDUCTIVE EFFECT

Let us consider acetic acid, propanoic acid, di, and tri methyl acetic acid for comparison. Methyl groups have +I effect (donate electrons). When a proton is lost it actually leaves behind the bonding electrons. Giving more electrons to the already negative carboxyl group will destabilize the carboxylate anion. Hence the acidity of trimethyl acetic acid (pivalic acid) is lower than acetic acid (Figure 12.3).

A more prominent effect can be clearly seen by the comparison of formic acid and acetic acid. Formic acid (no methyl group) pK_a is 3.75 and is a stronger acid than acetic acid whose pK_a is 4.8 (Figure 12.4).

When each of the protons in the methanol is replaced by a methyl group, the resulting alcohol pK_a increases as shown here. A more pronounced effect is seen when the pK_a of the methyl proton is compared (Figure 12.5).

Consider acetaldehyde, acetone, and ethyl acetate. There is 13-fold increase of pK_a between acetaldehyde and ethyl acetate. Similarly, the pK_a of the methylene proton in malonaldehyde, acetylacetone and ethyl acetoacetate is 5, 8.9, and 10.5, respectively (Figure 12.6).

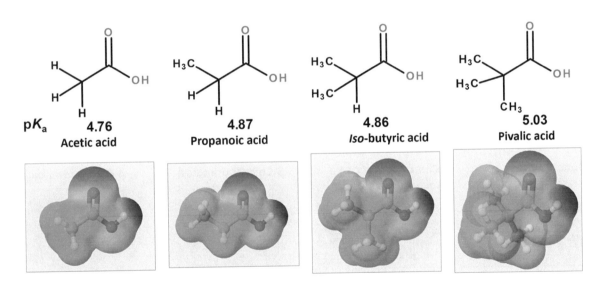

| pK_a | 4.76 | 4.87 | 4.86 | 5.03 |
| | Acetic acid | Propanoic acid | Iso-butyric acid | Pivalic acid |

FIGURE 12.3 Relationship between pK_a and Inductive effect.

| pK_a | 3.7 | 4.8 |

FIGURE 12.4 pK_a comparison of formic acid and acetic acid.

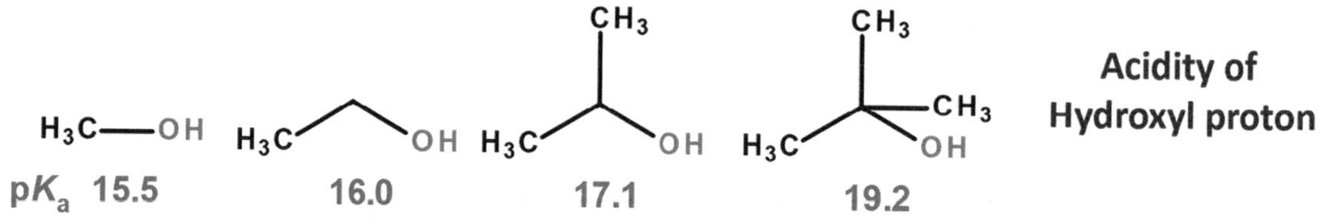

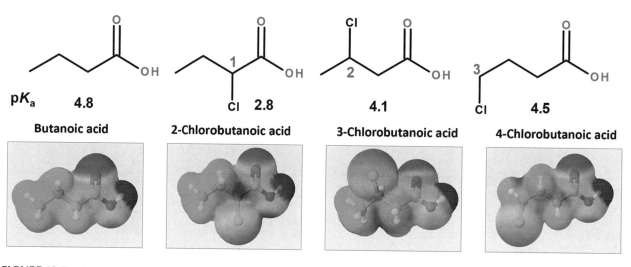

Acidity of Hydroxyl proton

$\text{H}_3\text{C}\!-\!\text{OH}$ $\text{H}_3\text{C} \diagdown \text{OH}$ $\text{H}_3\text{C} \diagup \text{OH}$ (with CH$_3$) $\text{H}_3\text{C} \diagup$ (with CH$_3$, CH$_3$, OH)

pK$_a$ 15.5   16.0   17.1   19.2

FIGURE 12.5   Comparison between acidity of hydroxyl protons.

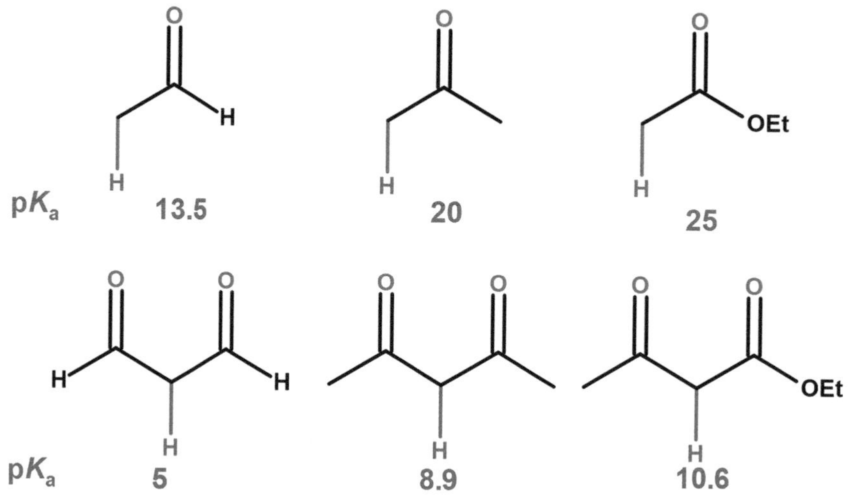

pK$_a$ 13.5   20   25

pK$_a$ 5   8.9   10.6

FIGURE 12.6   Comparison between pK$_a$ of hydrogen attached to carbonyl groups.

pK$_a$ 4.8   Cl 2.8   4.1   4.5

Butanoic acid   **2-Chlorobutanoic acid**   **3-Chlorobutanoic acid**   **4-Chlorobutanoic acid**

FIGURE 12.7   Comparison between pK$_a$ of 2- or 3- or 4-chlorobutanoic acid.

As far as the –I effect is considered, this will have the opposite effect compared with the +I effect of the methyl group. So, the presence of halogen will increase the acidity, and it is clearly seen for 2-chlorobutanoic acid (pK$_a$ 2.8) compared with butanoic acid (pK$_a$ 4.8). Since the inductive effect operates through the bond, its effect becomes weak after 2–3 bonds. 3-Chloro and 4-chlorobutanoic acids are only slightly more acidic than butanoic acid (Figure 12.7).

12.1.2   RELATIONSHIP BETWEEN pK$_A$ AND ELECTRONEGATIVITY

What is the effect of electronegativity on pK$_a$? When we move from left to right in a row in the periodic table electronegativity increases. That is clearly reflected in their respective pK$_a$ values. The pK$_a$ values of protons in methane, ammonia, water, and hydrogen fluoride decrease in the order of 48, 33, 16, and 3, respectively.

When we move down the column in the periodic table, the electronegativity decreases. The more electronegative fluorine gives the relatively acidic fluoroacetic acid with pK$_a$ 2.59 followed by less acidic chloroacetic acid (pK$_a$ 2.85), next comes bromoacetic acid (pK$_a$ 2.9), and then iodoacetic acid (pK$_a$ 3.17). Since the electronegative atom can stabilize the carboxylate anion, these acids are more acidic than acetic acid (pK$_a$ 4.8) (Figure 12.8).

When we extend the same with more electronegative substituents, the trend is clearly seen. Thus, the trichloroacetic acid is highly acidic compared with acetic acid (Figure 12.9).

12.1.3 RELATIONSHIP BETWEEN pK$_A$ AND RESONANCE STABILIZATION

What is the effect of resonance stabilization on pK$_a$ values? We saw electronegative atoms stabilize the carboxylate anion by pulling electrons toward themselves. Here we will look at the resonance stabilization of the carboxylate anion by the distribution of the negative charge. When perchloric acid (four oxygens) forms a perchlorate anion by loss of hydrogen, the negative charge on the oxygen enters into resonance with the remaining three oxygen atoms which stabilizes the perchlorate anion. Due to the multiple resonance structures of the conjugate base, the pK$_a$ of perchloric acid is −10 (Figure 12.10).

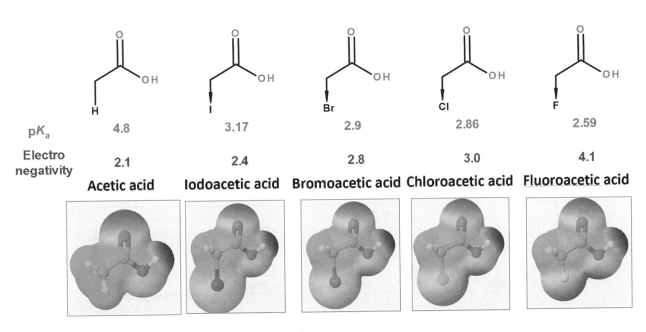

pK$_a$	4.8	3.17	2.9	2.86	2.59
Electro negativity	2.1	2.4	2.8	3.0	4.1
	Acetic acid	Iodoacetic acid	Bromoacetic acid	Chloroacetic acid	Fluoroacetic acid

FIGURE 12.8   Relationship between pK$_a$ and electronegativity.

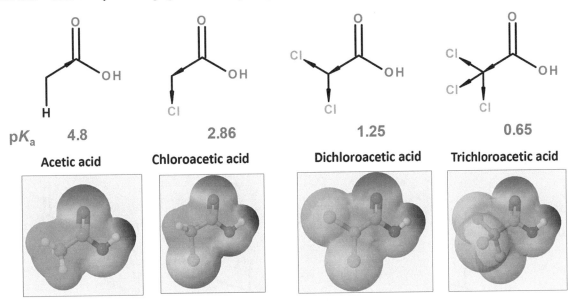

pK$_a$	4.8	2.86	1.25	0.65
	Acetic acid	Chloroacetic acid	Dichloroacetic acid	Trichloroacetic acid

FIGURE 12.9   Relationship between pK$_a$ and multiple halogens.

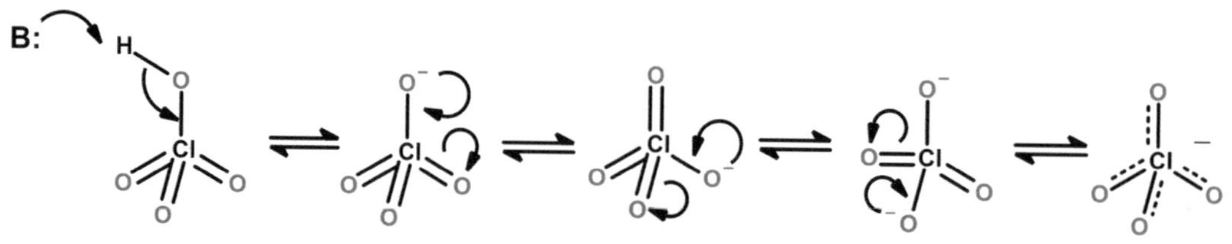

FIGURE 12.10   Relationship between pK$_a$ and resonance stabilization.

Let us compare the pK$_a$ of acetic acid (4.8) with ethanol (pK$_a$ =16). The huge difference in pKa values of this pair of compounds is attributed to the greater resonance stabilization of the carboxylate anion in acetic acid which is absent in ethanol (Figure 12.11).

Phenol has a pK$_a$ of 9.95. The corresponding resonance structures are shown here (Figure 12.12).

You may have some doubts regarding the acidity of phenol. Let me explain. We know more resonance structure means better stability. Better stability means a better conjugate base. Then why phenol is less acidic than acetic acid?

In acetic acid, (1) the resonance structures formed by the delocalization of negative charge have identical energy. (2) The negative charge is present on two oxygen atoms that are more electronegative than carbon.

But phenoxide ion has many high-energy carbanion resonance structures and there is only one electronegative

oxygen. There is one more reason. Generally, aromatic structures are more stable than non-aromatic ones. But in stabilization of the phenoxide ion, the aromaticity is lost. That is one of the reasons why the conjugate base of phenol is less stable than that of acetic acid. By this, we can easily visualize the importance of aromatic stability.

12.1.4   RELATIONSHIP BETWEEN PK$_A$ AND ELECTRON-WITHDRAWING EFFECT

Let us compare a couple of electron withdrawing group (EWG) and their pK$_a$ values. Acetyl, cyano, tri-N-methyl, and nitro. Since the nitro group is more EW than the rest of the groups given here, the nitroacetic acid is the most acidic one. The values of the pK$_a$ clearly tell the effect of EWG on pK$_a$ (Figure 12.13).

We know alcohols are neutral and their pK$_a$ is around 16. When the three hydrogens of methanol were replaced by the

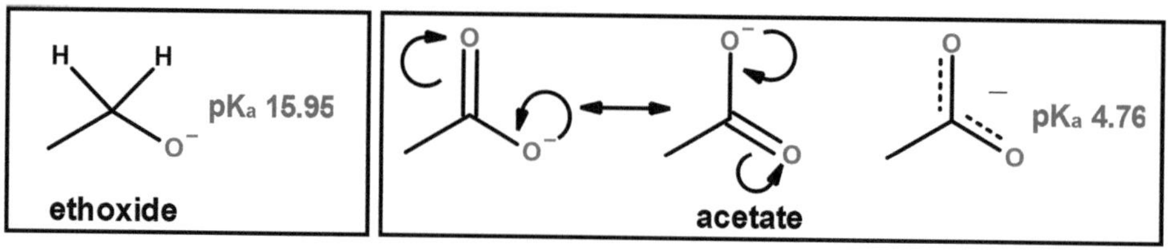

FIGURE 12.11   Resonance contribution in stabilizing carboxylate anion.

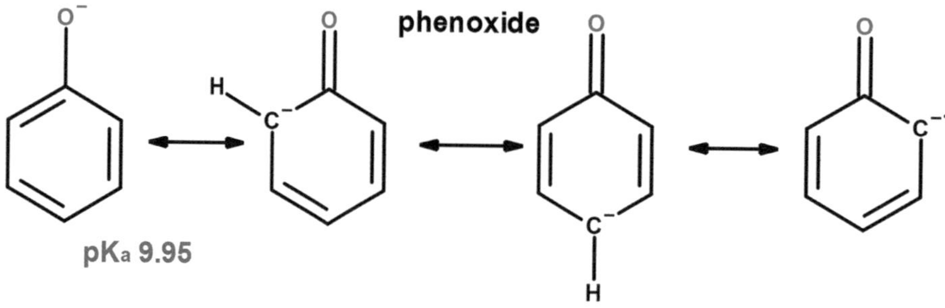

FIGURE 12.12   Resonance contribution in stabilizing phenoxide anion.

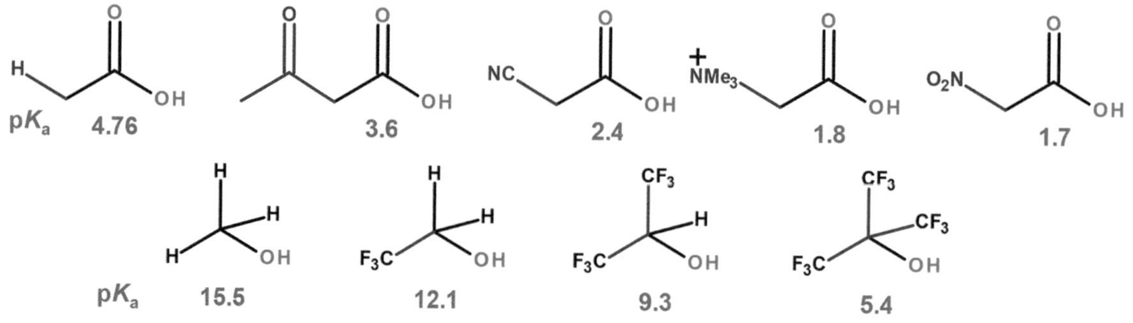

FIGURE 12.13   Relationship between pK$_a$ and electron-withdrawing effect.

EW CF$_3$ group one after another the pK$_a$ started to decrease. When we replace all three hydrogens with CF$_3$, we end up with an alcohol that is comparatively acidic similar to acetic acid.

12.1.5 Relationship between pK$_A$ and Hybridization

Here we will look at which hybridized orbitals can stabilize the negative charge efficiently. Between sp^3, sp^2, and sp, the s-character is more in sp-hybridized orbital (50%). The s orbitals are closer to the nucleus than p orbitals. The acidity or pK$_a$ of protons will be more pronounced in systems with sp hybrid orbitals. Thus the pK$_a$ for ethane, ethylene, and ethyne are 50, 44, and 25, respectively. From this, it is evident that ethyne can lose a hydrogen 10^{25} times faster than ethane hydrogen.

Next, we move on to alcohols and acids and see how hybridization affects their acidity. We have four alcohols propanol, allyl alcohol, benzyl alcohol, and prop-2-yn-1-ol. The hydroxyl carbon is directly attached to the sp/sp^2 center. Due to this, the pK$_a$ difference is not very high. The difference is to the magnitude of three or four (Figure 12.14).

When we talk about acids namely propanoic acid, acrylic acid, benzoic acid, and propynoic acid, the pK$_a$ difference is around magnitude of three (Figure 12.15).

Due to the presence of electronegative heteroatom namely oxygen, the anion is easily formed compared with alkanes where the negative charge was on carbon.

Alkynes (pK$_a$ 25) are stronger acids than typical alkanes (pK$_a$ > 50). But we know *alkyne C–H bonds are stronger than alkane C–H bonds because they have more s-character and now you are saying that the sp are more acidic and hence easier to break because they have more s-character.*

When we say *sp* bonds are stronger and difficult to break, we mean homolytic cleavage. But what we are doing here is a heterolytic cleavage, the formation of a proton and an anion. Stronger acid means the C-H bonds must be weak, then only it will easily dissociate to give the conjugate base. The acidity and the strength of C-H bonds are inversely proportional when we talk about heterolytic cleavage, less stable means strongly acidic. So, the stability we talk about is the stability with respect to the carbanion and not with respect to the C-H bonds. That is how hybridization influences the stability of carbanion.

12.1.6 Relationship between pK$_A$ and Attaining Aromaticity

If the anion can become aromatic, then it will be stable, and in turn, it will be ready to lose a proton, or we can say it will be more acidic. Let us consider the pK$_a$ of hydrogens represented in the 3 compounds cyclopentadiene, cycloheptatriene, and cyclopropane which are 15.5, 36, and 62, respectively. Cyclopentadiene has the lowest pK$_a$ because the resulting cyclopentadienyl anion is aromatic and hence easily formed.

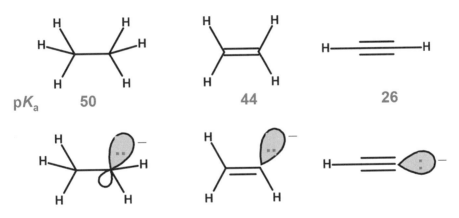

FIGURE 12.14 Relationship between pK$_a$ and hybridization.

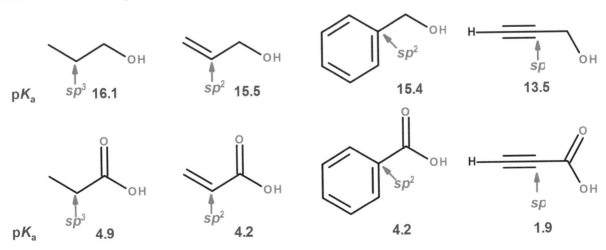

FIGURE 12.15 Relationship between pK$_a$, hybridization, and functional group.

Cycloheptatriene anion is an 8 π electron system and is non-planar. Aromatic stabilization does not work here. The 1,2,3-trimethylcyclopropenyl anion is 4 π electron system which is anti-aromatic, so it has the highest pK$_a$ (Figure 12.16).

Chloroform is more acidic than fluoroform which is opposite to their known electronegativity. "Strong electronegative atoms always do not increase acidity". Here we may have to consider the electrostatic repulsion between the negative charge and the electronegative atoms. As the size of the atom increases this electronic repulsion becomes less. That is the reason although bromine is less electronegative than fluorine shows greater acidity in this case. CF$_3$ anion is more destabilized. But in the case of chloroform and bromoform due to the presence of d orbital, there is little back bonding.

12.2 FACTORS INFLUENCING THE ACID STRENGTHS

How structure affects acid strength. (1) The strength of the bond (H-A from which the proton is lost). If the H-A bond is strong, then the acid is weak. (2) The electronegativity of the atom from which the proton is lost. In HA, if H is bonded to a more electronegative atom, the polarization $^{\delta+}$H—A$^{\delta-}$ becomes more pronounced and H is more easily lost as H$^+$. (3) Changes in electron delocalization on ionization. Stabilization of anion by electron delocalization increases the equilibrium constant for its formation.

When the two effects-bond strength and electronegativity-conflict, bond strength is more important when comparing elements in the same group of the periodic table as the pK$_a$s for the hydrogen halides show. Fluorine is the most electronegative and iodine is the least electronegative of the halogens, but HF is the weakest acid while HI is the strongest. Electronegativity is the more important factor when comparing elements in the same row of the periodic table. Inductive effects depend on the electronegativity of the substituent and the number of bonds between it and the affected site.

12.3 COMPARING THE ACID STRENGTH

How can we predict the possible site of reaction based on the pK$_a$ of protons in a molecule? Let us take the examples of aminoethanol and but-1-en-3-yne. In aminoethanol, NH and OH protons can be deprotonated. Their pK$_a$ differs by a factor of 19. Sodium hydroxide has a pK$_a$ of 15.7 and sodamide has a pK$_a$ of 35. Sodium hydroxide is a weaker base for removing the hydroxyl proton of aminoethanol. But it can be done by either ethoxide or *tert*-butoxide because they are comparatively stronger bases. But they cannot deprotonate the amino proton because its pK$_a$ is 35. So, to deprotonate the amino proton we need either sodium hydride, sodamide, or n-butyl Li. But the problem is it will also deprotonate the hydroxyl proton.

In the case of but-1-en-3-yne, the terminal alkyne proton can be deprotonated by sodium hydride or sodamide. However, the alkene proton can be deprotonated only by n-BuLi (Figure 12.17).

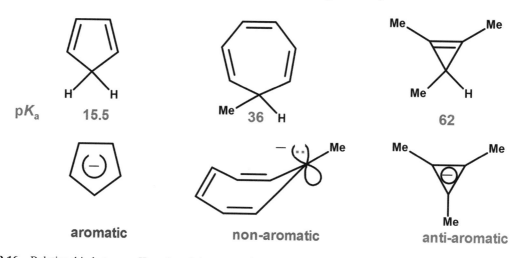

FIGURE 12.16 Relationship between pK$_a$ and attaining aromaticity.

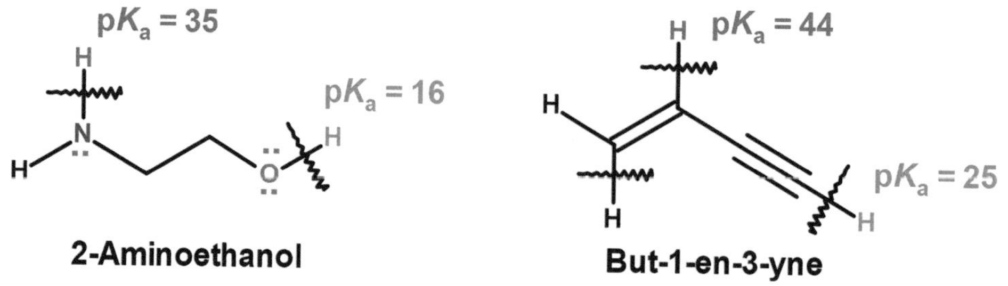

FIGURE 12.17 Comparing the acid strength.

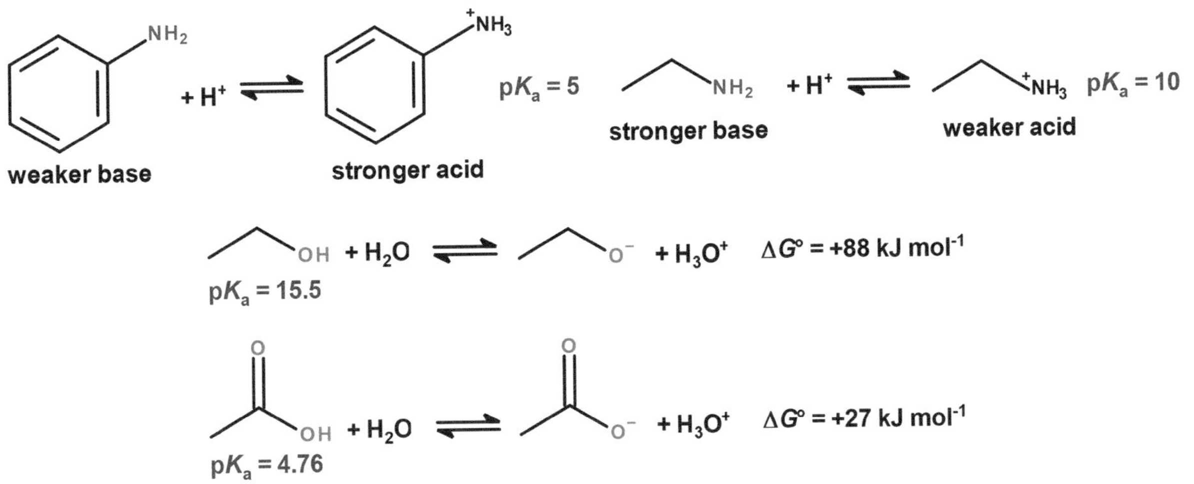

FIGURE 12.18 pK$_a$ for bases and conjugate acids.

For acidic proton lower pK$_a$ value means stronger acid. But for bases, higher pK$_a$ means stronger base. We can compare aliphatic and aromatic amino compounds. We know aniline is a weaker base compared with ethylamine. But how to prove it? One way to do this is to compare the pK$_a$ values. When we say pK$_a$ of a base what we really mean is the pK$_a$s of their conjugate acids. Aniline pK$_a$ is 5 and ethylamine pK$_a$ is 10. So, ethylamine is a stronger base than aniline (Figure 12.18).

Let us look at the relationship between pK$_a$ and Gibbs energy change. The sign and magnitude of pK$_a$ (ionization in H$_2$O) are directly proportional to the sign and magnitude of Gibbs energy change $\Delta G°$ (ionization). We know the $\Delta G°$ should be −ve for a spontaneous reaction. When we compare these two reactions, both of them are non-spontaneous. But the acetic acid reaction has a lower Gibbs energy change due to lower pK$_a$ compared with the ethanol reaction.

In an equilibrium reaction, the higher pK$_a$ component will be formed. In the reaction between benzoic acid and sodium hydroxide, the equilibrium shifts toward the sodium benzoate. This is because water has a higher pK$_a$ of 14 compared with benzoic acid (pK$_a$ 4.2).

Similarly, when phenol (pK$_a$ 9.94) is treated with sodium bicarbonate we get sodium phenoxide and carbonic acid (pK$_a$ 6.36). In this reaction, the equilibrium will lie toward phenol. In other words, phenol when treated with sodium bicarbonate does not give sodium phenoxide (Figure 12.19).

Benzoic acid can react with both sodium hydroxide as well as sodium bicarbonate to give the sodium salt. But phenol reacts with sodium hydroxide only to give the sodium salt and not with sodium bicarbonate. In the equilibrium reaction for benzoic acid, the pK$_a$ of benzoic acid is lower than the pK$_a$ of either water or carbonic acid. The reaction proceeds in the direction where the products have higher pK$_a$.

On the other hand, if you look at the reactions of phenol, the reaction with sodium hydroxide gives water whose pK$_a$ is higher than that of phenol, but the reaction with sodium bicarbonate does not produce a product that has higher pK$_a$ than phenol. So, the reaction lies toward phenol. This principle is employed in the separation of binary organic mixtures containing acids and phenol.

12.4 HARD AND SOFT ACIDS AND BASES

To understand the reactivity of carbanions, having an idea about the hard and soft acids and bases principles is important.

Molecules generally interact through attractive and repulsive forces. For a reaction to occur attractive forces play a vital role than the repulsive forces. In the attractive forces for bond formation or reaction to occur, we do not consider van Der Waals forces or hydrogen bonding. But Coulombic forces (electrostatic forces) play a vital role. The orientation of attraction (stereochemistry) between the reacting species, the energy of the reacting molecules (thermodynamics), and the pace with which they move (kinetics) all are important points to consider when a reaction takes place.

H–Acid: S–Base + S–Acid: H–Base ⇌ H–Acid: H–Base + S–Acid: S–Base

12.4.1 PEARSON'S HARD AND SOFT ACIDS AND BASES PRINCIPLE

Let us look at Pearson's hard and soft acids and bases principle (HSAB). Hard acids form strong bonds with hard bases, and soft acids form strong bonds with soft bases. The position of equilibrium depends on charge and orbital interaction. When two orbitals of unequal energy interact, the lowering in energy is less than when two orbitals of very similar energy interact. Knowledge of the HSAB principle helps us understand the kinetics of reactions.

The ability of a cloud of electron density to *deform/change* can affect its reactivity. A center that strongly

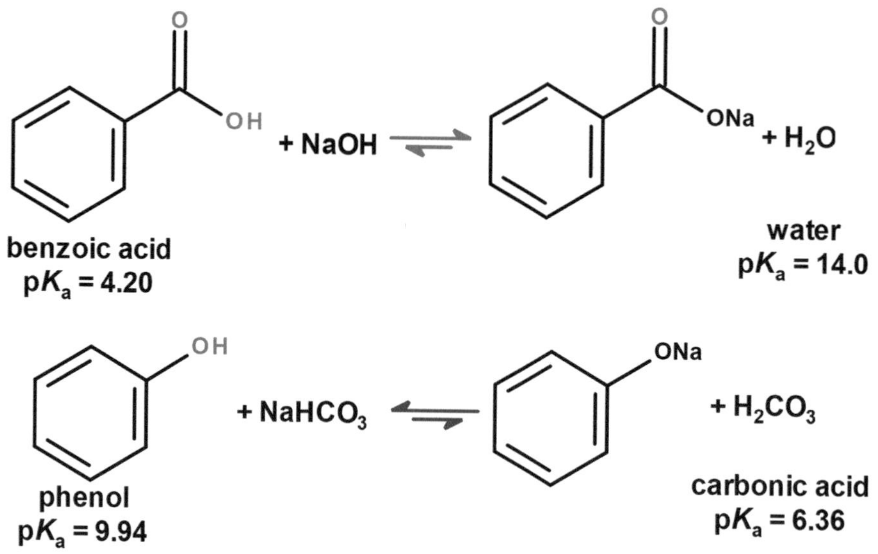

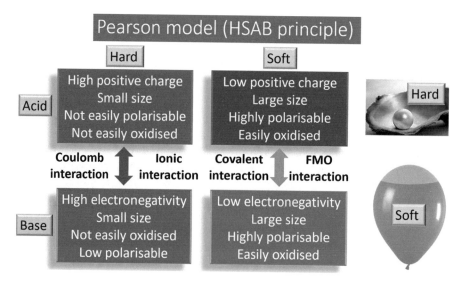

benzoic acid
$pK_a = 4.20$

water
$pK_a = 14.0$

phenol
$pK_a = 9.94$

carbonic acid
$pK_a = 6.36$

FIGURE 12.19 pK_a for reactions of phenol and benzoic acid with NaOH and NaHCO$_3$.

Pearson model (HSAB principle)

Hard	Soft
Acid High positive charge / Small size / Not easily polarisable / Not easily oxidised	Low positive charge / Large size / Highly polarisable / Easily oxidised
Coulomb interaction / Ionic interaction	Covalent interaction / FMO interaction
Base High electronegativity / Small size / Not easily oxidised / Low polarisable	Low electronegativity / Large size / Highly polarisable / Easily oxidised

FIGURE 12.20   Pearson's hard and soft acids and bases principle.

retains its electron density is said to be "hard" whereas a center whose electron density is more diffused is said to be "soft".

If we look at thiolate and alkoxide, they both have comparable basicity but thiolate is more nucleophilic and a better leaving group than alkoxide. This difference is attributed to the greater polarizability of thiolates.

The entire crux of the HSAB principle is given in the following diagram. A hard acid has a high positive charge, is of small size, is not easily polarizable, and is not easily oxidized. On the other hand, a soft low positive charge, large in size, is highly polarizable, and is easily oxidized (Figure 12.20).

A hard base has high electronegativity, is of small size, is not easily oxidized, and less polarizable. A soft base has low electronegativity, is of large size, is highly polarizable, and is easily oxidized.

Due to high charges the hard acid–hard base interaction is Coulombic in nature or we call this as ionic interaction. Whereas the soft acid soft base interaction is covalent in nature, we can explain this using FMO interaction.

Using a pearl and a balloon we can understand HSAB. Let me explain. What is the size of a pearl, it is small. Can we oxidize it? Very challenging. Can we break it? No. It is hard. Remember pearl, remember hard. Next look at the balloon what are its properties, large in size, easily be broken and it is soft.

In terms of HOMO/LUMO, ionization potential is the negative of the energy of HOMO, electron affinity is the negative of the energy of LUMO. The reaction takes place in the direction that increases the hardness. When a reaction takes place, HOMO/LUMO energies of the reactants have to be considered with regard to the symmetries and relative energies of the overlapping orbitals [whether they are atomic or molecular].

Hard acids and bases have the HOMO of the base and the LUMO of the acid far apart in energy. A soft acid bonds strongly to a soft base because the orbitals involved are close in energy. We get the maximum overlap for covalent bonding (Tables 12.1 and 12.2).

Sodium ion (Na^+) is *harder* than the silver ion Ag^+ (NaCN vs AgCN). You know Na is smaller than silver. Alkoxide ions, RO^-, are *harder* than thiolate anions, RS^-. Copper(II) ion, Cu^{2+}, is *harder* than copper(I) ion, Cu^+. The nitrogen anion end of the ambidentate cyanide ion, CN^-, is *harder* than the carbon anion end, NC^-. The ambidentate enolate ion has a hard oxyanion center while the carbanion center is softer and more nucleophilic.

Let us take a substrate that can undergo both hard and soft reactions. The attack can occur at *two* positions and nucleophiles exhibit regioselectivity. *Softer nucleophiles* like the carbon anion, NC^-, and the thiolate anion, RS^-, attack the β-alkyl carbon. *Harder* nucleophiles like alkoxide ion, RO^-, attack the acyl (carbonyl) carbon (Figure 12.21).

Nucleophilic substitution on a carbonyl or a phosphate ester (hard centers) typically leads to the formation of an intermediate; TS involves bond formation; and rate depends simply on the strength (basicity) of the (hard) nucleophile.

Nucleophilic substitution on a saturated carbon (a soft center) does not involve an intermediate; rather, there are five bonds around the carbon at the TS; large, polarizable (soft) nucleophiles react more quickly.

12.4.2 Luche Reduction

The selective 1,2-reduction of enones with sodium borohydride is achieved in combination with $CeCl_3$ in methanol. $CeCl_3$ is a selective Lewis acid catalyst for the methanolysis of sodium borohydride. The resulting reagents, various sodium methoxyborohydrides, are harder reducing agents (according to HSAB principles) and therefore effect a 1,2-reduction with higher selectivity (Figure 12.22).

TABLE 12.1
Characteristics of Acids with Example

Types of Acids	Characteristics	Examples
Hard Lewis acids	• Atomic centers of small ionic radius • High positive charge • No electron pairs in their valence shells • Low electron affinity • Likely to be strongly solvated • High-energy LUMO	H^+, Li^+, Na^+, K^+, Mg^{2+}, Ca^{2+}, Sn^{2+}, Al^{+3}, CO_2, RCO^+, NC^+, BF_3, N^{+3}, RPO_2^+, SO_3
Soft Lewis acids	• Large ionic radius • Low or partial δ+ positive charge • Electron pairs in their valence shells • Easy to polarize and oxidize • Low energy LUMOs, but large magnitude LUMO coefficients	Pd^{+2}, Pt^{+2}, Pt^{+4}, Cu^+, Ag^+, Au^+, Hg^{+2}, CH_2, carbene, HO^+, RO^+, Br^+, RS^+, O, Cl, BH_3

TABLE 12.2
Characteristics of Bases with Example

Types of Bases	Characteristics	Examples
Hard Lewis bases	• Small, highly solvated, electronegative atomic centers: 3.0–4.0 • Species are weakly polarizable • Difficult to oxidize • High-energy HOMO	NH_3, RNH_2, N_2H_4, H_2O, OH^-, O^{-2}, ROH, RO^-, R_2O, CO_3^{-2}, NO_3^-, PO_4^{-3}, SO_4^{-2}
Soft Lewis bases	• Large atoms of intermediate electronegativity: 2.5–3.0 • Easy to polarize and oxidize • Low energy HOMOs but large magnitude HOMO coefficients	H^-, R^-, C_6H_6, CN^-, RNC, CO, SCN^-, R_3P, R_2S, RSH, RS^-, I^-

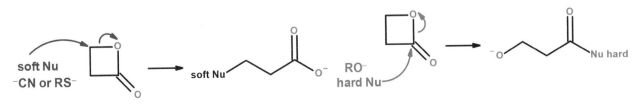

FIGURE 12.21 Example of a substrate that can undergo both hard and soft reactions.

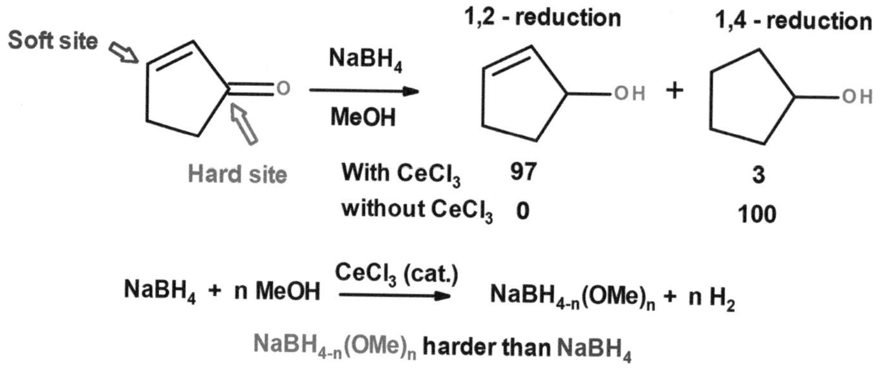
Soft site
Hard site
NaBH4
MeOH
1,2 - reduction
1,4 - reduction
OH
OH
With CeCl3 97 3
without CeCl3 0 100
NaBH4 + n MeOH
CeCl3 (cat.)
NaBH4-n(OMe)n + n H2
NaBH4-n(OMe)n harder than NaBH4

FIGURE 12.22 Luche reduction.

Let us compare the reactivity between carbonyl vs alkene reactivity. For easy understanding, we can assume that the carbonyl bond is a polar bond hence electron-rich (polar) species can attack the carbonyl bond.

A single bond between two atoms having different electronegativities breaks heterolytically. Similar to that, it is easier to break a C=O bond heterolytically and a C=C bond homolytically. Nucleophiles readily attack a carbonyl group but not an isolated C=C double bond. However, radicals readily attack C=C double bonds, and, although they can attack carbonyl groups, they do so less readily.

What is an acidic proton? How to identify them? Relationship between strength of the base and acidic proton (pK_a), How to remove the acidic proton? How to make a carbanion? If we know pK_a can we predict carbanion formation?

Sir James Black received the 1988 Nobel Prize for physiology or medicine "for their discoveries of important principles for drug treatment." He developed histamine receptors, H_2-receptors the first clinically useful H_2-receptor antagonist, cimetidine. Knowledge of pK_a is very important for deciding anion formation. We will look at the formation, structure, stability, and reactivity of carbanions.

12.5 STRUCTURE OF CARBANION

Carbocations (6 e^-) are sp^2 hybridized with a planar structure. But carbanions have eight electrons. Assuming the carbanion is trisubstituted, we can assume the structure will be similar to sp^3 hybridized orbitals (Figure 12.23).

There is some indirect evidence to show that carbanions are non-planar (trigonal pyramidal). We know bridgehead carbocations are not formed generally, because they need to be planar, but stable bridgehead carbanions are known. It also does not violate Bredt's Rule (double bonds cannot be formed at the bridgehead of a bridged ring system).

The $^-CH_3$ is isoelectronic with NH_3 and RNH_2. According to VSEPR theory, we can assume they all will have trigonal pyramidal structure. Carbanions are diamagnetic because all eight electrons are paired.

Specifically depending on the carbon at which the carbanion is generated, we can have different structures. Carbanions assume a trigonal pyramidal, bent, or linear geometry when the carbanionic carbon is bound to three (e.g., methyl anion), two (e.g., vinyl, phenyl anion), or one (e.g., acetylide anion) substituents, respectively (refer Figure 12.14).

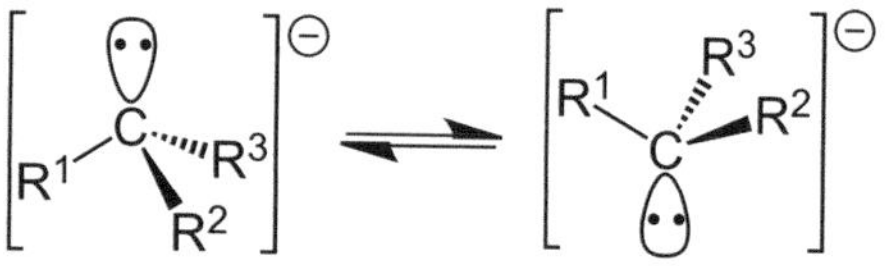
R1
R2
R3
C

FIGURE 12.23 Structure of carbanion.

12.6 CARBANION FORMATION

Let us look at the different ways of carbanion formation:

1. Deprotonation (inorganic bases, organic bases, organometallics).
2. Decarboxylation of carboxylate anion.
3. Elimination of formaldehyde from a primary alkoxide anion.

Organometallics include organolithium, alkoxides, alkylamides, organo tin, mercury, and magnesium derivatives. The second and third methods can be used to generate carbanions which otherwise cannot be formed by direct deprotonation and generally produce a single regioisomer.

Anions are colored. Sometimes it is possible to even visually see the color change during anion formation. Some examples are anions formed from nitro derivatives are yellow and triphenyl carbanion is blood red.

12.6.1 Carbanion Formation through Deprotonation

In many reactions, proton abstraction is an important initial and fast step. Here we have a list of various substrates. If you look at the bases that are used, for ketones and esters, which have one carbonyl group, we use stronger bases like amide or dialkyl amides. For dicarbonyl compounds or cyano esters, the methylene proton is somewhat more acidic so we can use an alkoxide base, and for nitro compounds, we can use hydroxide. As we had seen earlier, the different

Reactive Intermediates

acidic protons can be deprotonated using different bases which are matching with their pK$_a$s.

All these reactions that are mentioned below are equilibrium reactions. That means what we are discussing is the thermodynamics of the reaction. Due to the presence of stabilizing groups, that is, electronegative atoms in the vicinity, the kinetics are very fast. Only in the case of carbon acids, where there is no stabilization of anions, the reactions are slow and we focus on kinetics in those reactions (Figure 12.24).

12.7 ORGANOMETALLICS – LITHIUM BASES

The relative electronegativities of carbon and lithium suggest that the C-Li bond will be highly polar. Organolithium species aggregate, n-BuLi is tetrameric (in ether) or hexameric (in cyclohexane). The presence of Lewis bases such as tetrahydrofuran (THF), diethyl ether (Et$_2$O), tetramethylethylene diamine (TMEDA) or hexamethylphosphoramide (HMPA) influences the structure and reactivity of lithium bases.

We are using different bases for the carbanion generation. How do we decide which base is to be used? In all the equilibrium reactions we have 4 species, the substrate, carbanion, the base, and the conjugate acid of the base. Between the substrate and the conjugate acid of the base, the one with lower pK$_a$ or higher acidity will lose its proton. Due to complex side reactions, it is always preferable to use the base that is specific for the anion to be generated (Figure 12.25).

Now you will be wondering, why NaNH$_2$ is a stronger base than NaOH.

The pK$_a$ of H$_2$O (conjugate acid of ⁻OH) is 14, and the pK$_a$ of ammonia (conjugate acid of ⁻NH$_2$) is 35. That's a difference of 20, which means that sodamide is roughly 10^{19} times more basic than sodium hydroxide. Nitrogen is less electronegative than oxygen, so the ⁻NH$_2$ can readily give out the negative charge than ⁻OH during proton abstraction reactions. Hence acts as a strong base.

Alternatively, think like this, water is neutral whereas ammonia is a weak base. So, in the scale of base to neutral to acid, a weak base will have a strong conjugate acid. A moderate base will have a neutral conjugate acid and a strong base will have a weak conjugate acid. To get an idea of how strong sodamide is, we can look at its preparation. Sodamide is prepared by reacting sodium metal with ammonia gas or liquid ammonia. Sodamide is highly reactive, and it reacts violently with water to produce ammonia gas and caustic sodium hydroxide.

CH$_3$NH$_2$ is called methylamine, but NaNH$_2$ is called **sodamide** and not **sodamine** why?

Generally, amides (–CONH$_2$) give out NH$_3$ during hydrolysis with either acid or base. Sodamide is an ionic substance and readily forms NH$_3$ like other amides. But amines are covalent and do not readily give out ammonia. Hence NaNH$_2$ is called sodamide and not sodamine. This is because of the charge on nitrogen. You know Cl$_2$ is chlorine gas and Cl⁻ is chloride ion. Here it is ⁻NH$_2$ so it is amide. But CH$_3$NH$_2$ is called as methylamine (due to its neutral charge on nitrogen).

Let us compare the formation of the same anion by the removal of either hydrogen or halogen. We can consider triphenylmethane and triphenylmethyl chloride. The pK$_a$ of triphenylmethane is 33, so sodamide is required to remove

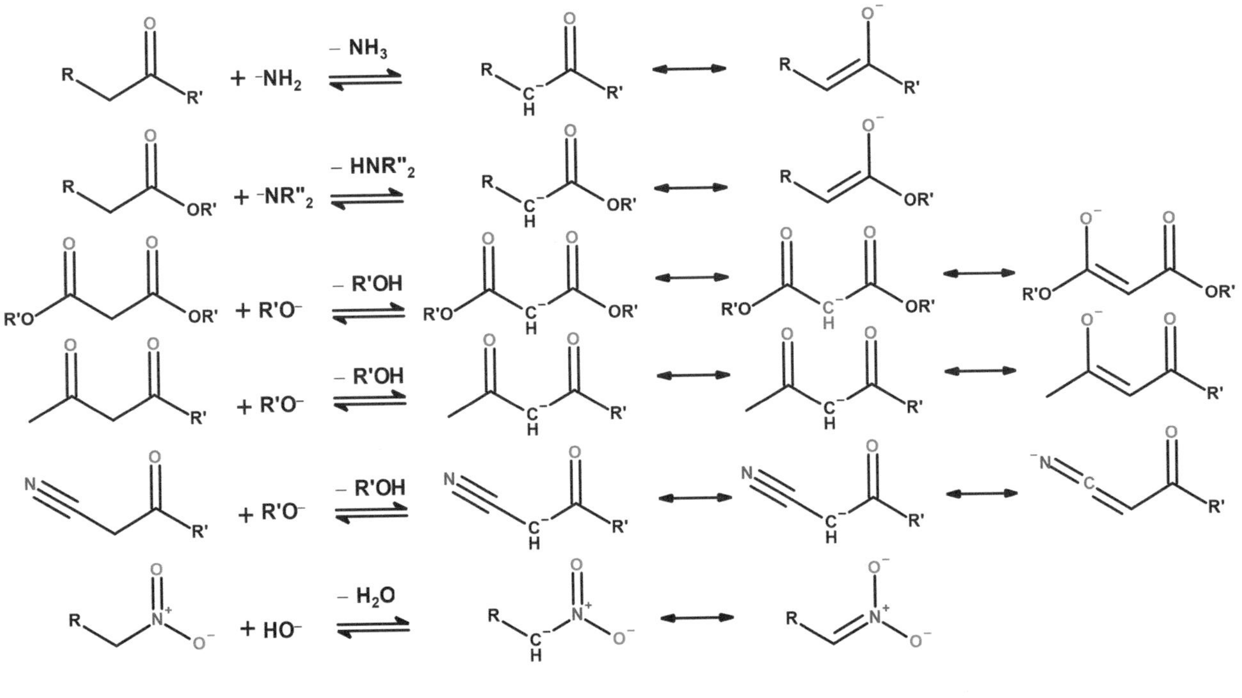

FIGURE 12.24 Formation of stabilized carbanions.

163

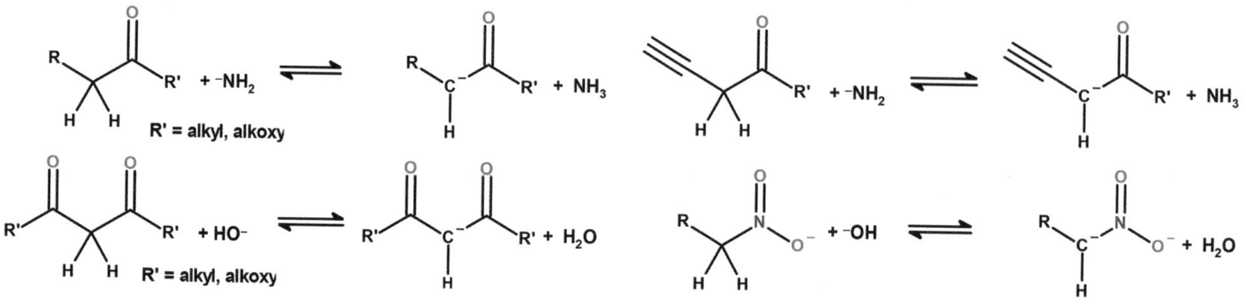

FIGURE 12.25 Carbanion generation using different bases.

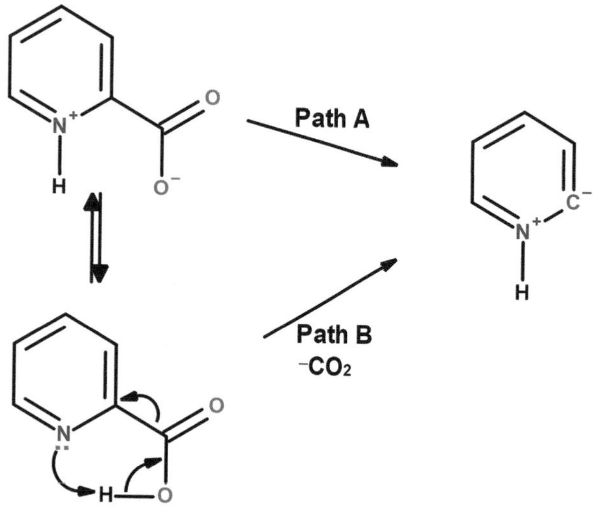

FIGURE 12.26 Decarboxylation of carboxylate anion.

the acidic proton. When the pK_a is lower, a weaker base (Na/Hg) can be used for the carbanion from the haloarene.

12.7.1 DECARBOXYLATION OF CARBOXYLATE ANION

The 2-pyridine carboxylate anion undergoes decarboxylation as given below. Proton transfer and decarboxylation occur in sequence (i.e., zwitterion decarboxylates, as in path A,) or in a concerted manner (i.e., neutral acid decarboxylates as in path B,) (Figure 12.26).

There are many places in organic chemistry where you will come across bases and nucleophile. Whether base and nucleophile are the same or different? Let us see the similarities first. Both base and nucleophile are electron-rich species and are either neutral or negatively charged.

Now let us look at the differences between them. Base abstracts proton (H-X), it is involved in elimination reactions, it is related to thermodynamics (stability), it is involved in equilibrium reactions. On the other hand, when

we talk about nucleophiles, they add to substrate (C-X), they are involved in substitution reactions and they are related to kinetics (rates of reaction).

Table 12.3 gives a comparison of common bases that are employed in organic synthesis.

From right to left in a row or bottom to top in the group in the periodic table, nucleophilicity and basicity increase. As the bulkiness of a group increases, its nucleophilicity decreases but basicity increases (provided we have electron-donating groups). Hence *tertiary* butoxide will act as a good base instead of nucleophile. When bulky groups are attached to carbon, many repulsion and steric hindrances will be there. In polar protic solvents (alcohol, water, etc.), nucleophilicity increases from bottom to top in groups in the periodic table. In polar aprotic solvents (DMSO, DMF, Acetone, etc.), nucleophilicity increases from top to bottom in the group in the periodic table. Electron-donating groups (+H groups) increase nucleophilic and basic strength, electron-withdrawing groups (-I groups) decrease the nucleophilic and basic strength.

12.8 SUBSTITUENT EFFECT ON STABILITY OF CARBANIONS

A carbanion is destabilized by an X substituent having +I effect (alkyl, methoxy). Inductive effects accumulate faster than resonance. The π effects of X substitutions are somewhat offset by the inductive effect of the electronegative X. Carbanion has a negative charge in $2p$ orbital. If there is a neighbor with a lone pair in higher p orbital, the repulsion will be less (e.g., -S) umpolung reaction. Halogen substitution favors or stabilizes carbanions. The enolate anion may be considered as an alkene with a powerful X-type substituent, the alkoxide oxygen.

Since the Z substituents are electron-withdrawing in nature, a carbanion is strongly stabilized by Z substituent. Phosphonium (R_3P^+) and sulfonium (R_2S^+) groups stabilize carbanionic centers, forming ylides. (Wittig reaction is one such example) The stabilization is due in part to

TABLE 12.3
Comparison of Common Bases

Strong Nucleophile Strong Base	Weak Nucleophile Strong Base	Strong Nucleophile Weak Base	Weak Nucleophile Weak Base
^-OH, RO^-, $^-NH_2$	DBN, DBU, KO-t-Bu, LDA, BuLi, NaH, KH	RCO_2^-, ^-CN, N_3^-, I^-, RNH_2, RSH, H_2S	H_2O, ROH, RCO_2H

electrostatic effects and in part to orbital interactions via σ* or 3d involvement. Trigonal boron can also stabilize carbanions. In the case of enolate anion, oxygen has more negative charge and carbon has more contribution from HOMO. Charged electrophiles (hard electrophiles) should preferentially add to the oxygen atom and neutral (soft) electrophiles would be expected to add to the carbon atom. This explains the reason for *C* vs *O* alkylation by different groups.

A carbanion is also stabilized by a C substituent. They have π electrons. These π electrons are involved in substantial delocalization of the negative charge (e.g., allyl).

12.9 FACTORS THAT AFFECT CARBANION STABILIZATION

Another important factor which stabilize carbanion is the reaction medium. In protic solvent, deprotonation of solvent competes with anion formation. Following is the stability order for different substituents that are present in carbanions in DMSO (aprotic solvent). $NO_2 > COR > CN \approx CO_2R > SO_2R > SOR > Ph \approx SR > H > R$. Hybridization, substitution, strain energy, polarization, and inductive effects may all contribute to stabilization and/or destabilization of the carbanions. Let us have a look at them.

1. Primary, secondary, and tertiary alkyl (inductive effect)
2. Allyl, benzyl (resonance)
3. Hetero atom (electronegativity/dipolar)
4. Electron-donating and electron-withdrawing groups (resonance)
5. Vinyl, phenyl, acetylide (hybridization)

12.9.1 PRIMARY, SECONDARY, TERTIARY ALKYL (INDUCTIVE EFFECT)

Addition of a methyl substituent to a carbanion center results in destabilization of the anion (all three anion types demonstrate a near linear increase in anion stabilization with substitution at the β-position, similar to what we saw in pK_a). Electron-donating groups (+H effect) destabilize negative charge.

If we look at the neopentyl carbanion, the destabilizing inductive effects of α-substitution are counteracted by the stabilization gained from β-substitution. The latter has been attributed to hyperconjugation, which involves a stabilizing interaction between the non-bonding electron pair of the anion and an anti-bonding σ orbital on the β-carbon. Increasing stabilization of the carbanion with increasing s-orbital character in the hybridization of the carbon.

12.9.2 ALLYL, BENZYL (RESONANCE)

Allylic or benzylic hydrogens are substantially more acidic than their saturated counterparts. Because loss of proton gives resonance stabilized allyl and benzyl carbanions. Other examples include cyclopentadienyl anion.

12.9.3 HETERO ATOM (ELECTRONEGATIVITY/DIPOLAR)

Alkoxide and alkylamide anions exhibit this effect. Electronegativity increases as we go across the periodic table. If you compare the carbanions having different substituents like C, N, O to F (across the periodic table), the stability of the negative charge will increase. The negative charge on carbon is stabilized by the neighboring electronegative groups.

12.9.4 ELECTRON-DONATING AND ELECTRON-WITHDRAWING GROUPS (RESONANCE)

Similar to +I effect, electron-donating groups destabilize carbanions. On the other hand, strong electron-withdrawing groups stabilize carbanions. An excellent example we have seen earlier is the aromatic nucleophilic substitution of nitro derivatives.

12.9.5 VINYL, PHENYL, ACETYLIDE (HYBRIDIZATION)

Alkenes (sp^2 hybridized carbons) like ethylene and benzene can be deprotonated by reaction with $^-NH_2$ in the gas phase. This is largely attributed to the sp^2 hybridization of the carbanion center. The stability of the acetylide carbanion is attributed to the significant s-character ($\cong 50\%$) of the charge-bearing sp-hybridized carbon.

The order of stability for various substituted carbanions is given here. How can you remember this? You can follow the order if you invoke the 's' character in other words hybridization.

Which R will bond with Li? On the left, we have alkyl group R, on the right we have alkyl group R' bonded with Li. The reasoning in these experiments is the R group that forms the more stable carbanion would be more likely to be bonded to lithium than to iodine (Figure 12.27).

Grignard reactions are generally carried out using an absolutely dry non-protic solvent like diethyl ether, THF, and so on. When the substrates contain -OH, −NH2 (that is groups containing active hydrogen), they undergo a reaction with the Grignard reagent. In fact, this feature is used in the Zerewitinoff/Zerevitinov determination of active hydrogens in a chemical substance.

Carbon is more electronegative than magnesium. Grignard reagents are highly polar. The alkyl group is not a full-fledged carbanion. However, the alkyl carbon carries a partial negative charge. RMgX being the salt of a very weak acid R-H, any reagent that is more acidic can easily cleave RMgX to an alkane. This is the reason RMgX is prepared

$RLi + R'I \rightleftharpoons RI + R'Li$	$R_2Mg + R'_2Hg \rightleftharpoons R_2Hg + R'_2Mg$
Vinyl> Phenyl > Cyclopropyl > Ethyl > n-Propyl > isobutyl > neopentyl > Cyclobutyl > Cyclopentyl	Phenyl > Vinyl > Cyclopropyl > Methyl > Ethyl> iso-Propyl

FIGURE 12.27 Comparison between lithiation and Grignard reaction.

in a dry condition by avoiding moisture. This is also the reason many acidic groups are sensitive to RMgX.

The ease of formation of anions is related to the number of resonance forms. Examples include enolate of ketones, esters, malonic ester anion, acetoacetic ester anion, cyanoacetic ester anion, and nitronate anion.

An increase in 's' character at carbanion carbon (hybridization) gives better stability. Carbanion on *sp* carbon is more stable than on sp^2 carbon compared with that on sp^3, that is alkynyl carbanions are more stable than that of alkenyl (vinyl/phenyl) compared with alkyl (3°, 2°, 1°). Let us look at the stabilization of carbanions by sulfur or phosphorus (ylide). This can be explained by the availability of *d*-orbitals on sulfur and phosphorus which help in stabilizing the anions. Let us move on to electron-withdrawing inductive effect (field effect). Acyl, cyano and nitro groups withdraw the electron density on the carbanion and thereby stabilize the carbanion. Conjugation of carbanion lone pair with polarized multiple bonds also stabilizes the carbanions (benzyl, allyl). Aromatization (cyclopentadienyl anion) also supports the stability of carbanions. Conjugation of the unshared pair with an unsaturated bond (resonance, allyl/benzyl) stabilizes carbanions.

12.10   CARBANION REACTIONS

We will be studying about few new reactions. Since many carbanions add to carbonyl compounds, we will not be studying aldol condensation, Dieckman condensation, or reactions of active methylene compounds.

We saw many factors that stabilize carbanion. Alkyl (+I effect) carbanions are destabilized hence they are highly reactive. We also know these alkane protons require a very strong base to deprotonate them. When we move into the stabilized category, we have resonance, hetero atom, and hybridization (Figure 12.28).

Let us see an example of reactions of hybridization stabilized carbanions. These alkyne protons require strong bases like BuLi, to deprotonate them. They react with carbonyl compounds and give additional products.

Similarly, the alkyne proton can be removed by Grignard reagents to form the carbanion. To this, we can add the carbonyl compound to get the addition product.

You may have a doubt. When the Grignard reagent reacts with carbonyl compound, we get alcohol. Going by that mechanism we are supposed to get ethyl-substituted alcohol and not alkynyl-substituted alcohols.

You are absolutely right. This is where the order of addition comes into the picture. For most of the synthetically useful carbanion chemistry, scientists prefer stepwise reaction. They first form the carbanion by reacting the substrate with the base at a low temperature. They allow sufficient time for the formation of the anion. After that, they will add the carbonyl compound. Although the reaction is done in one flask, it is a two-step reaction. Moreover, carbanions being highly reactive, were generally not isolated. They are generated *in situ*. If you look at the reaction temperature, generally the carbanion generation is done at very low temp like −78 °C or sometimes at 0 °C, and then the second substrate is added.

12.10.1   Manipulating Carbanion Reactivity

The carbonyl group can act as an acceptor (Grignard reaction) or as a radical partner. These two are common reactions with normal reactivity of the carbonyl group. In normal reactivity, the carbonyl carbon acts as the electrophile and is attacked by the anion. In umpolung reactivity, the anion is generated in the carbonyl group (Figure 12.29).

One way to introduce a deuterium or tritium atom into a specific location, mainly the most acidic proton, in a molecule is through the abstraction of that proton by a very strong base and treating the anion with D_2O or T_2O.

Types of carbanions			
Unstabilised	Stabilised by		
Produced by proton abstraction by strong base (*inductive effect*) BuLi, Mg Highly unstable (very reactive) R-Li, RMgX	Hybridisation Strong base Na, K pK_a 9-25 $N\equiv C^-Na^+$ $RC\equiv C^-Na^+$	Resonance Strong base Na, K pK_a 10-30 enolate	Heteroatom Strong base Na, K pK_a 10-30 ylide

FIGURE 12.28   Types of carbanions.

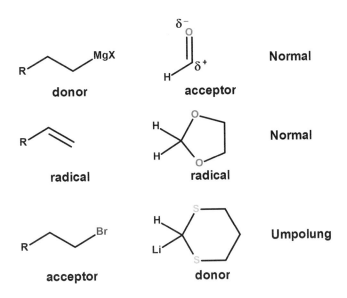

FIGURE 12.29 Carbanion reactivity, normal and umpolung.

12.11 ACYL ANIONS (CARBONYL EQUIVALENTS)

Following are the reactions we will be studying.

1. Cyanide ion
   - Benzoin condensation
2. Benzilic acid rearrangement
   - Rearrangement involving Oxygen
3. Favorskii Rearrangement
4. Umpolung (dithianes)
5. Nitronate anions
   - Henry reaction

12.11.1 Benzoin Condensation

It is a coupling reaction between two aldehydes that allows the preparation of α-hydroxyketones. Addition of the cyanide ion to create a cyanohydrin effect an umpolung of the normal carbonyl charge affinity, and the electrophilic aldehyde carbon becomes nucleophilic after deprotonation. A strong base is now able to deprotonate at the former carbonyl C-atom. A second equivalent of aldehyde reacts with this carbanion; elimination of the catalyst leads to the product.

The nucleophilic cyanide adds to the electropositive carbonyl carbon which results in the negative charge on the oxygen which is neutralized by abstraction of proton from the solvent to give the cyanohydrin. When the proton is picked up by the oxygen, a hydroxide anion is produced. This now acts like a base and abstracts a proton from the cyanohydrin. This leads to a carbanion. This attacks the second equivalence of the aldehyde which is similar to the aldol reaction leading to a diol. The diol then loses the cyanide during charge reversal to give the benzoin. Since the cyanide is regenerated in the last step, its role in this reaction is as a catalyst (Figure 12.30).

12.11.2 Benzilic Acid Rearrangement

The reversible addition of hydroxide ion to one of the benzil carbonyl groups produces an intermediate that undergoes a pinacol-like rearrangement. In contrast to the carbocation "pull" that initiates the pinacol rearrangement, this **benzilic acid rearrangement** complements a weak electrophilic pull by the adjacent carbonyl carbon with the "push" of the alkoxide anion. A rapid proton transfer then forms the relatively stable carboxylate anion. This is an example of 1,2-phenyl shifts. The driving force for the rearrangement is the conversion of a less stable anion into a more stable one (Figure 12.31).

12.12 REARRANGEMENT INVOLVING OXYGEN

12.12.1 Favorskii Rearrangement

When a nucleophilic base is treated with α-halogenated ketones, having acidic α-hydrogens, it undergoes a skeletal rearrangement known as the **Favorskii rearrangement**. This reaction is believed to proceed by way of a cyclopropanone intermediate. Facile conversion of cyclopropanones to hydrates and hemiacetals (relief of angle strain) occurs, and the cyclopropoxide conjugate base undergoes ring opening and protonation by the solvent. Alicyclic and heterocyclic halocycloalkanones containing 4–13 (but not 5) atoms in the rings undergo this reaction (Figure 12.32). When α-halocyclopentanones are treated with base, we get products of aldol, substitution and dehydrohalogenation reactions.

In general, the chloro compounds provide larger yields than the bromo and fluoro analogs, and branched-chain aliphatic alkoxides are claimed to be the most effective bases.

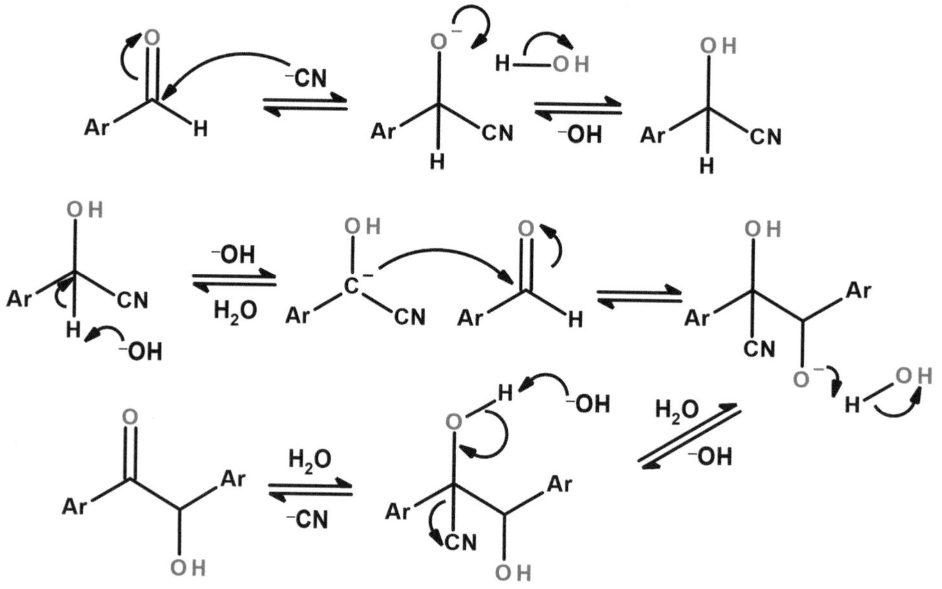

FIGURE 12.30   Mechanism of benzoin condensation.

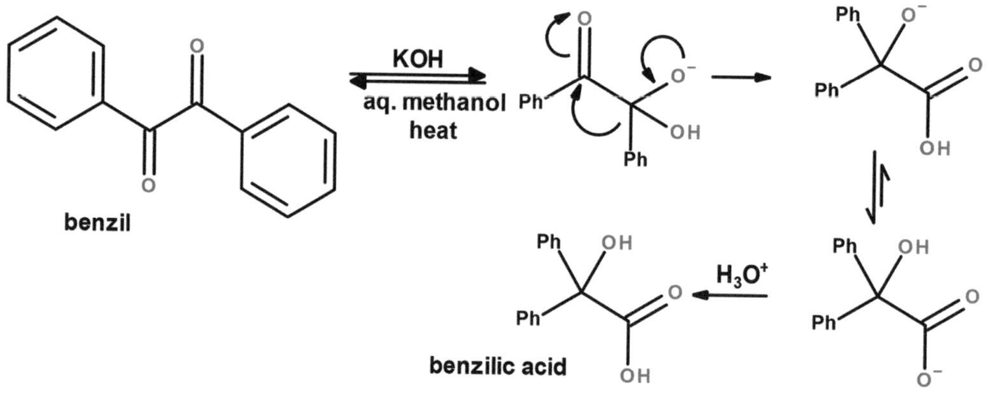

FIGURE 12.31   Mechanism of benzilic acid rearrangement.

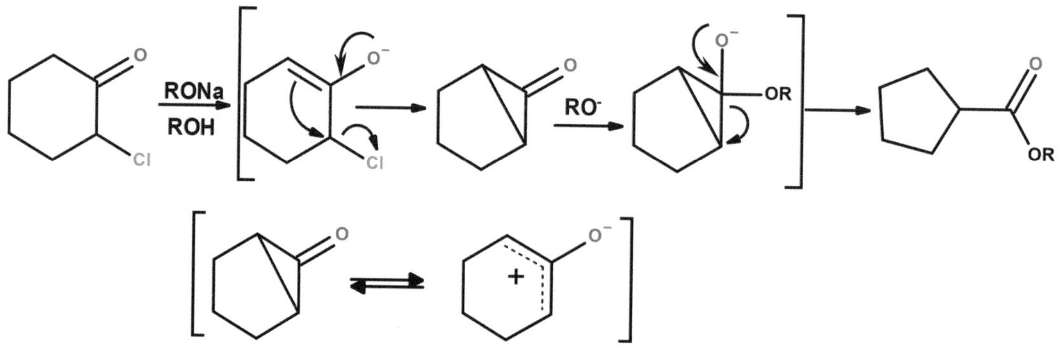

FIGURE 12.32   Mechanism of Favorskii rearrangement.

When stereoisomeric substrates were examined, the rearrangement proved to be stereospecific, ruling out a common zwitterionic intermediate.

12.13 UMPOLUNG-INVERSION OF REACTIVITY

Many organic reactions are classified as polar (acid–base, electrophile-nucleophile, oxidation-reduction, electron donor-acceptor) reactions. Charge separation occurs due to electronegativity difference (C, O, N, X). Umpolung is any *process* in which the donor and acceptor reactivity of the atoms are interchanged. It is also called masked acyl anion equivalents (Figure 12.33).

In aldehydes and ketones, the carbonyl carbon is electrophilic and hence is susceptible to attack by nucleophilic reagents. We can consider, the carbonyl group reacts as a formyl cation or as an acyl cation. A reversal of the positive polarity of the carbonyl group makes it act as a formyl or acyl anion would be synthetically very attractive. To achieve this, the carbonyl group is converted to a derivative whose carbon atom has a negative polarity. After its reaction with an electrophilic reagent, the carbonyl is regenerated. Although various acyl anion equivalents are known, we will be focusing only on the thioacetals or dithanes.

Let us look at the dithane formation. Here we are comparing the relative acidity of the 1,3-dioxane (acetal) or 1,3- thioacetal proton. Due to the favorable acidity, the thioacetal proton can be abstracted by a base. Of course,

you can also ask why specifically 1,3- why not 1,2- or even dimethyl? The reason is that 1,2-dithanes undergo a fragmentation reaction and dimethylthioacetals are susceptible to carbene formation (Figure 12.34).

12.13.1 FORMATION AND APPLICATION OF DITHANES

1,3-Dithianes are usually formed from corresponding aldehydes by thioacetalization. Lithium being a strong base removes the most acidic proton. This anion can now react with various electrophiles. Deprotection leads to the carbonyl compound (Figure 12.35).

12.13.2 REACTION OF DITHIANES

The various compounds that can be prepared from the lithiated dithane are given here. The reaction between alkyl halide and lithiated dithane gives ketones. Similarly, the reaction of ketones or imines with the lithiated dithane produces α-hydroxy or α-amino ketones respectively (Figure 12.36).

12.14 HENRY REACTION

When a nitro compound is treated with α, β- unsaturated carbonyl compound, it undergoes 1,4-addition to give the nitro ketone. The nitro group can be further transformed into ketone and that reaction is called the Nef reaction. The intermediate in this reaction is the carbanion as shown here (Figure 12.37).

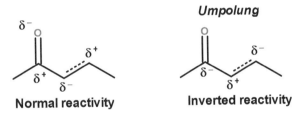

FIGURE 12.33 Umpolung-inversion of reactivity.

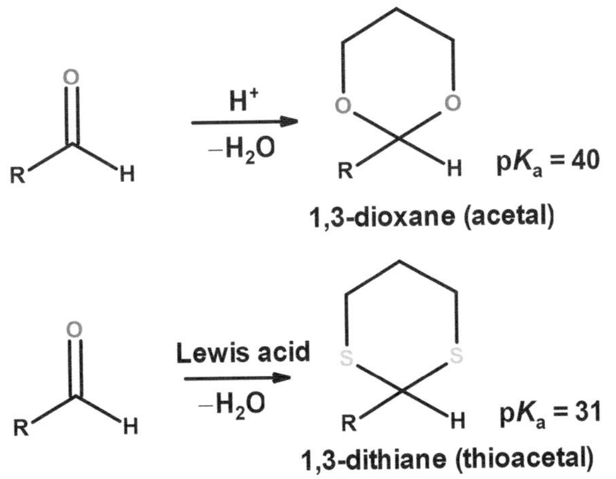

FIGURE 12.34 Dithane formation.

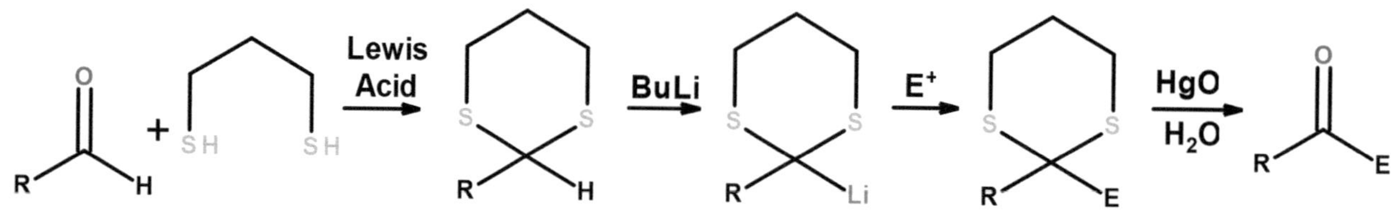

FIGURE 12.35 Formation and application of dithanes.

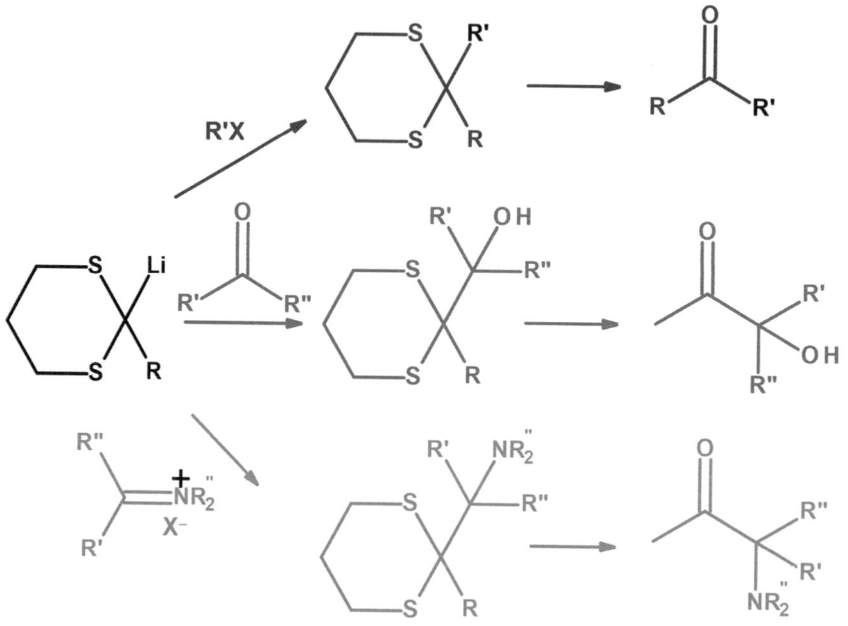

FIGURE 12.36 Reactions of dithanes.

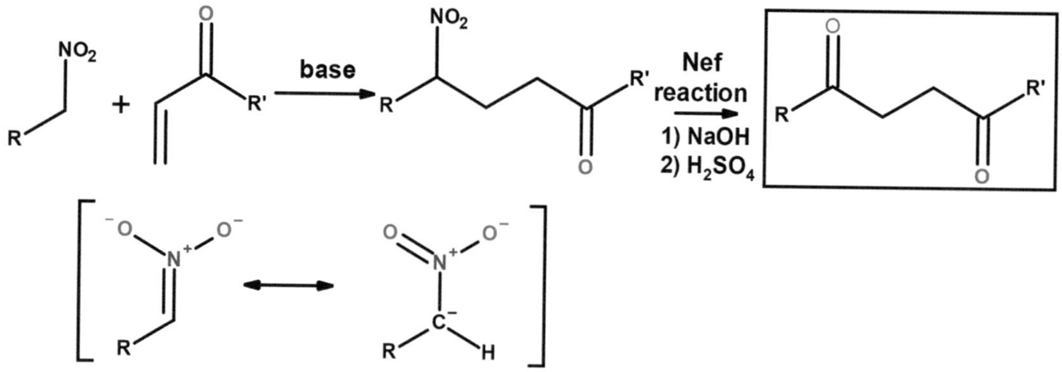

FIGURE 12.37 Henry reaction.

12.15 QUESTION

1. Write a short note on the structure of carbanions.
2. Mention a few methods by which carbanion can be formed through deprotonation.
3. Write a short note on the substituent effect on the stability of carbanions.
4. How resonance affects carbanion stabilization.
5. Why is it important to know the hybridization effect when discussing carbanion stabilization?
6. What is Henderson–Hasselbalch equation?
7. What are the applications of Henderson–Hasselbalch equation?
8. Briefly mention the factors that affect pKa.
9. What is the relationship between pKa and the electron-withdrawing effect?

10. Why it is important to know the relationship between pKa and resonance stabilization?

11. How by knowing the relationship between pKa and electronegativity, we can plan our experiments.

12. How do we decide the strength of the base to be used for proton abstraction.

13. With suitable examples explain why it is important to use the right base in a reaction.

14. What is a hard base? Give some examples.

15. How HSAB principle can be used to explain product formation in O vs C alkylation reaction.

16. What is Luche reduction?

17. What is the role of cyanide in Benzoin condensation? Explain using the mechanism.

18. How will you compare Pinacol rearrangement and Benzilic acid rearrangement?

19. What is Wittig rearrangement? How does it differ from Wittig's reaction?

20. Explain what happens when an unsymmetrical dibromo ketone is treated with nucleophilic base, sodium methoxide.

21. What is umpolung? Explain this with suitable examples.

BIBLIOGRAPHY

Ashenhurst, J. (2022). Orbital Hybridization And Bond Strengths in Bonding, Structure, and Resonance (https://www.masterorganicchemistry.com/2018/01/19/hybridization-and-bond-strengths/) (accessed, 15-Apr-2024)

Applequist, D. E., & O'Brien, D. F. (1963). Equilibria in halogen-lithium interconversions. *Journal of the American Chemical Society*, 85(6), 743–748.

Blanksby, Stephen J. & Bowie, John H. (2005). Carbanions: formation, structure and thermochemistry, 261–270. (https://ro.uow.edu.au/scipapers/3237) (accessed, 15-Apr-2024)

Carey, F. A., & Sundberg, R. J. *Advanced Organic Chemistry, Part B-Reactions and Synthesis*, 4th Ed. Kluwer Academic/Plenum Publishers, 2000, 1.

Clayden, J., Greeves, N., & Warren, S. *Organic Chemistry, Acidity, Basicity and pKa, Chapter 8*, 2nd Ed. Oxford University Press, 2014, 181, 237.

Dessy, R. E., Kitching, W., Psarras, T., Salinger, R., Chen, A., & Olivers, T. (1966). Organometallic electrochemistry. II. Carbanion stabilities. *Journal of the American Chemical Society*, 88, 460.

Dissociation Constants of Organic Acids And Bases (https://www.zirchrom.com/organic.htm) (for dissociation constants) (accessed, 15-Apr-2024)

Dunn, G. E. et al. (1972). Kinetics and mechanism of decarboxylation of some pyridinecarboxylic acids in aqueous solution. *Canadian Journal of Chemistry*, 50, 3017. (https://www.nrcresearchpress.com/doi/pdf/10.1139/v72-480) (accessed, 15-Apr-2024)

Edwards, J. O. (1954). Correlation of relative rates and equilibria with a double basicity scale. *Journal of the American Chemical Society*, 76, 1540–1547.

Favorskii Rearrangement in Reactions and Mechanism, Generated by Reaxys, (https://www.sciencedirect.com/topics/chemistry/favorskii-rearrangement) (accessed, 15-Apr-2024)

Flynn, A. B., & Amellal, D. G. (2016). Chemical information literacy: pKa values-where do students go wrong? *Journal of Chemical Education*, 93(1), 39–45.

It curdles if you don't stir it, 2013 (https://orgprepdaily.wordpress.com/2013/04/01/it-curdles-if-you-dont-stir-it/) (accessed, 15-Apr-2024)

Lee, A. C., & Crippen, G. M. (2009). Predicting pKa. *Journal of Chemical Information and Modeling*, 49(9), 2013–2033.

Mayr, H. Ludwig Maximilian University of Munich, Welcome to our Database of Nucleophilicities and Electrophilicities (https://www.cup.uni-muenchen.de/oc/mayr/DBintro.html) (accessed, 15-Apr-2024)

Mayr, H. Ludwig Maximilian University of Munich, A Quantitative Approach to Polar Organic Reactivity (https://www.arm-chemfront.com/2013/abstracts/ACF2013_P8_Mayr.pdf) (accessed, 15-Apr-2024)

Parr, R. G. & Pearson, R. G. (1983). Absolute hardness: companion parameter to absolute electronegativity. *Journal of the American Chemical Society*, 105, 7512. https://doi.org/10.1021/ja00364a005 (https://pubs.acs.org/doi/10.1021/ja00364a005) (accessed, 15-Apr-2024).

Pearson, R. G. (1985). Absolute electronegativity and absolute hardness of Lewis acids and bases, *Journal of the American Chemical Society*, 107, 6801. https://doi.org/10.1021/ja00310a009 (https://pubs.acs.org/doi/10.1021/ja00310a009) (accessed, 15-Apr-2024).

Pearson, R. G. (1986). Absolute electronegativity and hardness correlated with molecular orbital theory. *Proceedings of the National Academy of Sciences of the United States of America*, 83, 8440. (https://www.ncbi.nlm.nih.gov/pmc/articles/PMC386945/pdf/pnas00326-0013.pdf) (accessed, 15-Apr-2024).

Pearson, R. G., & Songstad, J. (1967). Application of the principle of hard and soft acids and bases to organic chemistry. *Journal of the American Chemical Society*, 89, 1827–1836. https://doi.org/10.1021/ja00984a014

Physical Organic Chemistry, 2004. (https://www.ch.ic.ac.uk/local/organic/tutorial/DB_Lecture_1.pdf) (accessed, 15-Apr-2024)

Po, H. N., & Senozan, N. M. (2001). The Henderson-Hasselbalch equation: Its history and limitations. *Journal of Chemical Education*, 78(11), 1499.

Rauk, A. *Orbital Interaction Theory of Organic Chemistry*, 2nd Ed. John Wiley & Sons, Inc., 2001, 108. Robert G. Parr and Ralph G. Pearson (1983). Absolute hardness: companion parameter to absolute electronegativity. *Journal of the American Chemical Society*, 105, 7512. https://doi.org/10.1021/ja00364a005 (https://pubs.acs.org/doi/10.1021/ja00364a005) (accessed, 15-Apr-2024)

Rossi, R. D. (2013). What does the acid ionization constant tell you? An organic chemistry student guide. *Journal of Chemical Education*, 90(2), 183–190.

Seebach, D. (1979). Methods of reactivity umpolung. *Angewandte Chemie International Edition*, 18, 239. (https://ethz.ch/content/dam/ethz/special-interest/chab/chab-dept/department/images/Emeriti/Seebach/PDFs/156-PULI_Ump-E.pdf) (umpolung) (accessed, 15-Apr-2024)

Smith, M. B., & March, J. *March's Advanced Organic Chemistry Reactions, Mechanisms and Structure, Chapter 5: Carbocations, Carbanions, Free Radicals, Carbenes, and Nitrenes*, 5th Ed. John Wiley & Sons, Inc., 2001, 227, 338.

Sykes, P. *A Guidebook to Mechanism in Organic Chemistry, Chapter 3: The Strengths of Acids and Bases, Chapter 10: Carbanions and their Reactions*, 6th Ed. John Wiley & Sons, Inc., 1985, 53, 270.

William D. Shipe, Princeton University, Umpolung: Carbonyl Synthons, 2004 (https://www.princeton.edu/~orggroup/supergroup_pdf/Umpolung2.pdf) (accessed, 15-Apr-2024)

William Reusch, Michigan State University, Ionization Constants of Heteroatom Organic Acids (https://www2.chemistry.msu.edu/faculty/reusch/virttxtjml/acidity2.htm) (for dissociation constants, and references given there) (accessed, 15-Apr-2024)

William Reusch, Michigan State University, Rearrangements Induced by Cationic or Electron Deficient Sites in Virtual Text of Organic Chemistry, 1999. (https://www2.chemistry.msu.edu/faculty/reusch/virttxtjml/rearang2.htm) (accessed, 15-Apr-2024)

13 Reactive Intermediates
Free Radicals

In this chapter, we will study about free radicals. Why are they important? How can we generate them? What are their types? What are their uses/applications?

In 2000, American chemical society celebrated, 100 years of organic free radical discovery by Moses Gomberg. In simple terms, we can say a free radical can be defined as any molecular species containing an unpaired electron in an atomic orbital and it is capable of independent existence even momentarily.

There is a brief history of the synthesis of free radicals. Gomberg first attempted the synthesis tetraphenylmethane.

Earlier Adolf Baeyer with help of August Kekulé tried to synthesize tetraphenylmethane. Even Victor Meyer attempted the synthesis of tetraphenylmethane. But no one was successful and it was believed that this compound could be extremely unstable. Since scientists never give up, at the University of Heidelberg in the lab of Meyer, Gomberg was trying to prepare tetraphenylmethane and was successful albeit in a very small amount.

He then tried to produce the next fully phenylated hydrocarbon—hexaphenylethane—by reacting triphenyl-methyl halides with sodium in benzene. When this failed, he repeated the experiment using silver instead of sodium. He obtained a white crystalline powder that reacted with oxygen in the air. To avoid this oxidation, Gomberg reacted triphenylmethyl chloride and zinc in an atmosphere of carbon dioxide. Sometimes, he allowed the experiment to run for weeks in a special airtight apparatus he had constructed. The reaction produced a thick, dark yellow syrup.

When Gomberg wanted to synthesize the next analog, hexaphenylethane, he failed. He tried various coupling reactions as shown here. Instead, he got the stable triphenylmethyl radical.

Different types of radicals based on charges are **neutral**, **+ve**, or **−ve**. Halogen (X), alkyl (R·), hydroxyl (OH·) are examples of neutral radicals. Molecular ion in mass (M^i), conducting polymers are examples of positive radicals and Birch reduction, the reaction of Ar compounds with alkali metal, superoxide ($O_2^{·-}$) are examples of anion radicals.

Our existence revolves around free radicals. The oxygen we breathe is a ground-state triplet molecule, which means it is a diradical. Oxygen is important for breathing, and cellular respiration, lack of oxygen leads to hypoxia. Oxygen is also related to aging. In fact, many cell cycle processes produce and consume reactive oxygen species. Finally, many of the foods like apples we eat undergo rapid air oxidation.

13.1 STRENGTH OF A BOND AND FORMATION OF RADICALS

13.1.1 BOND DISSOCIATION ENERGIES

The amount of energy required to homolytically cleave a particular bond is known as the **bond dissociation energy** (D). It is calculated by the difference in the enthalpies of formation of the products and reactants for homolysis.

$$A - B \rightarrow A + B$$

We cannot directly measure the BDE of a particular bond. Because, it requires two parameters: (1) the energy required to break the bond and (2) the energy difference between the starting material and the final product.

Let us compare bond dissociation energy and bond energy. We have a couple of different bond dissociation energies for methane. They are 104, 106, and 81 kcal/mol. But the average of these BDE is the bond energy. From this, we can clearly understand that it will not be accurate to equate bond energy with bond dissociation energy. Because radical stability has

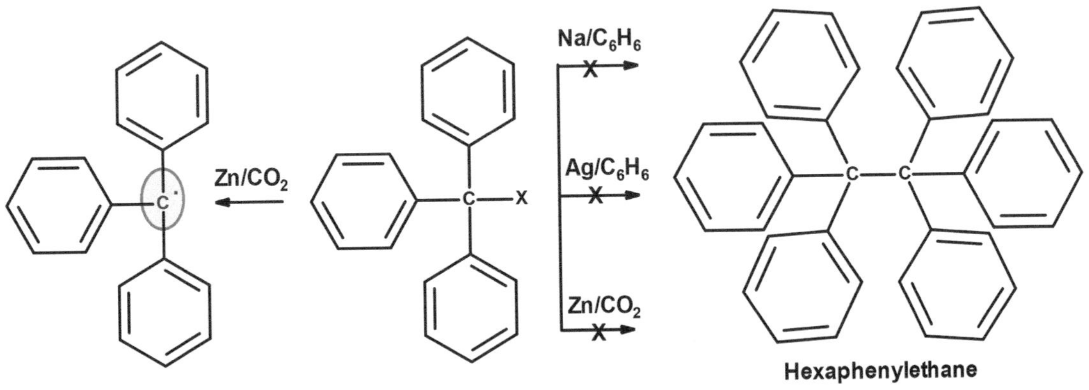

SCHEME 13.1 Synthesis of hexaphenylethane.

DOI: 10.1201/9781032631165-13

$$CH_4 \rightarrow CH_3\cdot + H\cdot \qquad D(CH_3-H) = 104 \text{ kcal/mol}$$
$$CH_3 \rightarrow CH_2\cdot + H\cdot \qquad D(CH_2-H) = 106 \text{ kcal/mol}$$
$$CH_2 \rightarrow CH\cdot + H\cdot \qquad D(CH-H) = 106 \text{ kcal/mol}$$
$$CH \rightarrow C\cdot + H\cdot \qquad D(C-H) = 81 \text{ kcal/mol}$$

$$\text{Bond energy, } E\,(C-H) = \frac{397}{4} \text{ kcal/mol} = 99 \text{ kcal/mol}$$

FIGURE 13.1   Comparison between bond dissociation energy and bond energy.

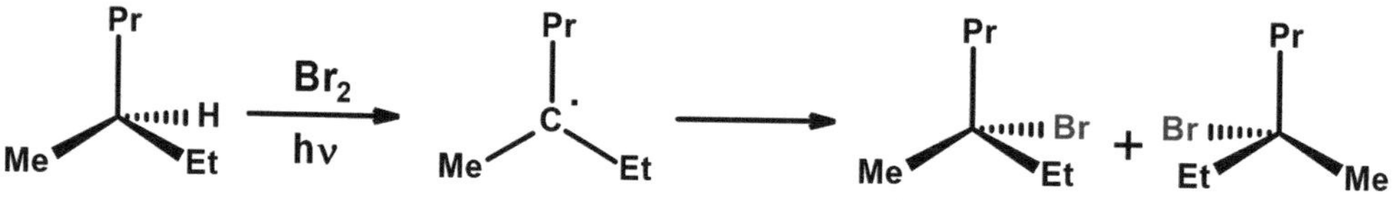

SCHEME 13.2   Radical bromination of 3R-methylhexane.

nothing to do with the enthalpy/energy difference between the starting material and the product, there is no relationship between BDE and radical stability (Figure 13.1).

Why does radical bromination of 3(R)-methylhexane give two brominated products?

We have to explain two observations: (1) type of the mechanism and (2) geometry of the product (which will take us to the intermediate). We cannot explain this reaction by ionic mechanism (heterolytic cleavage). Because we neither have any good leaving group nor the formation of a carbocation (by the loss of hydride ion). Now we are forced to think differently. Moreover, we get an equal mixture of (R)- and (S)-3-bromo-3-methyl-hexanes. This only tells us about the geometry but not about the intermediate itself. We will restrict our studies to homolytic cleavage of the C–H bond leading to radical intermediate. Using this mechanism, we can explain the product formation. Due to the planar nature of the radical, we get both the isomers in equal quantities.

13.2   FORMATION OF FREE RADICALS

Four methods of formation of free radicals are given here. (1) Thermal decomposition (from peroxide, persulphates, or azo derivatives) and (2) photolysis (AIBN – azobisiso-butyronitrile) are extensively used in organic synthesis. (3) Redox reactions (reduction of H_2O_2 by Fe) are more common in biochemical transformations, and (4) electrochemical methods (from monomer and electrolyte) are exploited by polymer and physical chemists.

13.2.1   THERMAL INITIATION

In thermal initiation, the initiator should contain a weak σ-bond (benzoyl peroxide and AIBN). The initiator has a lifetime ($t_{1/2}$) comparable to the time required for the completion of the reaction, 2,2'-Azobis(isobutyronitrile) gives 2-cyano-2-propyl radical and the reaction sequence is given here. The kinetics of its reaction is first order, regardless of the solvent or initiator concentration. The homolytic cleavage occurs between the C-N σ bond in AIBN.

In photochemical initiation, the initiator needs to absorb a photon of light and then use this energy to cleave a bond homolytically. This is suitable for substrates that are sensitive to heat, use of visible or long UV light is preferred (avoids exciting the reactant), AIBN (UV 345 nm), benzophenone (n → π*).

In chemical initiation, the initiator uses energy stored in bonds to form initiating radicals, for example, triethylboron reacts with O_2 giving ethyl radical.

13.2.2   KOLBE ELECTROLYSIS

In the electrochemical decarboxylation of carboxylic acid, we get alkyl radicals very similar to thermal decomposition to give dimers through radical intermediates. Two such intermediates combine to form a covalent bond. α-substituents such as carboxy, or cyano support the radical pathway, while alkyl, cycloalkyl, chloro, bromo, amino,

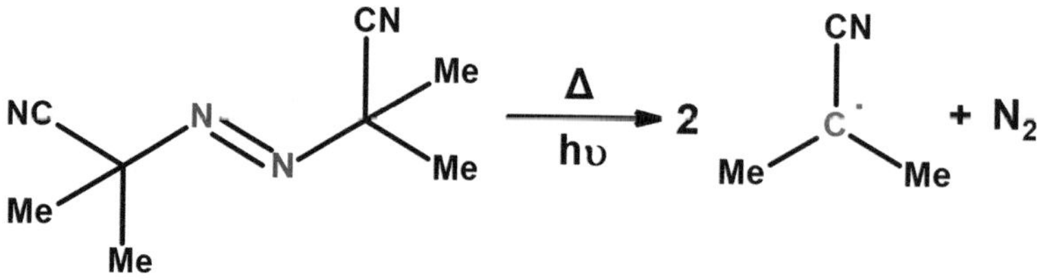

SCHEME 13.3   Radical formation through thermal initiation

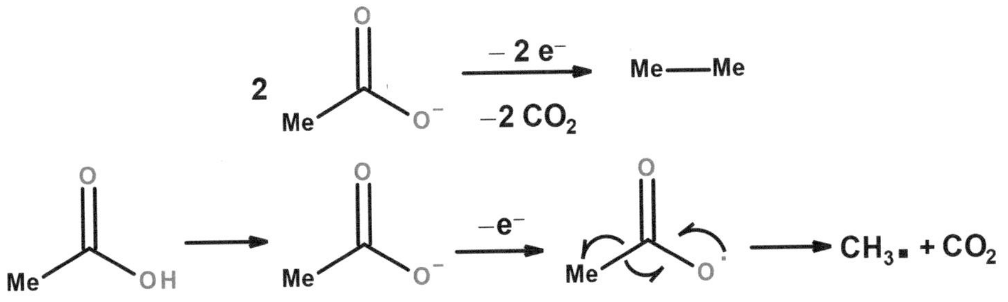

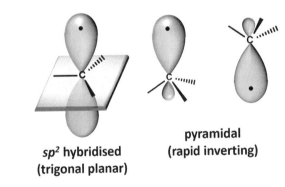

SCHEME 13.4 Kolbe electrolysis.

alkoxy, hydroxy, acyloxy, or aryl more or less favor the car-benium ions.

The main experimental factors that affect the yield in the Kolbe electrolysis are the current density, the pH of the electrolyte, ionic additives, the solvent, and the anode material.

13.2.3 BOUVEAULT–BLANC REDUCTION

Reduction of alkyl ester with sodium metal to give the corresponding alcohol is called the Bouveault–Blanc reduction.

13.3 STRUCTURE AND STEREOCHEMISTRY OF RADICALS

Here we will see the structure and stereochemistry of radicals. The methyl radical is **planar** and has D_{3h} symmetry. The planarity of the methyl radical has been attributed to steric repulsion between the H atoms. All other carbon-centered free radicals (with alkyl or heteroatom substituents) are best described as **shallow pyramids** (Figure 13.2).

When we say nearly planar and if the deviation is just 5°, in reality, it amounts to halfway between either planar or shallow pyramidal structure. The odd electron remains in the p orbital which is perpendicular to the sp^2 plane.

13.3.1 S AND P TYPES OF RADICALS

If a radical center has a *nearly planar arrangement* of attached atoms, the radical is described as being π-*type*. If a radical has a decidedly *pyramidal configuration*, the radical is described as being σ-*type*. Since in a π-type radical, the orbital in which the electron is centered is close to being a p orbital, this orbital often is referred to as being p type, i.e., one approaching that corresponding to sp^3 hybridization.

Radical centers with *electronegative atoms* (O, F) attached become more *pyramidal* (σ type). When no electronegative substituents are attached to the radical center, it will be planar (π-type).

FIGURE 13.2 Structure of carbon-centered radicals.

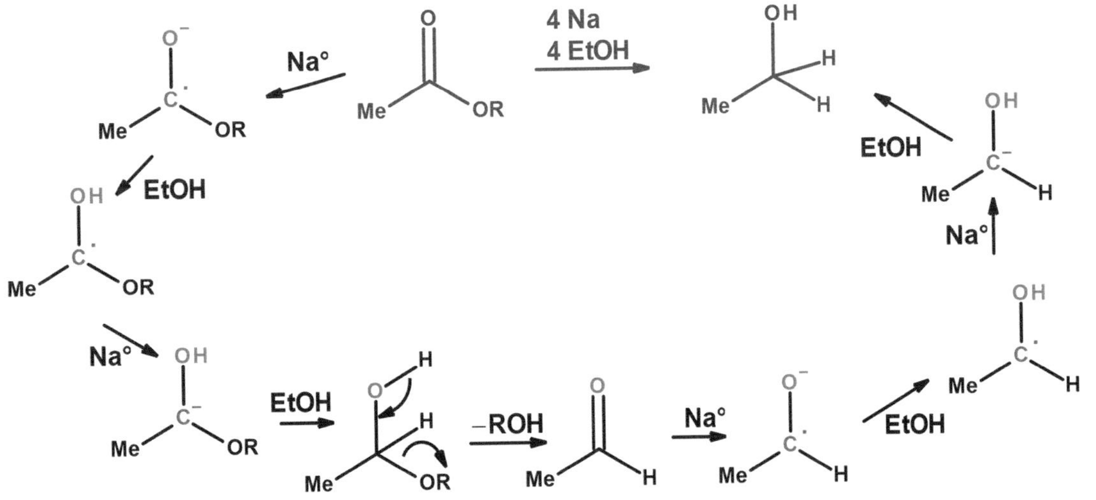

SCHEME 13.5 Bouveault–Blanc reduction.

When we look at alkyl radicals, methyl is planar whereas other alkyl radicals are pyramidal. *Tert*-butyl radical is pyramidal due to two effects, a torsional effect leading to staggered conformation and the other one is hyperconjugation.

In the pyramidal structure, the torsional effect is one in which the radical center tends to adopt a staggered conformation of the radical substituents. There is also a hyperconjugative interaction between the half-filled orbital and the hydrogen that is aligned with it. This hyperconjugation is stronger in the conformation in which the pyramidalization is in the same direction as to minimize eclipsing. In the case of group four hydrides, the delocalization of the unpaired electrons into the C–X bond increases. The eclipsed rotomer becomes the transitional structure for rotation (Figure 13.3).

We also have some indirect evidence that support the structure of the radicals. One such example is the study of product stereochemistry (acyclic system). A planar or rapidly inverting radical would lead to racemization when a chiral center is the reaction center. When a chiral center is the site of radical reaction, a racemic product is formed, indicating that alkyl radicals do not retain the tetrahedral geometry of their starting materials. Unlike carbocations, bridgehead radicals are known with pyramidal structures. Reactions involving cyclic systems or alkenes give mixture of stereoisomers; for example, reactions starting from pure *cis* or *trans* stereoisomers give mixtures of *cis* and *trans* products.

Next, we look at the nature of radicals. One or more electrons are unpaired, so there is a net magnetic moment and the species is paramagnetic. Electron spin resonance (ESR) or electron paramagnetic resonance (EPR) is a method used to study free radicals. The principle of ESR is similar to that of NMR, except that electron spin is involved rather than nuclear spin. Many times, the concentration may be too low for direct observation. In such cases, the spin-trapping technique can be used (think of amplifier, step-up transformer). Chemically induced dynamic nuclear polarization (CIDNP) is an NMR technique that is used to study chemical reactions that involve radicals.

13.4 STABILITY OF FREE RADICALS

We have two types of stability. One is thermodynamic stability, which is related to the substituent's electronic nature, and there is no persistent radical, which is related to the ground-state energy. The other one is kinetic stability. This is related to substituent's steric nature, persistent radical belongs to this category.

When the bond dissociation energy is lower, it is easy to undergo homolytic fission. Low BDE means a more stable radical. Moreover, a greater number of alkyl groups also increase the hyperconjugation. So *tert*-butyl radical is more stable than *iso*-propyl which in turn is more stable than ethyl radical. However, the increase in stability is not very pronounced like the carbocations (Figure 13.4).

Radicals can be stabilized by the substituents. A free radical is stabilized by an X: substituent through a two-orbital, three-electron π-type interaction. The nucleophilicity of the radical is greatly increased. The X: substituted free radicals are more easily oxidized. The π effects of X: substitution are somewhat augmented by the inductive effect of the electronegative X: in stabilizing the radical.

Since radicals lack octet configuration, they are electron deficient. So, electron-donating groups increase their stability, and thus we observe the following trend for the stability of alkyl radicals $CH_3^{\bullet} < CH_3CH_2^{\bullet} < (CH_3)_2CH^{\bullet} < (CH_3)_3C^{\bullet}$.

When we talk about the stability of the radicals, we talk about radical stabilization energy (RSE) similar to resonance stabilization energy for aromatic compounds, how do we determine or measure RSE? Are there some methods or procedures? Is it a theoretical one or an experimental one? Since the radicals are transient species, there are many ambiguities in some of the measurements. Generally, scientists use two types of measurements: one is theoretical, and the other one is experimental. In the theoretical one, RSEs have been calculated as the change in total energies of the species in the isodesmic reaction.

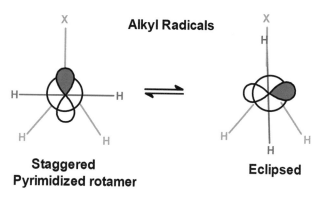

Alkyl Radicals

Staggered Pyrimidized rotamer   **Eclipsed**

FIGURE 13.3 Orbital rotation of radicals.

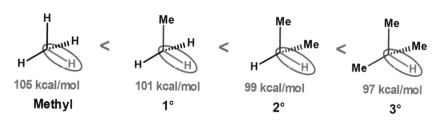

105 kcal/mol **Methyl** < 101 kcal/mol **1°** < 99 kcal/mol **2°** < 97 kcal/mol **3°**

FIGURE 13.4 Bond dissociation energy of substituted alkyl radicals.

13.4.1 ANOMALOUS BEHAVIOR OF HALO-SUBSTITUTED RADICALS

The halo-substituted radicals have different stabilities. The increasing stability of those radicals is given as $CF_3^{\bullet} < CF_3CH_2^{\bullet} < CH_3^{\bullet} < HCF_2^{\bullet} < FCH_2^{\bullet} < ClCH_2^{\bullet} < HCCl_2^{\bullet} < CCl_3^{\bullet}$. What we can infer from this series is that CF_3 is less stable compared with CCl_3.

Here we look at the notional stability with respect to several substituents. The trifluoromethyl radical is essentially tetrahedral, and difluoromethyl and monofluoro-methyl radicals are pyramidal, while the methyl radical is planar. The relative rate for the formation of methyl and fluoromethyl radicals decreases as follows: $HCF_2^{\bullet}$ (10.2) > $CH_2F^{\bullet}$ (9.0) > $CH_3^{\bullet}$ (1.0) > $CF_3^{\bullet}$ (0.08). C–H bond dissociation energy calculations for fluorinated methanes also suggest that the thermodynamic stability of fluoromethyl radicals decreases in a similar order: $CH_2F > HCF_2 > CH_3 > CF_3$. These data indicate that both mono- and difluorination stabilize the methyl radical, whereas trifluorination destabilizes the radical.

The fluorine exerts two types of effects. When it is in the α position, it inductively destabilizes the radical, and through resonance, it stabilizes the radical. But in the tri fluoro radical, the resonance stabilization is much lower, and hence the destabilizing effect predominates and this radical is more unstable (Figure 13.5).

Next, let us look at C substituents. A carbon-free radical is also stabilized by a C-substituent through π-type interactions which involve substantial delocalization into the substituent. The singly occupied molecular orbital (SOMO) energy is relatively unchanged, but the reactivity of the odd-electron center is reduced because the orbital coefficients are smaller.

Finally, what happens when there is a Z substituent? A carbon-free radical is stabilized by a Z substituent through the π-type interaction with the LUMO of the Z group. The SOMO is lowered in energy and the free radical is more electrophilic as a consequence.

We have four major factors that affect the stability of radicals. They are

1. Conjugation
2. Hybridization
3. Hyperconjugation
4. Captodative effects (carbon-centered radical)

13.4.2 CONJUGATION

In conjugation, we have π electrons involved in the stabilization of radicals. For allyl and benzyl radicals, we can apply resonance and aromaticity concepts. From bond dissociation energies and relative rotational energy barriers, we can say both these intermediates are stabilized by conjugation. Recent reports suggest that allyl radicals are relatively little bit more stable than benzyl radicals. Delocalization of π electrons is the reason for the observed stability (Figure 13.6).

The benzyl radicals and their derivatives are important reactive species that undergo changes during the oxidation of toluene and di/tri-methylbenzenes. In some instances, these are involved in the formation of polycyclic aromatic hydrocarbons (PAHs). Similarly, allylic radicals are involved in halogenation, addition, and in polymerizations reactions.

13.4.3 HYBRIDIZATION

When we look at hybridization, the values of heat of formation for the vinyl and phenyl radicals are high compared with the heat of formation of propyl radicals. High energy means not easily formed in other words they are not stable. So sp^3 radicals are stable compared with sp^2 which are stable compared with sp radicals. When the s character increases from sp^3 to sp^2 to sp, the electron is held closer to the nucleus than in the p orbital. This can be taken to mean that when s character increases the electronegativity of carbon also increases. This leads to more unstable radicals.

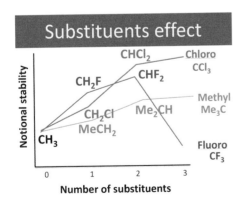

FIGURE 13.5 Substituent effect on halo-substituted alkyl radicals.

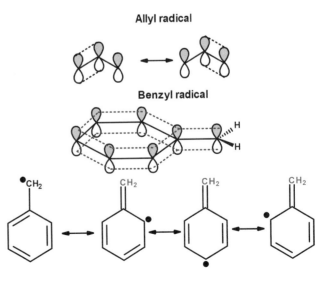

FIGURE 13.6 Effect of conjugation on the stability of free radicals.

13.4.4 HYPERCONJUGATION

In hyperconjugation, a C–H or C–C bond adjacent to an empty or partially filled non-bonding *p* orbital or *p* orbital (filled or antibonding) stabilizes it by hyperconjugation. The more the hyperconjugative structures, more the stability. So *tert*-butyl radical is more stable than secondary which in turn is more stable than primary or methyl radical.

13.4.5 CAPTODATIVE EFFECT

So far, we have seen a steric or electronic nature, which influences both the kinetic and thermodynamic radical stabilizations. Here we will look at a new type of stabilization called captodative effect. When a radical has both an electron-releasing (donor) substituent and an electron-withdrawing (captor) group on the radical carbon, it gets enhanced stabilization.

A word of caution: the captodative effect concept generally is studied and applied to ground-state electronic effects and not to the transition-state electronic effects for the reactions.

Radicals possess a single unpaired electron typically in a *p* orbital known as SOMO, is having high energy. Since low energy is related to thermodynamic stability, any factor that decreases this energy will give a radical higher stability. Since reactivity and stability are inversely related, higher stability means lower reactivity. This factor is more precisely presented by the electron-donating or -withdrawing groups attached to the radical carbon.

We will look at two aspects. One is the relative energy of the SOMO with respect to the other reacting molecule, and second one is which groups are responsible for the energy of SOMO.

We have two scenarios. SOMO has lower energy than LUMO, and SOMO has higher energy than HOMO. When the radical has electron donors, it will have lower energy, and when the radical has EWG on it, it will have higher energy (Figure 13.7).

If the SOMO is relatively low in energy, the principal interaction with other molecules will be with the occupied MOs (HOMO) (three-electron, two-orbital type). In this case, the radical is described as electrophilic and the HOMO of the incoming substrate is called the nucleophilic moiety.

If the SOMO is relatively high in energy, the principal interaction with other molecules may be with the unoccupied MOs LUMO (one-electron, two-orbital type). In this case, the radical is described as nucleophilic and the LUMO of the incoming substrate is called the electrophilic moiety the nucleophilic or electrophilic radical terminology is with respect to the unsubstituted radical. Substituents on the radical center will affect the electrophilicity or nucleophilicity of free radicals.

13.4.6 KINETIC STABILIZATION

The kinetic stabilization of radicals depends on the steric effects of the substituents. When the radical center is non-approachable, the radicals can exist for a longer time. Unlike thermodynamic stability where the energy is the main criterion for stability, here it is the reactivity that is the criteria for stability. If the radical center is easily approachable then that radical is less stable compared with a radical center that is difficult to approach.

Compared with *tert*-butyl radical, triphenylmethyl radical is more stable. One of the reasons is the resonance and delocalization of the "odd" electron in the aromatic nuclei (Figure 13.8).

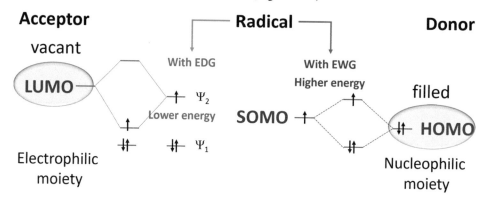

FIGURE 13.7 Energy relationship between SOMO, LUMO, and HOMO.

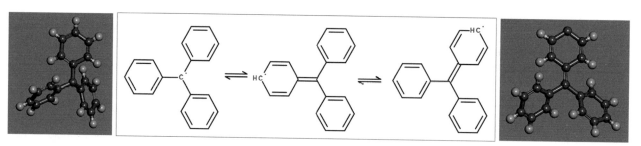

FIGURE 13.8 Delocalization of the "odd" electron in the aromatic nuclei.

When we look at the 3D structure, the triphenylmethyl radical is not planar. Similar to the propeller shape we saw for the carbocation, here also the central carbon radical is planar due to which there is delocalization.

When we compare both triphenylmethyl and hexaphenylethane radicals, we have some interesting facts. The energy of two moles of the triphenylmethyl radical or the hexaphenylethane radical are very similar. It is 35 kcal/mol. But their expected stabilization is 80 kcal/mol. Both are not lower in energy due to two different effects.

In the case of triphenylmethyl radical, it is partially twisted out of plane due to *ortho* hydrogen interactions, and hence, there is partial rotation. That leads to loss of resonance and hence it is higher in energy. In the case of hexaphenylethane, there is frontal strain, similar to ethane conformation. This leads to higher energy, and hence this molecule has higher energy. The stability of triphenylmethyl radical is attributed to the steric hindrance for dimerization (Figure 13.9).

13.5 FACTORS THAT AFFECT RADICAL STABILITY

The following factors influence radical stability (Figure 13.10):

1. Free radical's stability increases as we go down the periodic table (larger size)
2. Adjacent electron-withdrawing groups decrease the stability of free radicals
3. Free radical's stability increases as the electronegativity of the atom decreases
4. Free radicals are stabilized by resonance
5. Free radicals are stabilized by adjacent atoms with lone pairs
6. Free radicals decrease in stability as we go from sp^3 to sp^2 to sp hybridization
7. Stability increases in the order methyl <primary <secondary <tertiary

13.5.1 PARAMETERS THAT INFLUENCE REACTIVITY OF FREE RADICALS

We have two major types of radical reactivity. One is a unimolecular reaction, that is, reactivity of a radical by itself (fragmentation, rearrangements, etc.).

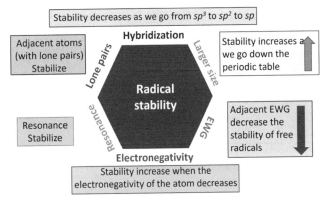

FIGURE 13.10 Factors that influence the radical stability.

The second one is radical interaction with other species. If it reacts with another radical, then it can undergo disproportionation or dimerization. If the radical reacts with another molecule, we have addition, displacement, and atom abstraction (Figure 13.11).

13.5.2 STEPS INVOLVED IN FREE RADICAL CHAIN REACTIONS

Three crucial processes occur in a free radical reaction. The initiation or the formation of the radicals is the first step. We have already seen this. The next is the propagation, and the last one is termination (Figure 13.12).

13.5.3 THE POLAR NATURE OF RADICALS

Since radical philicity appears to have an important role in radical reactivity, it is valuable to be able to determine easily whether a particular radical is electrophilic or nucleophilic (look at the structure of the radical). One is based on atom electronegativity and the other on cation and anion stability. By increasing the electron-withdrawing group, the radical can be turned into an electrophilic radical.

13.5.4 ATOM ELECTRONEGATIVITY AND PHILICITY OF RADICALS

Radicals such as RO•, RCO₂• R₂N•, Cl•, and F• where the radical center is on an atom more electronegative than carbon, are considered to be electrophilic, and those with the radical centered on an atom less electronegative than carbon (e.g., R₃Ge•, R₃Sn•, and R₃Si•) are classified as nucleophilic.

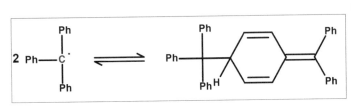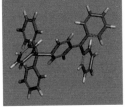

FIGURE 13.9 Strain in hexaphenylethane.

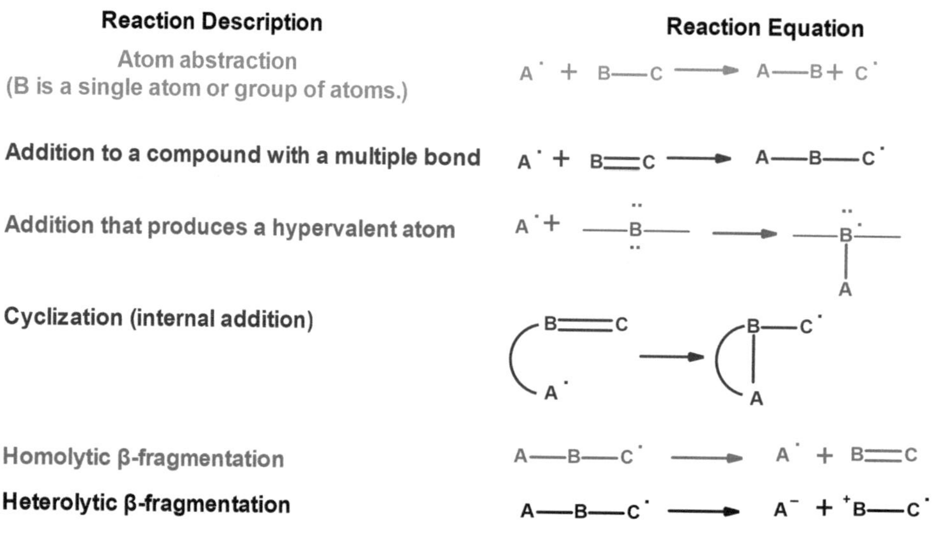

Reaction Description   **Reaction Equation**

Atom abstraction
(B is a single atom or group of atoms.)

$$A^\bullet + B-C \longrightarrow A-B + C^\bullet$$

Addition to a compound with a multiple bond

$$A^\bullet + B{=}C \longrightarrow A-B-C^\bullet$$

Addition that produces a hypervalent atom

$$A^\bullet + \ -\!\!\overset{..}{\underset{..}{B}}\!\!- \ \longrightarrow \ -\!\!\overset{..}{\underset{|}{B}}\!\!\overset{\bullet}{\ } \ $$
                                                                    A

Cyclization (internal addition)

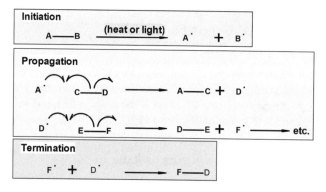

Homolytic β-fragmentation

$$A-B-C^\bullet \longrightarrow A^\bullet + B{=}C$$

Heterolytic β-fragmentation

$$A-B-C^\bullet \longrightarrow A^- + {}^+B-C^\bullet$$

FIGURE 13.11   Common reactions of radicals.

Initiation

$$A-B \xrightarrow{\text{(heat or light)}} A^\bullet + B^\bullet$$

Propagation

$$A^\bullet \quad C-D \longrightarrow A-C + D^\bullet$$

$$D^\bullet \quad E-F \longrightarrow D-E + F^\bullet \longrightarrow \text{etc.}$$

Termination

$$F^\bullet + D^\bullet \longrightarrow F-D$$

FIGURE 13.12   Steps involved in free radical chain reactions.

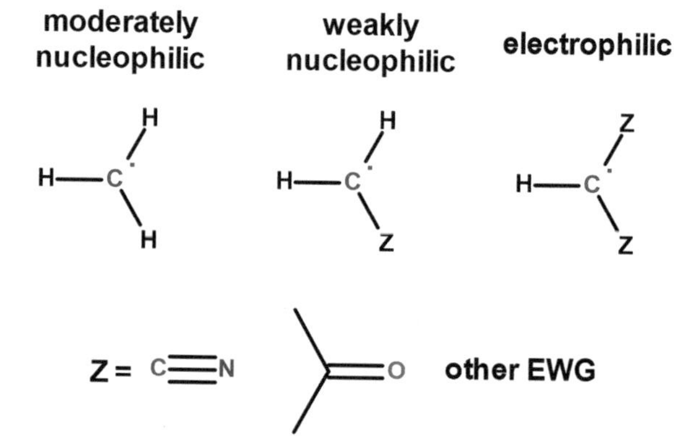

moderately nucleophilic

$$H-\overset{H}{\underset{H}{C}}{}^\bullet$$

weakly nucleophilic

$$H-\overset{H}{\underset{Z}{C}}{}^\bullet$$

electrophilic

$$H-\overset{Z}{\underset{Z}{C}}{}^\bullet$$

$$Z = \ C{\equiv}N \qquad \backslash\!\!=\!\!O \qquad \text{other EWG}$$

FIGURE 13.13   Examples of radical philicity.

13.5.5   CATION AND ANION STABILITY

Alkyl radicals and carbon-centered radicals bonded to an atom that can donate electron density to the radical center (e.g., an oxygen or nitrogen atom with an unshared pair of electrons, e.g.,·CH$_2$OH) are taken to be nucleophilic.

Carbon-centered radicals themselves are considered to be electrophilic if the carbon atom bearing the radical center also contains at least one powerful electron-withdrawing group (e.g., –NO$_2$, –CN, –CO$_2$R, –COR).

Another approach to assigning radical philicity depends upon knowledge of ion stability. According to this method, a nucleophilic radical leads more easily to a cation by electron loss than to an anion by electron gain.

Radicals such as RO$^\bullet$, RCO$_2^\bullet$, R$_2$N$^\bullet$, Cl$^\bullet$, F$^\bullet$, and RS$^\bullet$, which form anions by accepting electrons more easily, are designated as electrophilic.

If a hydrogen atom attached to a carbon-centered radical is replaced by an electron-withdrawing substituent (e.g., a cyano or carbonyl group), the resulting radical becomes more electrophilic (Figure 13.13).

13.6   FREE RADICAL REACTIONS

13.6.1   ATOM ABSTRACTION

We have two types of atom abstraction. One is hydrogen atom abstraction, and the other one is halogen-atom abstraction.

The ease of hydrogen atom abstraction is controlled by thermodynamic, polar, stereoelectronic, and geometric effects. From the BDEs we can predict alkoxyl, aminyl or conjugated groups like aryl and vinyl radicals can facilitate efficient hydrogen atom transfers.

Similarly, based on the strength of the C–H bond, we can also predict which proton will be abstracted, that is, the regiochemistry of intermolecular hydrogen atom abstraction. Generally, the weakest C–H bond is broken.

There are instances where the polar effect also plays a role. In the reaction of *tert*-butoxyl radical with γ-butyrolactone, the proton is abstracted from the position adjacent to the ether oxygen. Since the *tert*-butoxyl radical is electrophilic, we can expect the polar effect to operate. On the other hand, the reaction with the nucleophilic aminyl-boryl radical proceeds exclusively at the position adjacent to the carbonyl group.

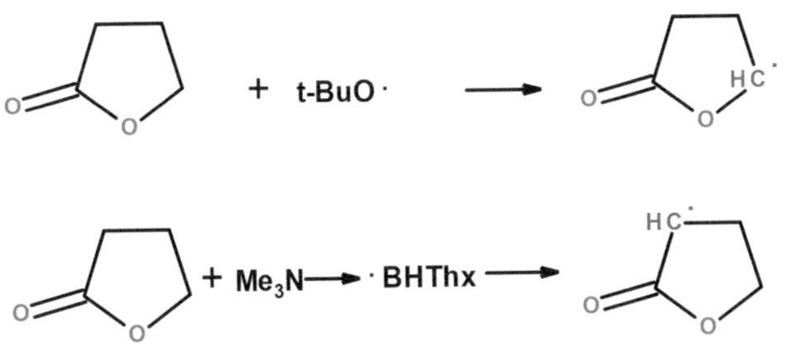

Thx = thexyl = 1,1,2-trimethylpropyl

SCHEME 13.6 Examples of hydrogen atom abstraction.

In the following reactions, we can see how the polar effect influences the radical formation.

Stereoelectronic effects also have a strong influence on the rate of hydrogen transfer. For example, the cleavage by the *tert*-butoxyl radical of an axial C–H bond is 14 times faster than that of an equatorial C–H bond. This is rationalized by interactions between lone pair on oxygen from the axial side and the bond being broken (Figure 13.14).

When we talk about intramolecular hydrogen atom abstraction, efficient 1,4-, 1,5-, 1,6- and 1,7-hydrogen atom transfers (= HT) have been observed. But the 1,5-process is by far the most common reaction.

The preference for 1,5-hydrogen transfer results from a more favorable entropy of activation of the six-membered transition state and not from the enthalpy of activation. The calculated transition state for a 1,5-hydrogen atom abstraction in butoxyl radical is shown here. It resembles a five-membered ring having an envelope shape like that of cyclopentane, but with one long bond (2.5 Å) between the carbon bearing the hydrogen atom and the oxygen of the alkoxyl radical.

The O–C–H bond is reasonably close to linear (153°) and far away from the 109° expected for chair-like transition states resembling a cyclohexane ring (Figure 13.15).

13.6.2 Intermolecular Hydrogen Abstraction

Lead tetraacetate is an efficient reagent for the generation of radicals from alcohols. This reagent is used to convert an alcohol to a macrolide. Lead tetraacetate is also used with iodine. This combination is very efficient for the preparation of tetrahydrofuran derivatives. The iodo alcohol can undergo two types of reactions as shown here. It can either form a tetrahydrofuran ring or can be converted to lactone by further reactions.

13.6.3 Bromination

Free radical bromination can be caried out by N-bromo-succinimde (NBS) or Br_2. Both reactions require a radical initiator. Wohl–Ziegler bromination uses NBS. The radical initiator reacts with the alkane to produce the alkyl radical and HBr. This HBr reacts with NBS to produce Br_2. Now the previously formed alkyl radical reacts with bromine to give the alkyl bromide and generates the bromine radical.

In the reaction with bromine, the initially formed bromine radical abstracts hydrogen from the hydrocarbon to produce the alkyl radical and HBr. This alkyl radical can react with either HBr or Br_2. It is shown that the reaction of the alkyl radical with bromine is faster than the reaction of the alkyl radical with HBr. Although bromine radical is produced in both these reactions, the other products are alkyl bromide and alkane, respectively.

The study based on relative rates of brominations of the halogenated alkanes is shown in Figure 13.16. Except for

k_{abst} (rel.) **14 : 1**

FIGURE 13.14 The cleavage of axial vs equatorial C–H bond.

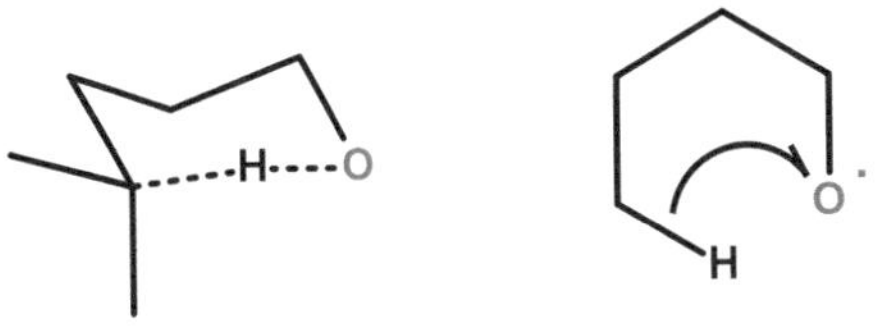

FIGURE 13.15 1,5-Hydrogen atom abstraction.

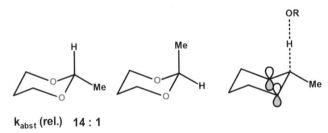

X = I or OAc

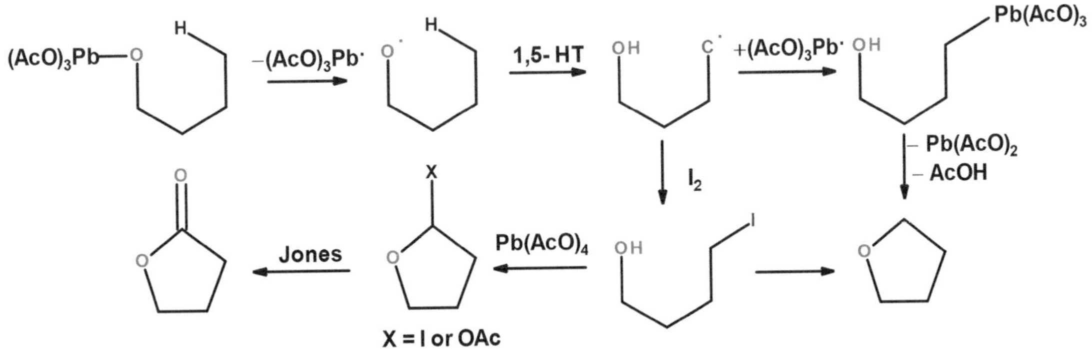

SCHEME 13.7 Intermolecular hydrogen abstraction.

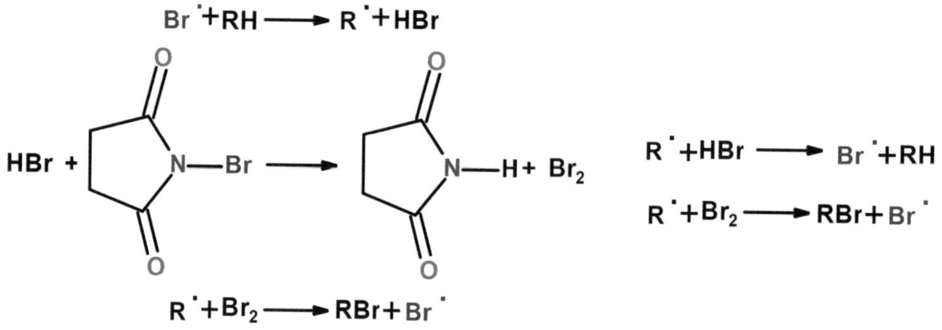

$$Br^{\cdot} + RH \longrightarrow R^{\cdot} + HBr$$

$$R^{\cdot} + HBr \longrightarrow Br^{\cdot} + RH$$

$$R^{\cdot} + Br_2 \longrightarrow RBr + Br^{\cdot}$$

$$R^{\cdot} + Br_2 \longrightarrow RBr + Br^{\cdot}$$

SCHEME 13.8   Bromination.

	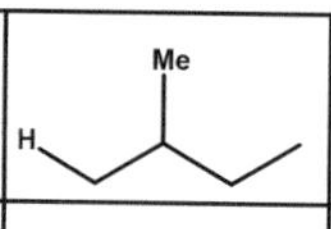 Me (H)	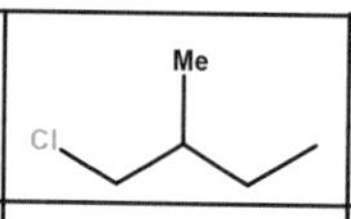 Me (Cl)	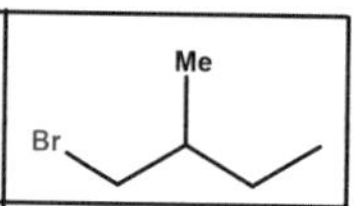 Me (Br)
Br₂/hυ	1.0	1.9	16.3
NBS/hυ MeCN	1.0	0.29	1.61
NBS/AIBN MeCN	1.0	0.27	1.61

	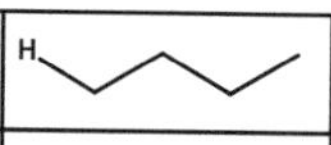 H	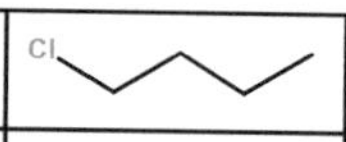 Cl	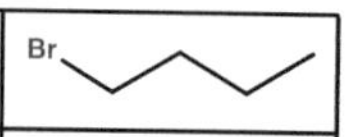 Br
Br₂/hυ	1.0	0.74	3.4
NBS/AIBN MeCN	1.0	0.31	0.58

FIGURE 13.16   Relative rates of bromination.

1-chlorobutane, all the halogenated alkanes react at a faster rate than their unsubstituted parent hydrocarbons.

When we apply the microscopic reversibility principle, the nonpolar carbon-centered radical will react with polar hydrogen bromide and will go back to the starting alkane. Only when it reacts with the bromine radical, the product will be formed. This is what happens in the NBS reaction.

On the other hand, when the alkyl radical reacts with nonpolar bromine, the product is formed. Even if the alkyl radical reacts with bromine radical still, the bromo product is formed. So this is one of the reasons, and we get more products of bromination when we use bromine compared with NBS.

When we have terminal carbon (6H) and internal carbons (2H), they will react at a 6:2 (3:1) ratio. But in reality, 8% of 1-bromopropane and 92% of 2-bromopropane are obtained in the reaction. This can be explained based on the relative bond energies of the C–H bonds that are broken. We can also invoke the stability of the resulting radical after hydrogen abstraction. We can also invoke kinetics vs thermodynamic control. Here both favor the 2-bromo derivative (Figure 13.17).

13.6.4   PHILICITY OF A RADICAL

How the philicity of the radical affects the hydrogen abstraction. We have two scenarios, either nucleophilic (methyl) or electrophilic (chlorine) radical abstracting a hydrogen.

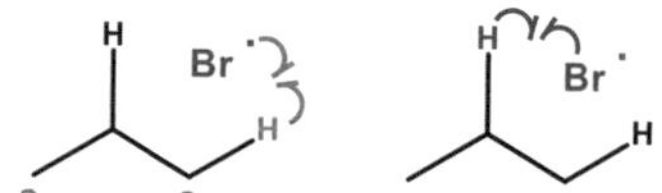

Number of hydrogens	3	3	2
% of Product (expected)		75	25
% of Product (obtained)		8	92

Explanation: 1) Based on bond energy (100.38 vs 98.7 kcal/mol)
2) Based on resulting radical stability

FIGURE 13.17   Relative rates of bromination and hydrogen abstraction.

These two radicals abstract hydrogen from a *tertiary* substrate. We have two *tertiary* compounds: one has three methyl substituents, that is,, tertiary butane, and the other has three chlorines, that is, chloroform. It was seen that the electrophilic chlorine radical abstracts the hydrogen from chloroform more readily than from the *tertiary* butane. On the other hand, nucleophilic methyl radical abstracts the hydrogen from the *tertiary* butane more readily than from the chloroform (Figure 13.18).

Moreover, when we compare the α and β proton abstraction from carboxylic acids, the methyl radical abstracts the α hydrogen, whereas the chloro radical abstracts the β hydrogen (Figure 13.19).

13.7   RADICAL ADDITION

We have two major types of radical addition reactions. One is intermolecular reaction, and the other one is an intramolecular

$$X^{\cdot} + H{-}G \longrightarrow H{-}X + G^{\cdot}$$

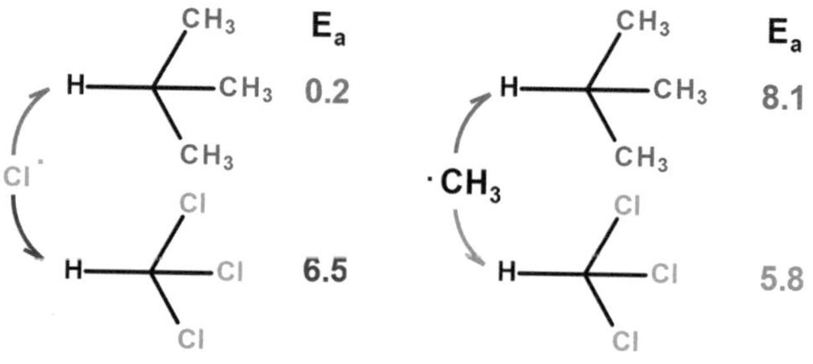

FIGURE 13.18   Relationship between philicity of the radical and hydrogen abstraction.

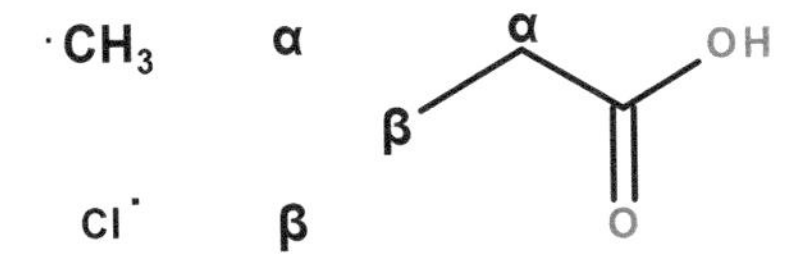

FIGURE 13.19 Relationship between α and β proton abstraction from carboxylic acids.

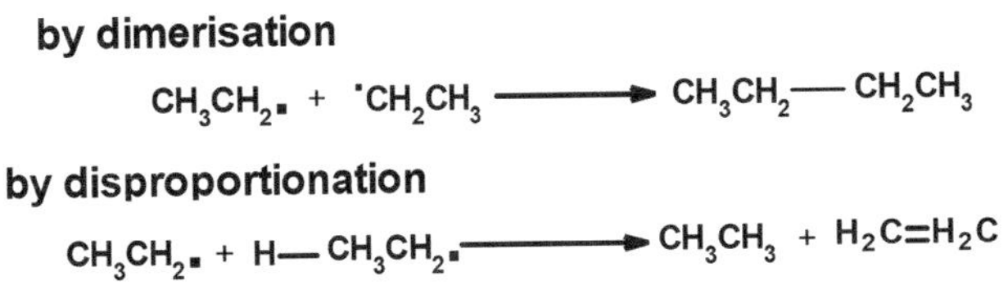

reaction (radical cyclization). Intermolecular reactions include an addition to a compound with multiple bonds and an addition that forms a radical with a hypervalent atom.

13.7.1 FRAGMENTATION REACTIONS

Homolytic β-fragmentation of a radical is an elementary reaction that cleaves a bond adjacent to a radical center (Figure 13.20).

When a radical is centered on a carbon atom that has an effective leaving group attached to a neighboring carbon atom, the radical cation is readily formed. The bond from the neighboring carbon atom to the leaving group need not cleave homolytically with ease; otherwise, β-fragmentation-producing ionic intermediates could be preempted by homolytic fragmentation. Heterolytic β-fragmentation occurs in the reaction shown.

α-Fragmentation is an elementary reaction in which a bond attached to a radical center cleaves homolytically. This reaction is rare because it requires the energy-demanding step of bond breaking without the energetic compensation of bond formation.

13.7.2 SINGLE ELECTRON TRANSFER

Single electron transfer produces radicals. Here we have a metal or its ion which undergoes oxidation by the loss of an electron. This electron is captured by the substrate to produce an activated complex, which is an anion radical complex. This complex then loses an anion to form the radical.

13.7.3 DISPROPORTIONATION REACTIONS

The disproportionation reaction is also called dismutation. When one of a pair of encountering radicals has a hydrogen atom β to the radical center, two stable molecules are produced when this hydrogen atom is abstracted by the other

by dimerisation

by disproportionation

SCHEME 13.9 Disproportionation reactions.

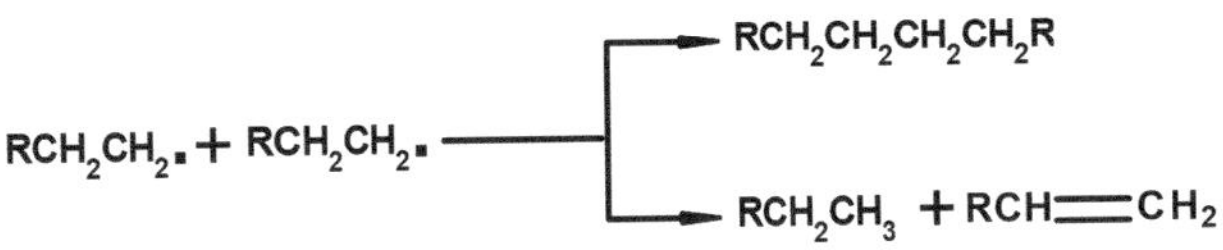

SCHEME 13.10 Alkyl radicals disproportionation to give an alkene and an alkane.

radical. One product is an unsaturated compound and the other one is a saturated compound.

When two alkyl radicals are disproportionate, we get an alkene and an alkane. They may also combine to give the dimer. Since the disproportionation rates are similar in magnitude to recombination rates, it led to the inevitable conclusion that the transition state for the two reactions must be identical.

Alkyl radical reaction with aromatic systems gives two different sets of products. One has having aromatic compound and an alkane, and other one is cyclohexadiene system. It is observed that no cyclohexadienyl radical is formed in any of these reactions. Because if they are formed, they will be less stable compared with an aromatic product. This must be reflecting, in turn, an energy barrier for the process.

13.8 RADICAL CYCLIZATION-BALDWIN'S RULES

This rule talks about how a ring closure will take place when an alicyclic system is formed. We have three different parameters: (1) the size of the ring that is formed, (2) the type of ring closure whether it is *endo* or *exo*, and (3) the nature of the carbon.

This is applicable to the formation of 3–7-membered rings. Two types of ring closures can occur, *exo* and *endo*, and it is based on the bond being broken during the ring closure.

The radical may add to three types of carbon geometry, namely, *tet*rahedral (sp^3 carbon), *trig*onal (sp^2 carbon), and *dig*onal (sp carbon) (Figure 13.21)

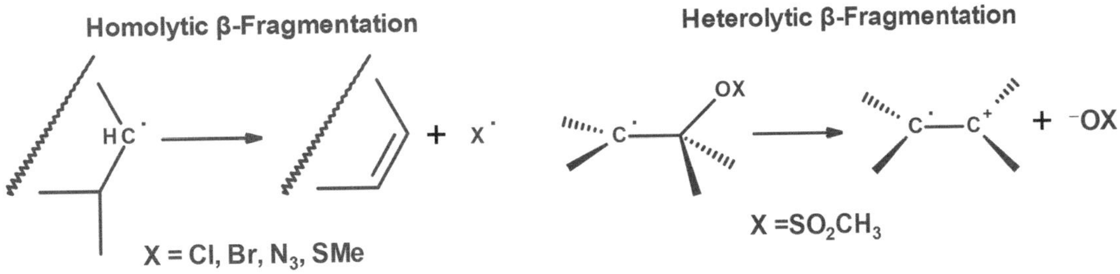

Homolytic β-Fragmentation

X = Cl, Br, N$_3$, SMe

Heterolytic β-Fragmentation

X = SO$_2$CH$_3$

FIGURE 13.20 Homolytic and heterolytic β-fragmentation.

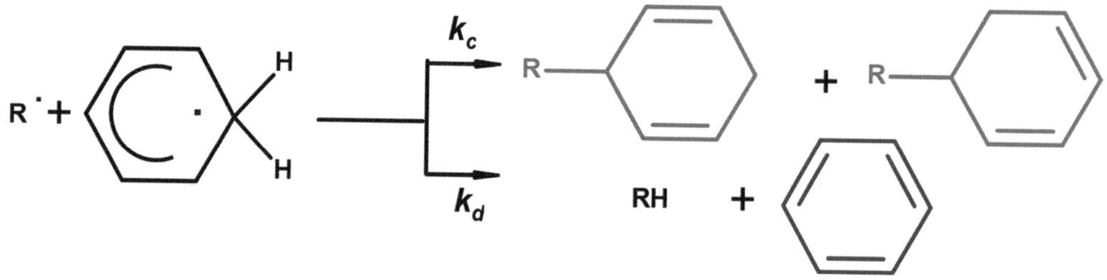

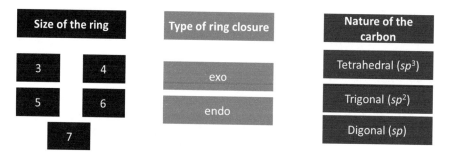

SCHEME 13.11 Alkyl radical reaction with aromatic systems.

Size of the ring	Type of ring closure	Nature of the carbon
3 4	exo	Tetrahedral (*sp³*)
5 6	endo	Trigonal (*sp²*)
7		Digonal (*sp*)

FIGURE 13.21 Factors that affect radical cyclization.

We can consider *exo* and *endo* cyclization. In the case of *exo* cyclization, the substituent is not part of the ring. In the *endo* cyclization, the substituent is part of the cyclic system. 5-*exo*-tet means, the resulting product is a five-membered ring, the radical attacked an *sp³* carbon to form the ring and the substituent is outside the cyclic system.

Except for the small rings three- and four-membered rings other ring members up to seven, *exo* cyclization is allowed for all the geometries. For alkenes and alkynes undergoing cyclization forming six- or seven-membered rings, all *exo* and *endo* are allowed. Due to the lower angle of attack for the *sp* cyclization, three- and four-membered *exo* cyclization is forbidden. With *tetrahedral* geometry, all *endo* cyclization (five, six, and seven) are forbidden (Figure 13.22).

Similar to an addition reaction, cyclization also requires radical generation, cyclization, and selective radical removal. Each step must be more rapid than the loss of radicals by (nonselective) radical/radical or radical/solvent

reactions, and the method that is chosen must convert the cyclic radical, but not the initial radical, to a stable product. Unlike addition reactions which require two components to react, the cyclization is intramolecular; hence, it is very fast. Moreover, entropy changes also favor cyclizations. Some of the potential problems include: (1) reactions of radicals with other radicals or solvent and (2) removal of initial radical before cyclization. Due to these issues, controlling radical reactions is a challenge. *Exo* cyclizations are typically favored kinetically over *endo* cyclizations (Table 13.1).

TABLE 13.1

Baldwin Rule for Radical Cyclization

Type	3 exo	3 endo	4 exo	4 endo	5 exo	5 endo	6 exo	6 endo	7 exo	7 endo
Tet	✓		✓		✓	✗	✓	✗	✓	✗
Trig	✓	✗	✓	✗	✓	✗	✓	✓	✓	✓
Dig	✗	✓	✗	✓	✓	✓	✓	✓	✓	✓

exo

endo

5-exo-tet   **6-endo-trig**   **6-exo-dig**

6-endo-tet

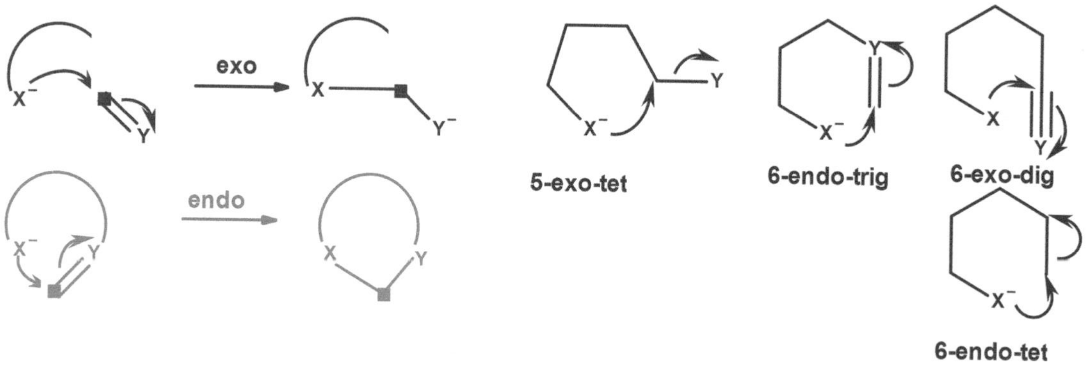

FIGURE 13.22 Types of radical cyclization.

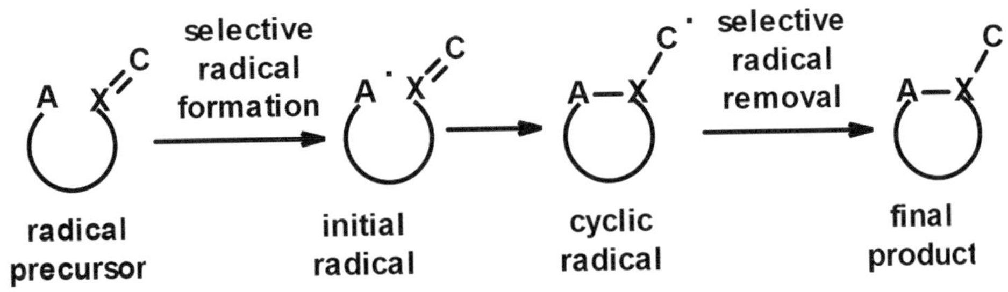

SCHEME 13.12   5-Hexenyl radical cyclization.

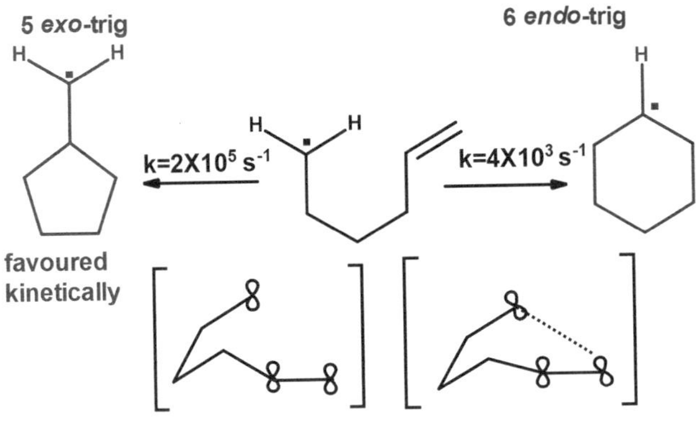

SCHEME 13.13   5 *exo* vs 6 *endo* cyclization.

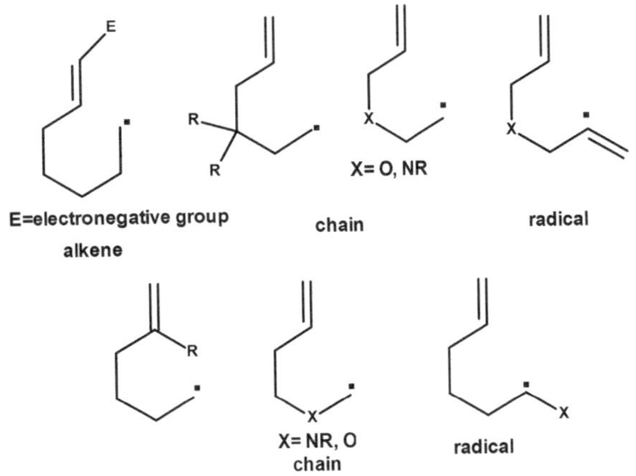

13.9   RADICAL CYCLIZATION

5-Hexenyl radical cyclization is an important cyclization, because it is well understood and also constitutes the most successful ring formation reaction in organic chemistry. There are six carbons. We can very well get both five-membered ring through *exo* and six-membered ring through *endo* cyclizations. This cyclization is highly exothermic, and strictly kinetically controlled. Entropy favors cyclization over addition.

You may be wondering, why five-membered cyclization is preferred over six-membered in radical cyclizations?. According to conformational analysis, we know six-membered rings are more stable than five-membered rings.

This can be explained based on the transition-state model. If it is 5 *exo* cyclization, the interacting orbitals are in close proximity compared with 6 *endo* cyclization.

Accelerating substituents, Decelerating substituents

Most substituents accelerate 5-*exo* cyclizations. Substitutions of oxygen or nitrogen for C-3 are powerfully accelerating because they provide better overlap in the 5-*exo* transition state. Most alkyl substituents on the radical carbon (C-l) have relatively little effect on the rate: primary, secondary, and tertiary alkyl radicals all cyclize at about the same rate. Vinyl and phenyl radicals are much more reactive in cyclizations than alkyl radicals. Acyl radicals are also excellent substrates (Figure 13.23).

The most common reason that 6-*endo* products are observed in radical cyclizations is because 5-*exo* cyclizations are decelerated by substituent effects. The substitution of any group for hydrogen at the alkene carbon (C-5) slows 5-*exo* cyclization.

FIGURE 13.23   Examples of accelerating substituents, decelerating substituents.

13.9.1   ACTIVATION OF 6-ENDO CYCLIZATIONS

The appropriate placement of activating groups can also be used to accelerate 6-*endo* cyclizations.

We will study about mechanism of the following reactions: Giese reaction, Hofmann–Löffler–Freytag reaction, Barton decarboxylation, Barton–McCombie deoxygenation, Birch reduction, and Sandmeyer reaction.

13.9.2   ANTI-MARKOVNIKOV ADDITION

In anti-Markovnikov addition, the first step is the addition of bromine radical to the alkene. This will result in an alkyl radical. The stability of the resulting radical is the deciding factor for the initial bromine radical addition. This is a reversible reaction. Hence, it is a thermodynamically controlled reaction. Unlike electrophilic addition where the initial addition is the addition of proton, in the radical reaction initial addition is the addition of bromine radical.

13.10   GIESE REACTION

Alkyl halides undergo C–C-coupling on reaction with tributyltin hydride in the presence of electron-deficient alkenes. Carbon-free radicals (generated from organic halides, Barton esters, etc.) are nucleophilic and can be trapped with various electrophiles. In particular, the trapping of free radicals with electron-deficient alkenes to form carbon–carbon bonds is called the Giese reaction.

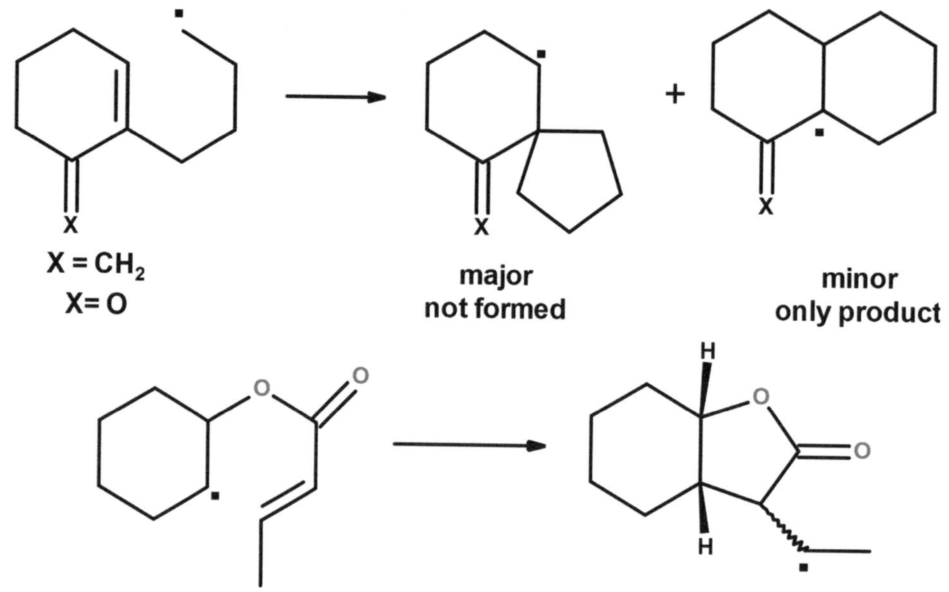

X = CH₂
X = O

major
not formed

+

minor
only product

SCHEME 13.14   Activation of 6-endo cyclizations.

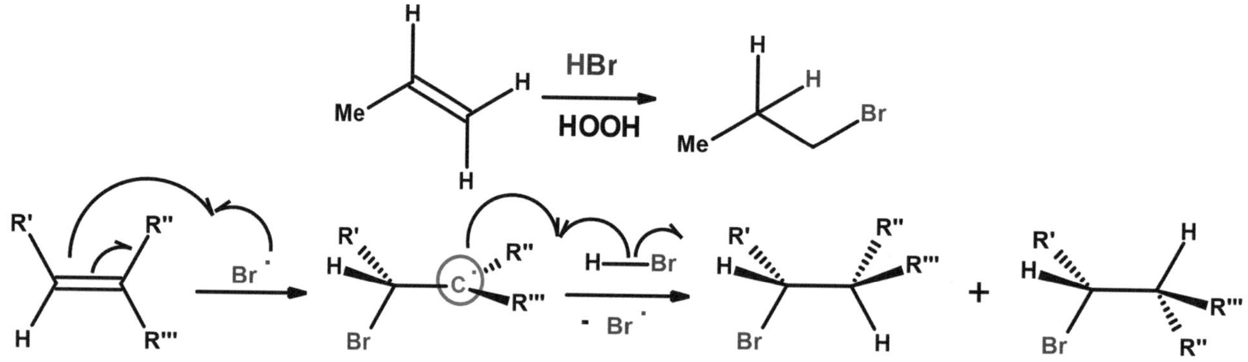

SCHEME 13.15   Anti-Markovnikov addition.

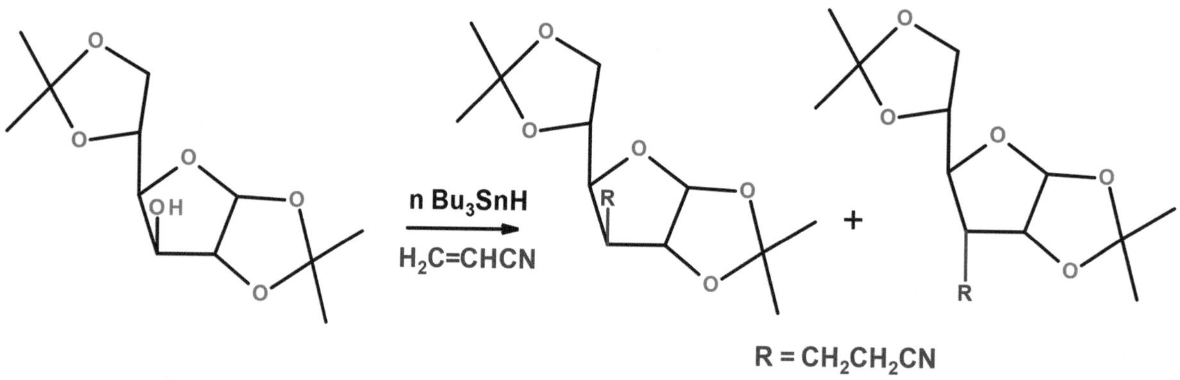

R = CH₂CH₂CN

SCHEME 13.16   Giese reaction.

The newly formed α-carbon radicals can be reacted further with the second electrophile in tandem fashion.

With alkyl halide and electron-deficient alkenes as reactants, a radical chain reaction leads, via the intermediary radicals, to the formation of the product. In the case of alkyl iodides, the reactions could be carried out with substoichiometric amounts of organotin compounds because the butyltin hydride could be regenerated *in situ* from the tin halide with NaBH₄, if it is used. The photochemically initiated reaction of the alkyl iodides (X=I) in ethanol at 25 °C with 0.2 molar equivalents of n-Bu₃SnCl and 1.3 molar equivalents of NaBH₄ in the presence of the alkenes afforded the product.

This C–C-coupling reaction could also be extended to thioacylated alcohols, analogously to the reduction method of the Barton method. One of the issues with this reaction is the reduction of the alkyl halide or alcohol to the corresponding alkane.

It is known that in the case of free radical copolymerizations, the rate of addition of polymeric benzyl radicals to monomers increases when the alkenes have an electron-withdrawing substituent. The rate of addition of radicals increases with the electron-withdrawing ability of the substituent Z the alkene. For example, the rate of addition of the cyclohexyl radical at 20 °C increases by almost 10^4 when the reaction is carried out using acrolein instead of 1-hexene (Figure 13.24).

Similarly, it is also observed that an increase in the rate of addition of cyclohexyl radical is caused by the presence of electron-withdrawing substituents Z in the substituted styrenes, methyl acrylates, and acrylonitriles.

Substituents at that carbon atom of the alkene which is attacked (α-substituents) exert both polar and steric effects on the rate of addition of free radicals. The polar α-effect is somewhat smaller than the polar β-effect. Alkene-stabilizing substituents exert no additional effect on the rate of addition.

Substituents at the carbon atom of the alkene that is not attacked (β-substituents) exert, on the whole, only polar effects on the rate of addition of free radicals. Radical stabilizing and space-filling β-substituents influence the rate of addition only slightly.

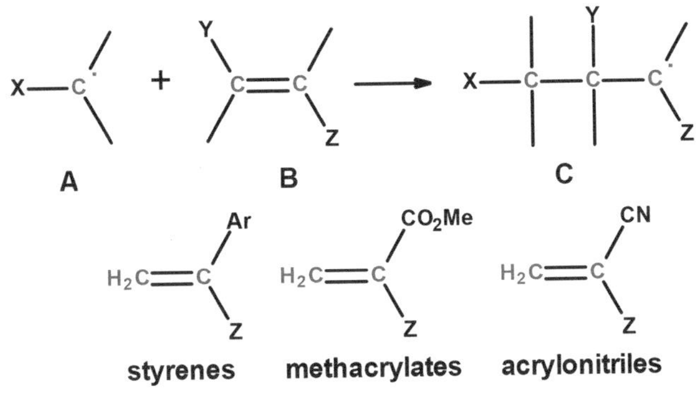

13.11 HOFMANN–LÖFFLER–FREYTAG REACTION

Next, we will see the Hofmann–Löffler–Freytag reaction. In this reaction, a cyclic amine (pyrrolidine or, in some cases, piperidine) is generated by thermal or photochemical decomposition of N-halogenated amine in the presence of a strong acid (concentrated H_2SO_4 or concentrated CF_3CO_2H). Intramolecular hydrogen atom is transferred to a nitrogen-centered radical.

When N-halo-amines are treated with concentrated H_2SO_4, the amino radical cation is produced. Hydrogen atom transfer via a six-membered transition state generates the ammonium radical, which reacts with the protonated form of amine (ammonium salt) to give an intermediate that is neutralized, the amine displaces the primary halide to give a hexahydroindolizine.

13.12 BARTON DECARBOXYLATION

Sir Derik Barton got the Nobel Prize in chemistry in 1969. The mechanism by which the Barton decarboxylation occurs is through a radical intermediate. Here, a carboxylic acid is initially converted to a thiohydroxamate ester (commonly referred to as a Barton ester) by treatment with thiohydroxamic acid. This intermediate ester product was not isolated. It was directly decarboxylated by treatment with Bu_3SnH (hydrogen donor) by heating. Use of a radical initiator like AIBN initiates the reaction. The initially formed radical on reaction with thiohydroxamate ester produces carboxylate radical. Unlike Kolbe electrolysis, here the alkyl radical reacts with hydrogen donor to give the alkane and not the dimer. This is an example of reductive decarboxylation.

The first step is the formation of the radical from the radical initiator, azobisisobutyronitrile (AIBN). This then reacts with Bu_3SnH to produce the tributyltin radical in the second step. Third step is the reaction between the tributyltin radical and the thiocarbonyl. This leads to the formation of a single bond between the sulfur and tin. In the next step, the N–O bond cleaves homolytically resulting in alkylcarboxy radical. In the last step, the loss of CO_2 from alkylcarboxyl results in alkyl radical which is quenched to form the alkane. The driving force for the reaction is the S–Sn bond formation.

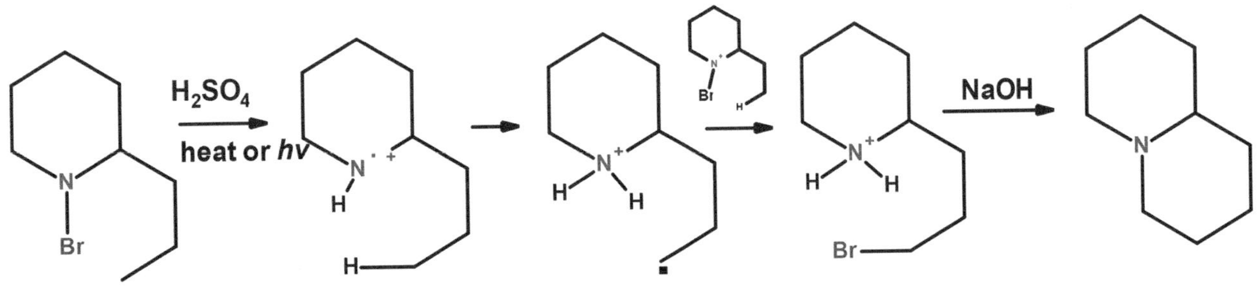

FIGURE 13.24 Free radical copolymerizations.

SCHEME 13.17 Hofmann–Löffler reaction.

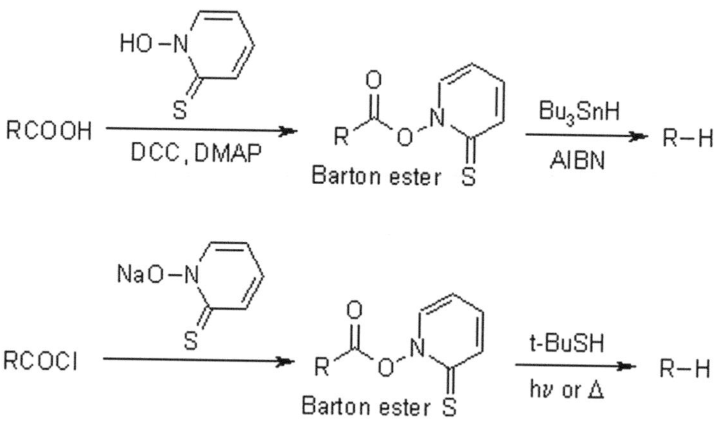

SCHEME 13.18 Barton decarboxylation.

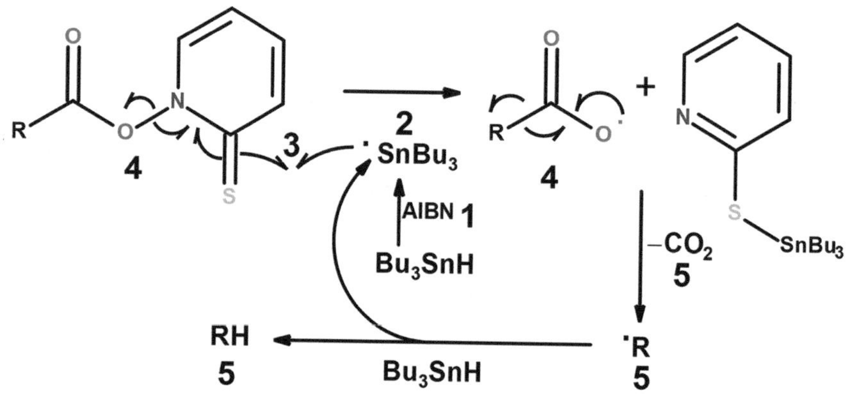

SCHEME 13.19 Mechanism of Barton decarboxylation.

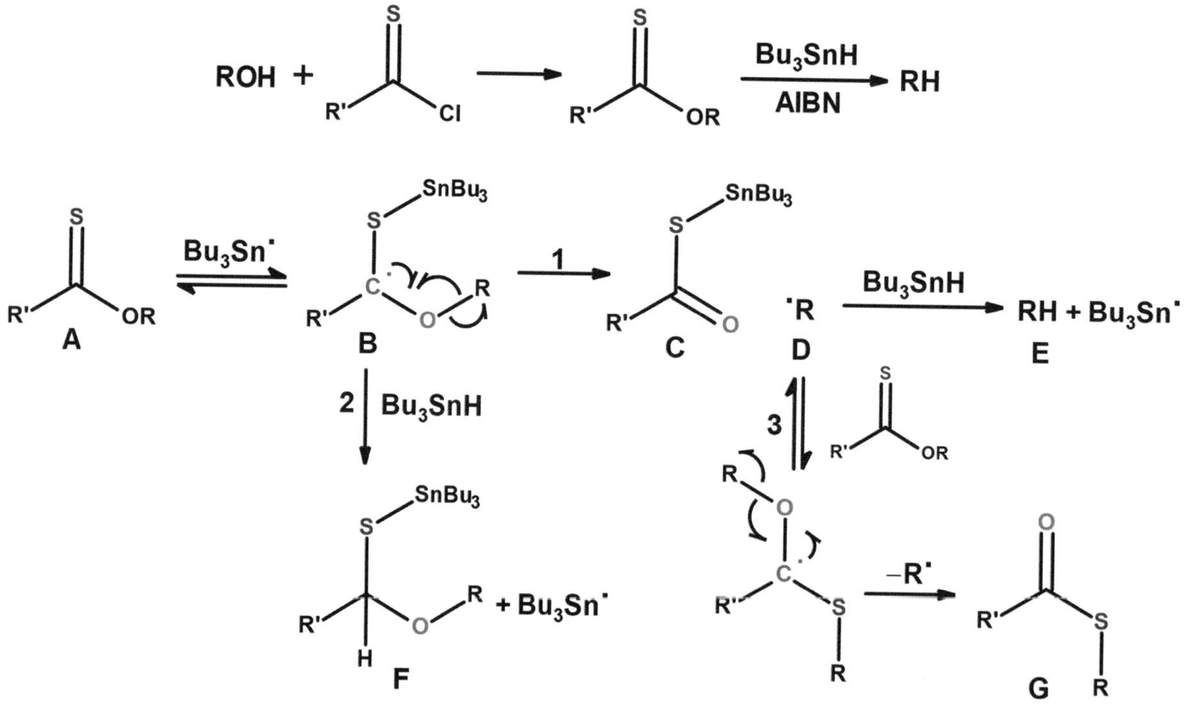

SCHEME 13.20 Barton–McCombie deoxygenation.

13.13 BARTON–MCCOMBIE DEOXYGENATION

The Barton–McCombie deoxygenation is the conversion of an alcohol into an alkane. The alcohol is treated with thioacyl chloride to give the corresponding alkoxythioacyl ester. This is reduced by tributyltin hydride to the alkane. The driving force for the reaction is similar to the Barton decarboxylation, that is, the S–Sn bond formation.

The individual steps are given in Scheme 13.20. Initiation produces small amounts of the tributyltin radical, which adds across the carbon-sulfur double bond at sulfur, yielding resonance-stabilized adduct radical B. Irreversible reduction of this intermediate by tributyltin hydride to form F via pathway (2) is sometimes a problematic side reaction, depending on the nature of R' and the reaction conditions. Fragmentation of B via pathway (1) affords co-product C and carbon-centered radical D, which abstracts a hydrogen atom from tributyl tin hydride to yield the product and propagate the radical chain. D also may add reversibly to starting material A via pathway (3), which in some cases leads to thioester G.

13.14 WOHL–ZIEGLER BROMINATION

According to Wohl–Ziegler bromination, allylic or benzylic substrates undergo bromination at allylic or benzylic positions when reacted with N-bromosuccinimide, respectively. This reaction uses a radical initiator.

13.15 BIRCH REDUCTION

Here an aromatic compound having a benzenoid ring is converted into a product, 1,4-cyclohexadienes, that is, an unconjugated diene. It is the reduction of aromatic rings in liquid ammonia with sodium, lithium, or potassium and an alcohol, such as ethanol and *tert*-butanol. This reaction is quite different from catalytic hydrogenation, where the aromatic ring is completely reduced to give a cyclohexane. Now this is very similar to the Bouveault–Blanc reduction. There it was esters, and here it is an aromatic system.

In the mechanism of Birch reduction, initially an electron adds to the aromatic system. This leads to two different anion formations. One is *ortho* to the substituent, and the other one is *meta* to the substituent. The carbanion abstracts a proton from the solvent. Then the second electron is added which produces the second carbanion. This also abstracts a proton from the solvent and the reduction is complete.

13.16 SANDMEYER REACTION

Aryl halides are produced by the reaction of aryldiazonium salts and copper halide salts, and this reaction is called the Sandmeyer reaction. It is an example of a radical-nucleophilic aromatic substitution. Using Sandmeyer reaction, several benzene derivatives containing functional groups like halo, cyano, trifluoromethyl, hydroxyl, and so on can be prepared.

Here loss of nitrogen gas produces a carbon radical. This is attacked by the nucleophile, and an electron is generated. This reduces the Cu^{+2} ion to Cu^{+1} ion.

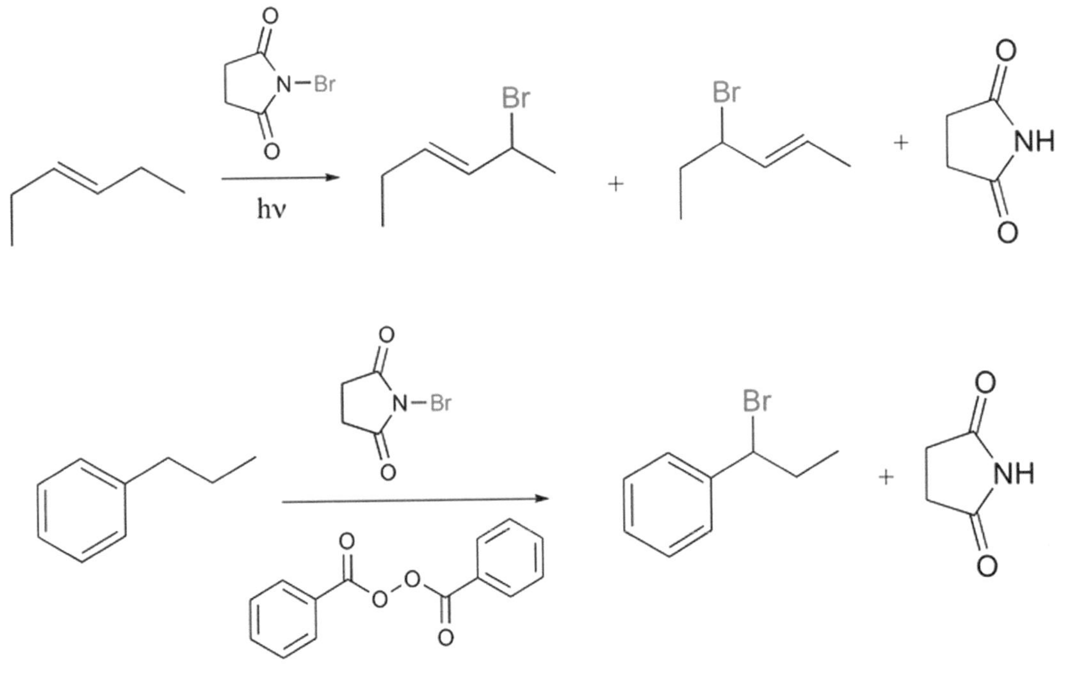

SCHEME 13.21   Wohl–Ziegler bromination.

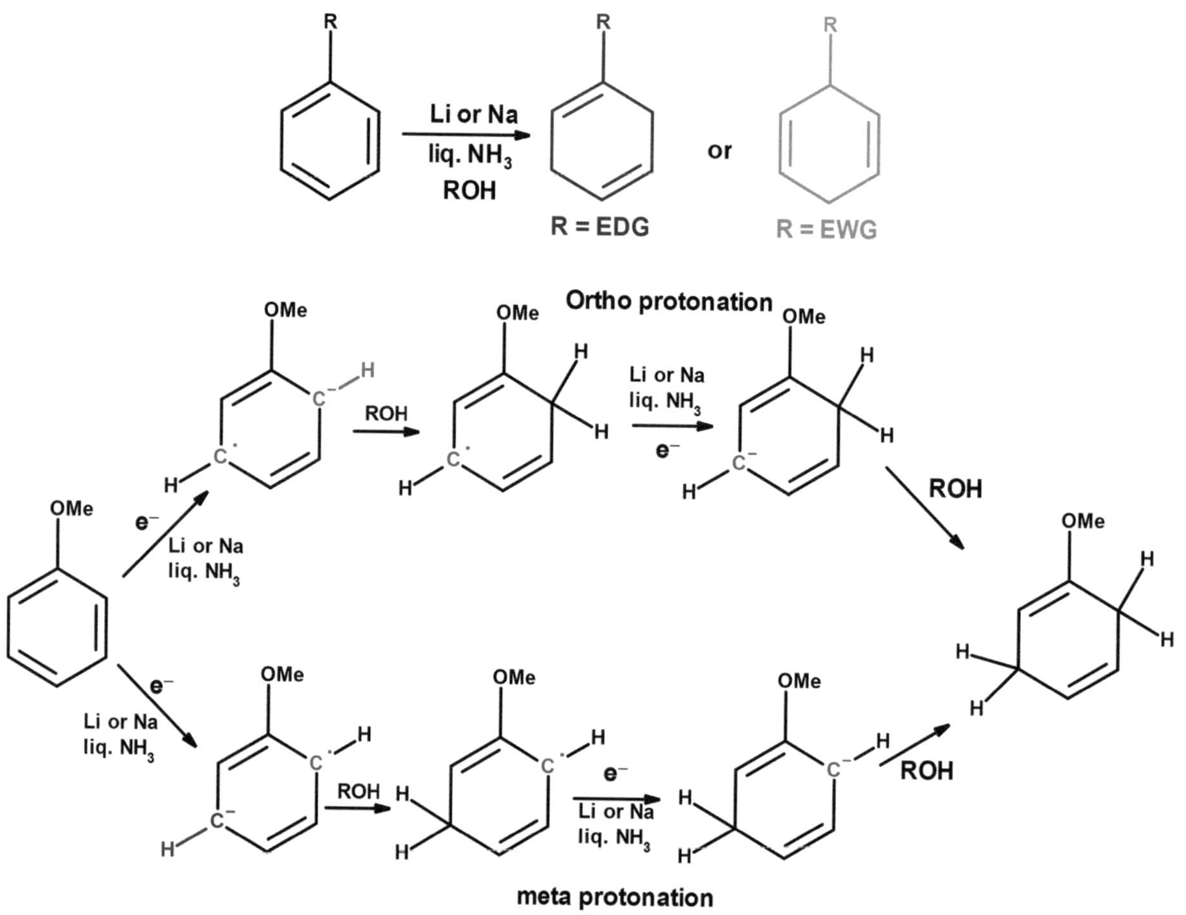

SCHEME 13.22   Birch reduction.

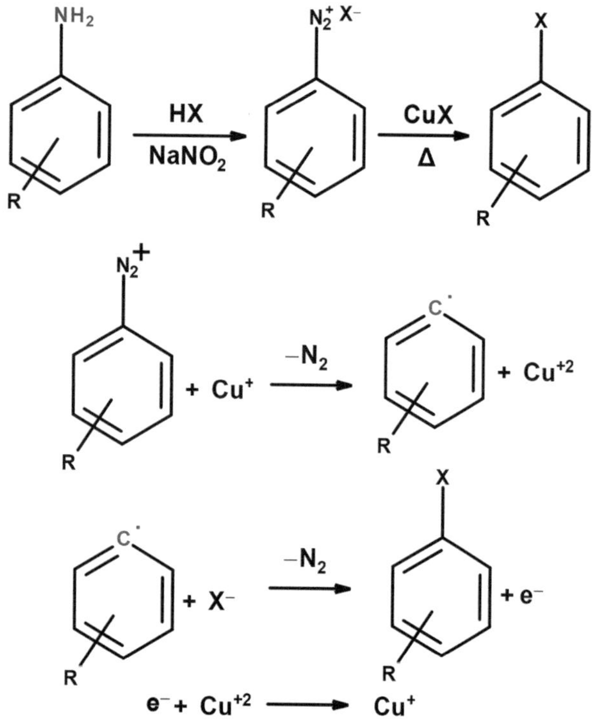

SCHEME 13.23   Sandmeyer reaction.

13.17 QUESTIONS

1. How do you compare between bond dissociation energy and bond energy.
2. What is Fenton reaction?
3. What is Kolbe electrolysis?
4. What is Bouveault–Blanc reduction? Write the mechanism for this reaction.
5. Write a brief note on alkyl radical structures.
6. Write a short note on how conjugation stabilizes free radicals.
7. What is the captodative effect?
8. Using suitable examples explain the kinetic stabilization of radicals.
9. What are the different factors that affect radical stability?
10. Briefly mention the various steps involved in free radical chain reactions.
11. Write a short note on atom abstraction in radical chemistry.
12. What are the different types of hydrogen abstraction? Explain the same with suitable examples.
13. How philicity of a radical is determined?
14. What are radical disproportionation reactions? Give some examples. How it differs from combination reactions?

15. Explain the mechanism of disproportionation reaction in radical chemistry.
16. Write a short note on anti-Markovnikov addition.
17. What groups favor Kolbe electrolysis.
18. Explain the mechanism of Bouveault–Blanc reduction.
19. What is Giese reaction? Explain using a suitable example.

BIBLIOGRAPHY

Ashenhurst, J. (2022). Bond Strengths and Radical Stability in Free Radical Reactions. (https://www.masterorganicchemistry.com/2013/08/14/bond-strengths-radical-stability/) (accessed, 15-Apr-2024)

Balaji Rao, R. 2.1: Baldwin's Rule for Ring Closure Reactions (https://chem.libretexts.org/Bookshelves/Organic_Chemistry/Book%3A_Logic_of_Organic_Synthesis_(Rao)/02._Rules_and_Guidelines_Governing_Organic_Synthesis/Baldwin%E2%80%99s_Rule_for_Ring_Closure_Reactions) (accessed, 15-Apr-2024)

Carey, F. A., & Sundberg, R. J. *Advanced Organic Chemistry, Part B-Reactions and Synthesis*, 4th Ed. Kluwer Academic/Plenum Publishers, 2000, 663.

Curran, D. P. (1991). 4.2 radical cyclizations and sequential radical reactions. *Comprehensive Organic Synthesis*, 4, 779–831.

Dziubla, T., & Butterfield, D. A. *Oxidative Stress and Biomaterials, Chapter One - A Free Radical Primer*. Academic Press, 2016, 1–33.

Feray, L., Kuznetsov, N., & Renaud, P. *Hydrogen Atom Abstraction. Radicals in Organic Synthesis*, Philippe, R., & Mukund P. (Ed.) Sibi Wiley-VCH Verlag GmbH, 2001, 246–278. https://doi.org/10.1002/9783527618293.ch39

Gibian, M. J., & Corley, R. C. (1973). Organic radical-radical reactions. Disproportionation vs. combination. *Chemical Reviews*, 73(5), 441-464.

Giese, B. (1983). Formation of CC bonds by addition of free radicals to alkenes. *Angewandte Chemie International Edition in English*, 22, 753–764. https://doi.org/10.1002/anie.198307531

Giese, B., et al. (1984). The scope of radical CC–Coupling by the "Tin method". *Angewandte Chemie International Edition in English*, 23, 69. https://doi.org/10.1002/anie.198400691

Giese, B., Kopping, B., Göbel, T., Dickhaut, J., Thoma, G., Kulicke, K. J., & Trach, F. (1996). Radical cyclization reactions. *Organic Reactions*, 753–856.

Giese Radical Addition (https://en.chem-station.com/reactions-2/2015/11/giese-radical-addition.html)

Henry Rzepa, The mechanism of the Birch reduction. Part 1: reduction of anisole (https://www.ch.imperial.ac.uk/rzepa/blog/?p=8452) (accessed, 15-Apr-2024)

Jiang, X., Li, X., & Wang, K. (1989). Reversal of the nature of substituent effect by changing the number of the. alpha.-substituent. Relative ease of formation of the three α-fluoromethyl radicals. *The Journal of Organic Chemistry*, 54(24), 5648–5650.

Kuang S. Chen and Jay K. Kochi (1974). The effects of a-halogen substitution on the configuration of alkyl radicals in solution by electron spin resonance. *Canadian Journal of Chemistry*, 52, 3529. (https://www.nrcresearchpress.com/doi/pdf/10.1139/v74-524) (accessed, 15-Apr-2024)

Ni, C. & Hu, J. (2016). The unique fluorine effects in organic reactions: recent facts and insights into fluoroalkylations. *Chemical Society Reviews*, 45, 5441. (https://pubs.rsc.org/en/content/articlepdf/2016/cs/c6cs00351f?page=search) (accessed, 15-Apr-2024).

Rauk, A. *Orbital Interaction Theory of Organic Chemistry*, 2nd Ed. John Wiley & Sons, Inc., 2001, 110.

Sandmeyer Reaction (https://www.organic-chemistry.org/namedreactions/sandmeyer-reaction.shtm)

Schäfer, H. J. (1991). Nonstabilized carbanion equivalents. *Comprehensive Organic Synthesis*, 3, 633–658.

Smith, M. B., & March, J. *March's Advanced Organic Chemistry Reactions, Mechanisms and Structure, Chapter 5: Carbocations, Carbanions, Free Radicals, Carbenes, and Nitrenes*, 5th Ed. John Wiley & Sons, Inc., 2001, 238.

Stuart W. McCombie, William B. Motherwell, Matthew J. Tozer (2012). The Barton-McCombie Reaction in Organic reactions (https://onlinelibrary.wiley.com/doi/10.1002/0471264180.or077.02) (accessed, 15-Apr-2024)

Sykes, P. *A Guidebook to Mechanism in Organic Chemistry, Chapter 11, Radicals and their Reactions*, 6th Ed. John Wiley & Sons, Inc., 1985, 299, 315.

Tanner, D. D., et al. (1973). Polar radicals. VI. Bromination reactions with molecular bromine and N-bromosuccinimide. Apparent anomalies and similarities. *Journal of the American Chemical Society*, 95(14), 4705–4711.

Van Hoomissen, D. J., & Vyas, S. (2017). Impact of conjugation and hyperconjugation on the radical stability of allylic and benzylic systems: A theoretical study. *Journal of Organic Chemistry*, 82, 5731–5742. https://pubs.acs.org/doi/pdfplus/10.1021/acs.joc.7b00549

Viehe, H. G., Janousek, Z., Merenyi, R., & Stella, L. (1985). The captodative effect. *Accounts of Chemical Research*, 18(5), 148-154.

Ballester, M. (1967). Inert carbon free radicals. *Pure and Applied Chemistry*, 15(1), 123–152.

14 Reactive Intermediates
Carbenes and Nitrenes

In this chapter, we will look at another reactive intermediate called carbenes. Carbenes are neutral species. The carbene carbon atoms have six valence electrons. So, they are divalent; they are directly bonded to only two other atoms and have no multiple bonds. Another term we will encounter is carbenoid. It is a carbene-like species which is bound to transition metals.

Here you can look at their electronic configuration and their hybridization. Carbocations have six electrons, carbon radical has seven electrons, and carbanion has eight electrons. So, if we compare the number of electrons, carbocations, and carbenes have six electrons. So, what is the difference between them?

Carbocations and carbenes have both six electrons. But carbocation is trivalent and carbene is divalent. We also see here other differences between carbocation and carbenes. Carbocations are charged species whereas carbenes are neutral. Carbenes have spin multiplicity, singlet, and triplet. Carbocations are trigonal planar, whereas carbenes are bent or linear. As you know carbocation acts like an electrophile whereas carbene can act either like electrophile or nucleophile depending on the substituents on the carbene carbon.

Singlet carbene has paired electrons whereas triplet carbene has unpaired electrons. In triplet carbene, the two electrons are kept in the orthogonal 'p' orbitals.

Two electrons are distributed in two nonbonding molecular orbitals (NBMOs) of the carbene and can have anti-parallel spins (singlet state) or parallel spins (triplet state). If the carbene is linear, the two NBMOs are degenerate (p_x, p_y), and the carbon center adopts sp hybridization.

Bending the molecule breaks such degeneracy and stabilizes the p_x orbital that acquires some 's' character, causing the carbon center to adopt sp^2 hybridization, whereas the p_y orbital remains almost unchanged (Figure 14.1).

14.1 TYPES OF CARBENES

14.1.1 ELECTRONIC CONFIGURATION OF CARBENES

Here we are looking at the ground and excited state carbenes. In the ground state, singlet carbene is paired, whereas in the excited state, it is not paired. On the contrary, the triplet ground state carbene is unpaired, and in the excited state, it is paired.

14.1.2 COMPARISON BETWEEN SINGLET AND TRIPLET CARBENES

In singlet carbene, the H-C-H bond angle is 103°, and the structure is similar to sp^2-hybridized carbon (bent). It has two unshared electrons in sp^2 orbital and the p orbital is unoccupied, it is diamagnetic, it reacts like electrophilic species. This undergoes concerted reaction. It reacts with *cis*-olefin to give *cis*-cyclopropane.

In triplet carbene, the H-C-H bond angle is 136°, and the structure is similar to sp hybridized carbon (linear). Unpaired electrons are in the two orthogonal 'p' orbitals, paramagnetic. It reacts like diradicals. This undergoes stepwise reaction. It reacts with *cis*-olefin to give *cis*- and *trans*- cyclopropane.

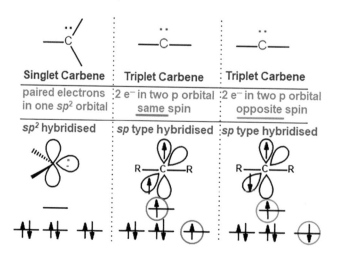

FIGURE 14.1 Differences between carbocation and carbenes.

DOI: 10.1201/9781032631165-14

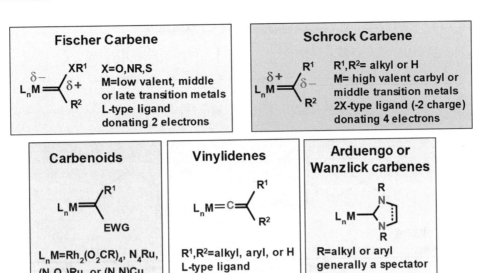

FIGURE 14.2 Types of carbene complexes.

14.1.3 TYPES OF CARBENE COMPLEXES

Carbenes can be stabilized by complexation with transition metals. Two extremes are known.

Carbene complexes of low valent/low oxidation state 18 e⁻ metals are electrophilic at carbon and are called Fischer carbenes (often behave like a glorified carbonyl group). Fischer carbenes, in which the carbene is bonded to a metal that bears an electron-withdrawing group (usually a carbonyl). In such cases, the carbenoid carbon is mildly electrophilic.

Carbene complexes of high valent/high oxidation state < 18 e⁻ metals are nucleophilic at carbon and are called Schrock carbenes and are bonded to a metal having an electron-donating group. They resemble Wittig reagent (which are not considered carbene derivatives).

Persistent carbenes, also known as Arduengo or Wanzlick carbenes include the class of *N*-heterocyclic carbenes (NHCs) and often are used as ancillary ligands in organometallic chemistry. Such carbenes are spectator ligands of low reactivity.

We also have carbenoids which are synthetically useful species (Figure 14.2).

14.2 SUBSTITUENT EFFECT ON CARBENES

Let us look at some of the substituent's effects on carbenes. Unstable singlet carbenes usually decompose/rearrange by alkene dimerization, 1,2-shifts, C–H insertion, fragmentation, or [2+1] cycloaddition to alkenes.

Attachment of π-donor and σ-acceptor groups (e.g., halogen atoms) stabilize the singlet state by raising the energy of the *p* orbital of the carbene, while the σ orbital remains basically unperturbed. In contrast, π-acceptor or conjugating groups (e.g., aryl groups) have the opposite effect, by lowering the energy of the *p* orbital of the carbene. Alkyl groups stabilize the singlet through hyperconjugative interactions with the empty $2p$ orbital. For bulky alkyl groups, this effect is less pronounced.

In the case of triplet carbene, there are two α-phenyl groups (they are diradical, highly reactive, and difficult to isolate). The triplet carbenes do not react with carbon–halogen bonds.

14.3 GENERATION OF CARBENES

In α-elimination, α-carbon loses a group without its electron pair, usually a proton, and then α-group with its pair, usually a halide ion: The most common example is the formation of dichlorocarbene by treatment of chloroform with a base and geminal alkyl dihalides with Me_3Sn^-, but many other examples are known.

14.3.1 FACTORS FAVORING 1,1 α-ELIMINATION

a. By powerful EWG (this increases the acidity of α proton and also stabilizes the –ve charge on the α carbon). In fact, when α bromine or iodine is present, they may not produce carbene as opposed to chlorine.

b. Use of strong bases.

c. The absence of β hydrogen (may or may not be). When chloroform is hydrolyzed by a strong base it produces dichlorocarbene. This undergoes a reaction with say, *cis*-but-2-ene to give the cyclopropane derivative.

Dihalocarbenes are formed when trihalomethanes are treated with a strong base, such as potassium *tert*-butoxide. The trihalomethyl anion produced on proton abstraction dissociates to a dihalocarbene and a halide anion.

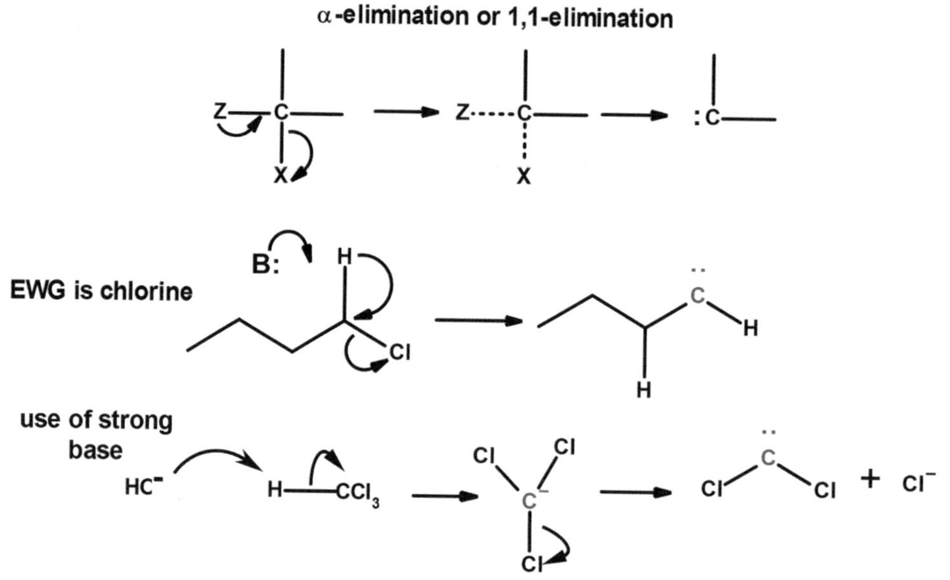

α-elimination or 1,1-elimination

EWG is chlorine

use of strong base

SCHEME 14.1 Generation of carbenes.

dibromomethylbenzene

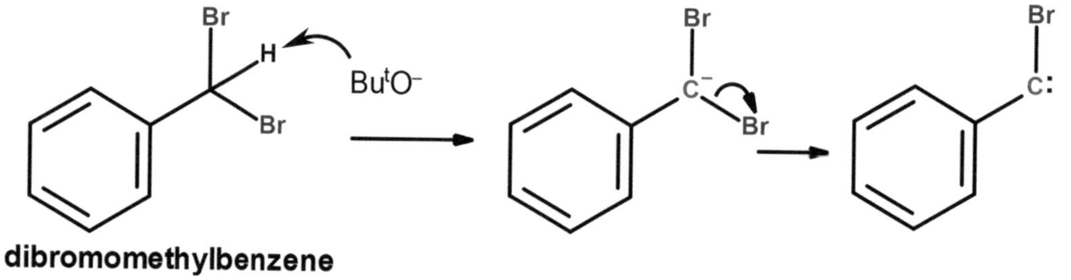

SCHEME 14.2 Generation of carbenes from dibromomethylbenzene.

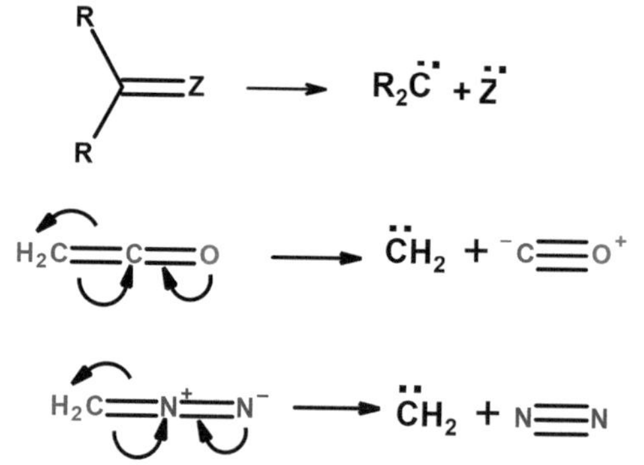

SCHEME 14.3 Disintegration of certain double bonds to form carbenes.

A carbenoid is formed when potassium *tert*-butoxide reacts with dibromomethylbenzene. This is similar to the earlier reaction. Instead of 3-halo substituents, we have one phenyl and two bromo groups.

Disintegration of compounds containing certain double bonds. The most important one is diazo decomposition.

Here we have some examples of diazo decomposition. In both cases the carbene carbon has either a phenyl group or an ester group. Since many of these carbenes are short-lived, they are never isolated. They are generated *in situ* and will be subjected to further reactions. When transition metals like Rh or Cu are used, we are getting carbenoids. They generally undergo C–H insertion or cyclopropanation reactions.

14.4 STABILITY OF CARBENES

Dialkyl groups stabilize carbenes. Electron donor substituents stabilize carbenes. Delocalization of the electron pair into 'p' orbital is shown by a double bond formed between the donor and the electron-deficient carbon center.

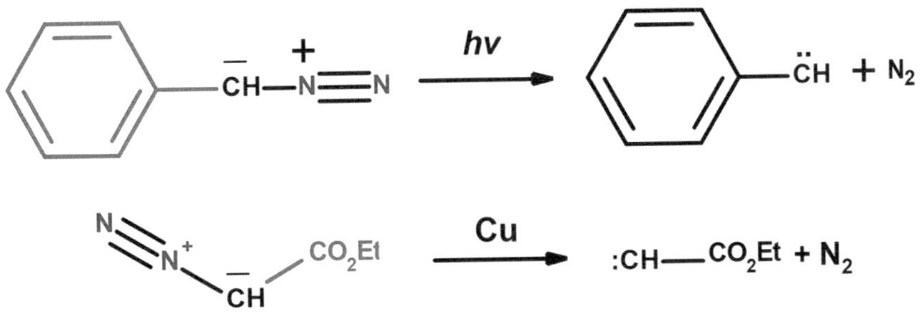

SCHEME 14.4 Diazo decomposition to form carbenes.

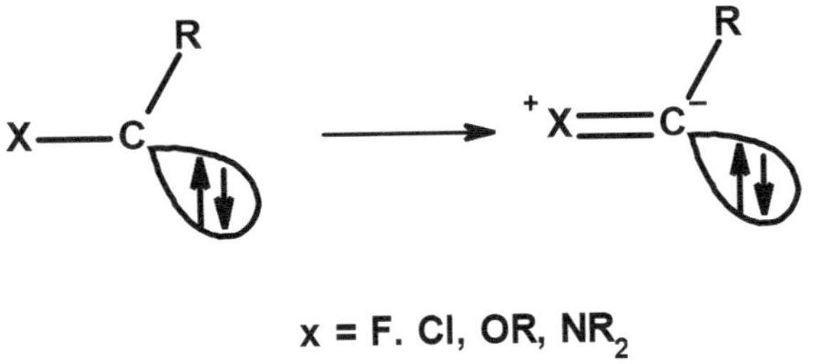

x = F. Cl, OR, NR$_2$

SCHEME 14.5 Stability of carbenes.

With dihalocarbenes, we have the interesting case of a species that resembles both a carbanion (unshared pair of electrons on carbon) and a carbocation (empty p orbital). Various questions come to our mind. Which structural feature controls its reactivity? Does its empty p orbital cause it to react as an electrophile? Whether its unshared pair make it nucleophilic?

Bonding in dihalocarbenes is based on the sp^2 hybridization of carbon. Two of carbon's sp^2 hybrid orbitals are involved in σ-bonds to the halogens. The third sp^2 orbital contains the unshared electron pair, and the unhybridized $2p$ orbital is vacant. From the electrostatic potential map, we can find out that the highest negative character is concentrated in the region of the lone pair orbital, and the region of the highest positive charge is situated above and below the plane of the molecule.

14.5 REACTIVITY OF CARBENES

Singlet carbenes react like electrophilic or nucleophilic species depending on the substituents on the carbene carbon atom. Whereas the triplet carbene behaves like a diradical.

The observation of stereospecific or nonstereospecific addition remains one of the most used experimental tests of singlet or triplet reactivity.

One thing you have to remember is, not all reactions are carried out in the ground state, because carbene generation requires, heating or photolysis. So, we may get excited states of both singlet and triplet carbenes.

14.5.1 STEREOCHEMISTRY OF CARBENE REACTIONS

If the singlet species adds to *cis*-2-butene, the resulting cyclopropane should be the *cis*-isomer since the movements of the two pairs of electrons should occur either simultaneously or with one rapidly succeeding another. However, if the attack is by a triplet species, the two unpaired electrons cannot both go into a new covalent bond, since by Hund's rule they have parallel spins. So, one of the unpaired electrons will form a bond with the electron from the double bond that has the opposite spin, leaving two unpaired electrons that have the same spin, and therefore cannot form a bond at once but must wait until, by some collision process, one of the electrons can reverse its spin. During this time, there is free rotation about the C—C bond, and a mixture of *cis*- and *trans*-1,2-dimethylcyclopropanes should result.

Depending upon the method of generation, some carbenes react as triplets, some as singlets, and others as singlets or triplets.

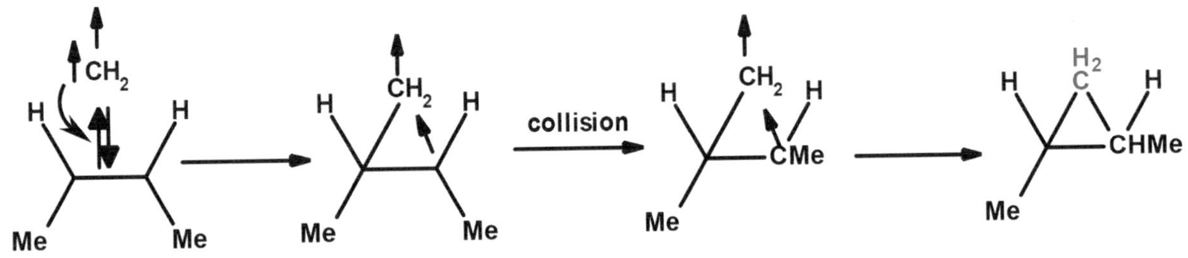

SCHEME 14.6 Stereochemistry of carbene reactions.

14.5.2 Simmons–Smith Reaction

When diiodomethane is treated with zinc–copper couple, a carbenoid is formed. This reacts with alkenes to give cyclopropanes.

14.5.3 Electrophilic Carbene

By comparing the rate of reaction of CBr_2 toward a series of alkenes with that of typical electrophiles toward the same alkenes (Table 14.1), we see that the reactivity of CBr_2 parallels that of typical electrophilic reagents such as Br_2 and peroxy acids. Therefore, dibromocarbene is electrophilic, and it is reasonable to conclude that electrons flow from the

π system of the alkene to the empty p orbital of the carbene in the rate-determining step of cyclopropane formation.

In insertion reactions, the carbene inserts itself between two atoms. Some of the examples of insertions include C-H, C-C, C-X, N-H, O-H, S-S, S-H, and M-C bonds. This insertion proceeds via a concerted mechanism. According to this mechanism, the carbene and the two atoms between which it is inserted can be visualized to have a three-atom centered transition state.

14.5.4 Nonstereospecific Addition of a Carbene

Note that the arrows used to show the flow of unpaired electrons (radicals) have only a halfhead. Moreover, the

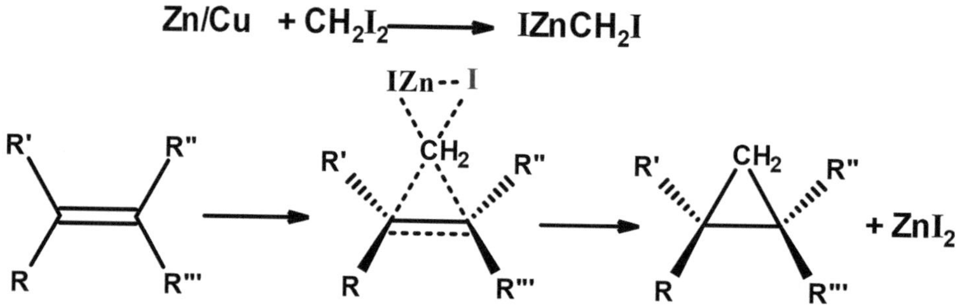

SCHEME 14.7 Simmons–Smith reaction.

TABLE 14.1
Reactivity of Electrophilic Carbenes

Alkene	CBr_2	Br_2	Epoxidation
$Me_2C = CMe_2$	3.5	2.5	Very fast
$Me_2C = CHMe$	3.2	1.9	13.5
$Me_2C = CH_2$	1.00	1.00	1.00
$(C_4H_9)CH = CH_2$	0.07	0.36	0.05

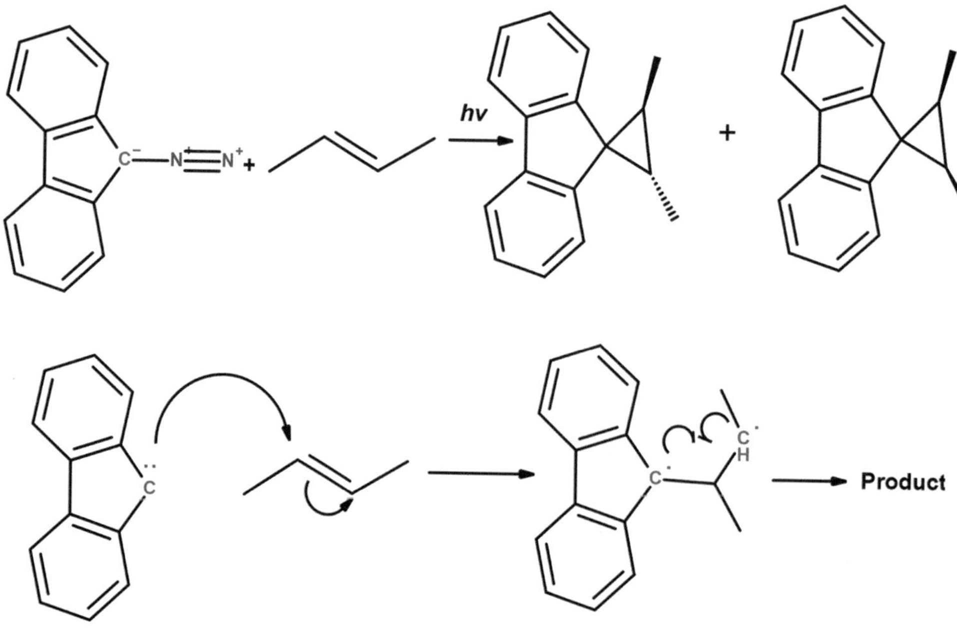

SCHEME 14.8 Nonstereospecific addition of a carbene.

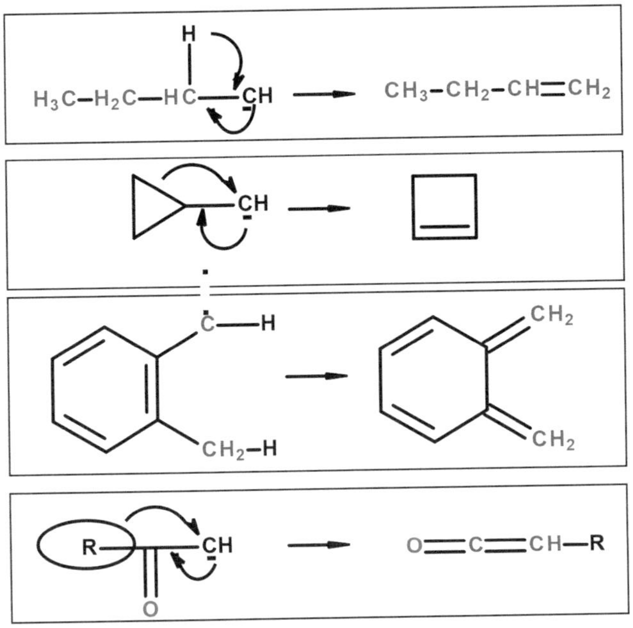

FIGURE 14.3   Few rearrangement reactions of carbenes.

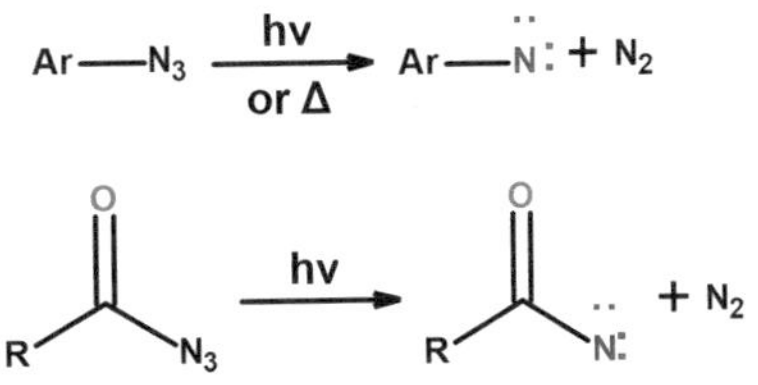

SCHEME 14.9   Formation of nitrenes.

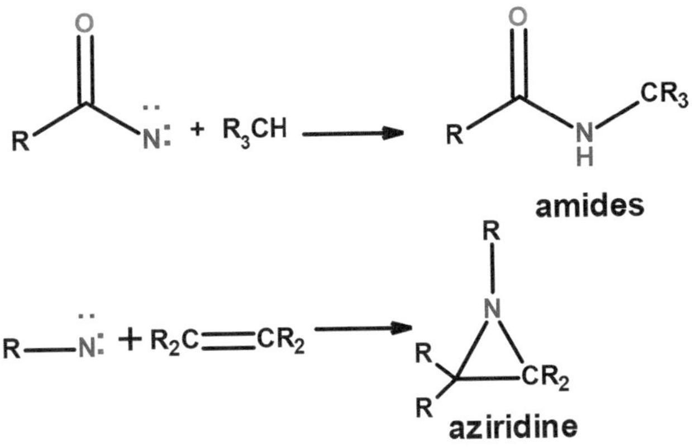

SCHEME 14.10   Few reactions of nitrenes.

intermediate in the reaction is a diradical. (Radicals are discussed in more detail in the previous chapter.) Rotation about the highlighted single bond takes place fast enough that the stereochemistry of the starting olefin is lost.

14.5.5   REARRANGEMENTS

They are (1) alkene formation, (2) ring expansion, and (3) Arndt Eistert synthesis. Here formation of the ketene is the important step. This ketene can be reacted with nucleophiles like water, amines, alcohols to give acids, amides, and esters, respectively (Figure 14.3).

14.6   NITRENES

Nitrenes are neutral species that contain a mono-coordinated nitrogen atom formally with four valence electrons.

14.6.1   FORMATION OF NITRENES

Common methods for generating nitrene intermediates are the photolysis and thermolysis of azides. Generation of nitrenes from acylazides can only be effected photochemically; thermolysis of an acylazide gives the corresponding isocyanate.

When we look at the stabilization of nitrenes based on substituents we have the following scenarios. Some X: substituents (i.e., halogens) are axially symmetric and do not lift the degeneracy of the $2p$ orbitals. Alkyl substituents are also approximately C_3 symmetric and probably would not lift the degeneracy enough to make the singlet state more stable.

Nitrene is very strongly basic and nucleophilic. The relatively small HOMO-LUMO gap suggests that such nitrenes may be colored and have a strong tendency to dimerize.

Z substituents also may yield nitrenes with singlet ground states, in this case by lowering the energy of the p system. 'C' substituents will not alter the $2p_\pi - 2p_\sigma$ gap appreciably.

Here a few reactions of nitrenes are given, they are insertion, elimination, and addition to C=C double bonds. When an acyl nitrene adds to a C–H bond we get amides. Similarly, when a nitrene is added to an alkene we get aziridines.

14.6.2   WOLFF REARRANGEMENT

It is generally accepted that a free carbene is an intermediate in the Wolff rearrangement. Here a diazoketone rearranges to a highly reactive ketene. The versatile ketene intermediate then reacts to give the carboxylic acid, ester, amide, and so on by reaction with water, alcohol, and amine, respectively.

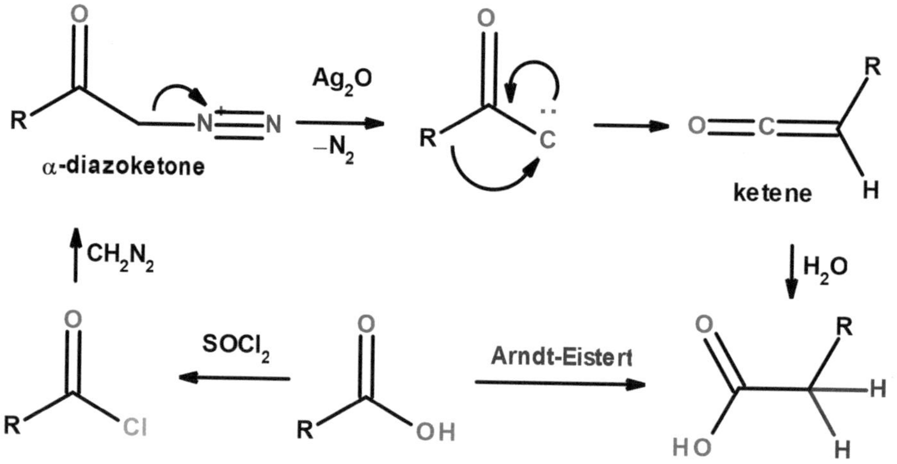

SCHEME 14.11 Wolff rearrangement.

14.7 QUESTIONS

1. Write a brief note on electronic configuration of carbenes.
2. Give a detailed comparison between singlet and triplet carbenes.
3. Mention a few methods by which carbenes can be generated.
4. Write a note on the stability of carbenes.
5. How stereochemistry is determined in the reactions of carbene.

BIBLIOGRAPHY

de Fremont (2008). "Synthesis of Well-Defined N-Heterocyclic Carbene (NHC) Complexes of Late Transition Metals". University of New Orleans Theses and Dissertations. 829. https://scholarworks.uno.edu/td/829 (accessed, 15-Apr-2024)

de Frémont, P., Marion, N., & Nolan, S. P. (2009). Carbenes: Synthesis, properties, and organometallic chemistry. *Coordination Chemistry Reviews*, 253, 862–892.

Grubbs, R. H., Trnka, T. M., & Sanford, M. S. (2003). Transition metal-carbene complexes in olefin metathesis and related reactions. *Fundamentals of Molecular Catalysis*, 34, 187–231. https://doi.org/10.1016/s1873-0418(03)80006-4.

Savin, K. A. (2014). Reactions involving acids and other electrophiles. *Writing Reaction Mechanisms in Organic Chemistry*, 2000, 161–235. https://doi.org/10.1016/b978-0-12-411475-3.00004-x

Smith, M. B., & March, J. *March's Advanced Organic Chemistry Reactions, Mechanisms and Structure, Chapter 5: Carbocations, Carbanions, Free Radicals, Carbenes, and Nitrenes.* John Wiley & Sons, Inc., 2001, 248.

Vega, E. M. Carbenes and Nitrenes in Reactive Matrices, Ph. D. thesis, RUHR Universitat Bochum, 2018, https://d-nb.info/1160442363/34

15 Carbon–Carbon Bond Formation Using Carbon Nucleophiles
Enolates, Enamines, and Enol Ethers

We will study nucleophilic additions involving carbonyl addition reactions. They are important building blocks in complex organic synthesis. We will be studying the addition of water, alcohol, thiols, amines, carbanions, and so on to carbonyl compounds and their mechanism of addition.

Why carbonyl groups are important? What are their synthetic utilities? How can we manipulate them? Since the carbonyl group has both nucleophilic and electrophilic sites, they can be easily converted to various other functional groups.

There are a few questions to be answered. What is the driving force? What are the factors that influence these reactions? What are kinetic and thermodynamic controls?

What is the driving force? Here we have an example of an amine reacting with a carbonyl compound to give an addition product. Look at the pK_a of the protons and heteroatoms. When carbonyl addition takes place in the presence of general acid-base catalysis two major changes drive the reaction namely change in (1) pK_a and (2) bonding. The amine proton has a pK_a of around 30 and the carbonyl oxygen around −4. In the intermediate that is formed after the addition, the pK_a of the proton attached to the positive nitrogen is around 8 and that of the oxide anion is nine. In the case of amino proton, the acidity increases, and in the case of carbonyl oxygen, the basicity increases. These two changes are huge, that is, the change in amino proton is around 22 pK_a units, and in the case of oxygen, it is 13 pK_a units. A sudden change of large pK_a leads to unstable intermediates or transition states, which cannot exist for a long time. Now, the acid or the base catalyst helps in either trapping the intermediate or pushing the reaction toward the product formation.

We have two major types of carbonyl addition reactions: (1) the nucleophile can add to the electropositive carbonyl carbon, and (2) the proton abstraction from the carbon adjacent to the carbonyl group.

In the first process, the carbonyl molecule acts as a Lewis acid, and in the second (when the molecule has hydrogen in the neighboring carbon), it acts as a Brønsted acid. In the first one, the sp^2-hybridized carbonyl carbon is converted to a sp^3-hybridized carbon after the nucleophilic addition, whereas in the second one, the carbonyl carbon retains its hybridization. Because of the presence of electronegative oxygen, the C=O double bond is polarized toward oxygen. This results in the development of a partial positive charge on the carbonyl carbon. It was observed that out of the two reactions, the second one is readily reversible and the first one is reversible in many instances. The presence of Lewis acid or acid catalyst favors coordination with carbonyl oxygen and hence makes it electrophilic. This, in turn, facilitates the nucleophilic attack on the carbonyl carbon or removal of proton from the neighboring carbon.

When we say catalysis by acids or bases, that does not mean that addition will not take place if acid or base is not present. In this chapter, our focus will be on this reaction. Even radical addition can happen where we have no acid or base.

15.1 COMPARISON OF C=C AND C=X (X = O, N, S)

Consider four types of double bonds: (1) C=C, (2) C=O, (3) C=N, and (4) C=S. Their approximate bond energies and geometries are given here (Figure 15.1).

The C=O bond is generally considered to be very strong, presumably because of its dipolar character, but compared with C=C, it is not very strong.

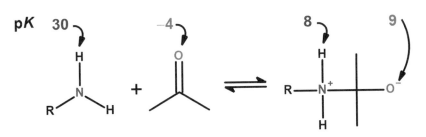

SCHEME 15.1 Amine reacts with a carbonyl compound.

DOI: 10.1201/9781032631165-15

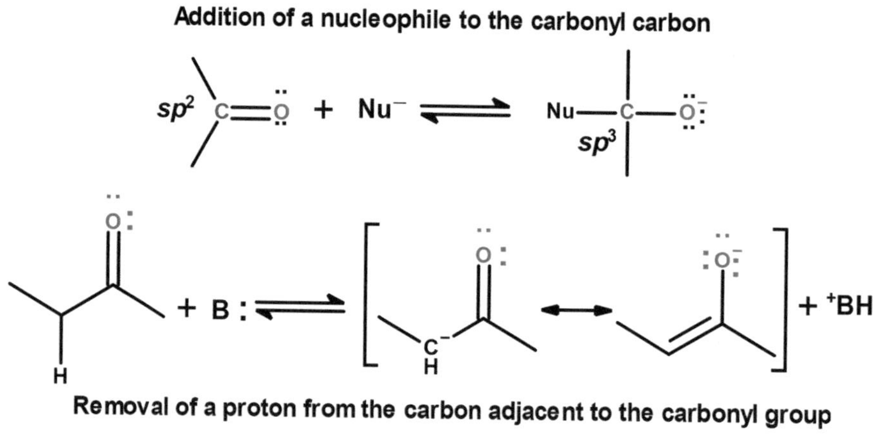

Addition of a nucleophile to the carbonyl carbon

Removal of a proton from the carbon adjacent to the carbonyl group

SCHEME 15.2 Addition of nucleophile and abstraction of hydrogen.

Type	BDE (kcal/mol)	Geometry
• C=C	• C=C 163-170	• C=C E/Z
• C=O	• C=O 173-181	• C=O does not exist
• C=N	• C=N 140-145	• C=N problem E or Z
• C=S	• C=S 135-140	• C=S does not exist

FIGURE 15.1 Comparison of C=C and C=X (X=O, N, S).

The striking difference in reactivity between C=C and C=O can be explained based on the electronegativity difference. In the case of alkenes, we see electrophilic additions occurring very readily, whereas in the case of carbonyls, nucleophilic additions occur more readily.

Due to the electronegativity difference between carbon and oxygen, nitrogen, or sulfur, C=O, C=N, and C=S bonds are always polarized resulting in electropositive carbon. This results in nucleophilic attacks on carbon and electrophilic attacks on oxygen, nitrogen, or sulfur.

Oxygen and sulfur belong to the same group. Based on the electronegativity of O and S, we can say the charge on the thiocarbonyl carbon is considerably smaller than that of the carbonyl carbon. In other words, the thiocarbonyl carbon will have reduced electrophilicity. Hence, its reduced reactivity toward nucleophiles.

15.2 NUCLEOPHILIC ADDITION VS SUBSTITUTION OF CARBONYL GROUPS

The principal question remaining in carbonyl addition is which attacks first, the nucleophile or electrophile. As in electricity, the electron flow takes place from the high potential to the lower potential, the nucleophilic addition to the carbonyl carbon is the first. The addition of proton in the last step is the electrophilic addition.

When A and X are H, R, or Ar, the substrate is an aldehyde or ketone. These substrates rarely undergo substitution. Because, H, R, and Ar are not good leaving groups. In the case of X if it is a heteroatom like OH, OR, NH_2, and so on due to the higher ability of these groups to leave, the nucleophilic addition does not take place very readily, rather substitution takes place. In short, we can say that the nature of X determines whether a nucleophilic attack at a carbon-hetero atom multiple bonds will lead to substitution or addition.

Competitive reactions:

* If A and/or X are electron-withdrawing substituents rates of nucleophilic attack increase and if they are electron-donating groups, then the rates of nucleophilic attack decrease.

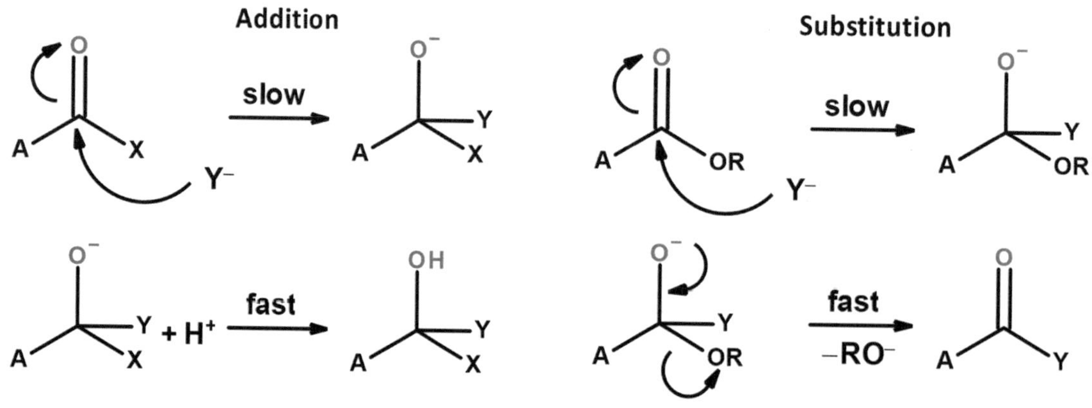

SCHEME 15.3 Addition vs substitution at the carbonyl group.

- Between aryl and alkyl groups, due to loss of resonance (that stabilizes the substrate molecule), on-going to the intermediate Aryl groups less reactive.
- Similar to aryl groups, double bonds in conjugation with the C=O also lower nucleophilic addition. There is also competition from 1,4 addition (Michael addition).
- Steric factor is responsible for the decreased reactivity of ketones compared with aldehydes.
- The initially formed product can undergo further elimination (H_2O) to give an alkene.
- The initially formed product can be attacked further by other nucleophiles.

15.3 TYPES OF CARBONYL ADDITION REACTIONS

We can broadly classify the carbonyl addition reactions into two major categories. One is hetero atom nucleophilic addition. and the other is carbon nucleophilic addition.

15.3.1 CLASS E AND CLASS N REACTIONS

In a class n reaction, the catalyst can abstract a proton from the nucleophilic reagent, and in a class e reaction, the catalyst donates a proton to the electrophilic reagent (in this case the aldehyde or ketone) in the first step and then removes it in the subsequent step. Thus, the base-catalyzed process is a class n reaction (removal of proton from nucleophile), while the acid-catalyzed process here is a class e reaction (donation of proton to electrophile).

In mechanism 1, water is the nucleophile. The catalyst (base, B) here abstracts a proton from the water molecule and makes ^-OH which attacks the carbonyl carbon. The initial hydrogen bonding between the catalyst B and the water molecule makes abstraction of the proton easier by the catalyst.

In mechanism 2, the catalyst BH is weakly hydrogen bonded to the carbonyl oxygen. This facilitates the transfer of the proton to the oxygen when the nucleophilic water attacks the carbonyl carbon. In this way, B and HB accelerate the reaction even beyond the extent that they form ^-OH or H_3O^+ by reaction with water.

15.3.2 MECHANISM OF BASE-CATALYZED HYDRATION

The base-catalyzed mechanism is a two-step process. In the first step, the nucleophilic hydroxide attacks the carbonyl group and is also the rate-determining step. The nucleophile is attached to the substrate which results in alkoxide ion as the product. The second step is the fast abstraction of a proton from the solvent by the alkoxide ion resulting in a gem diol.

Neutral water molecule can very easily attack a full-fledged carbocation compared with the partially electropositive carbon in the carbonyl group. The use of base catalyst which abstracts a proton from the neutral water molecule to form ^-OH enhances the nucleophilic strength as well as the rate of addition (Figure 15.2).

The following table gives the K_{hyd} values for few carbonyl compounds.

Increasing stabilization of the carbonyl group leads to decreasing K for hydration. From the table, we can find that the hydration occurs very readily for formaldehyde and hexafluoroacetone. The highly electron-withdrawing trifluoromethyl groups pull the electrons away from the carbonyl group which destabilizes the carbonyl group. Due to this the carbonyl group is prone to addition because of greater equilibrium constant.

This observation can be explained by steric and electronic parameters. Equilibrium constants correlate with Taft's inductive and steric parameter (Chapter 10.7).

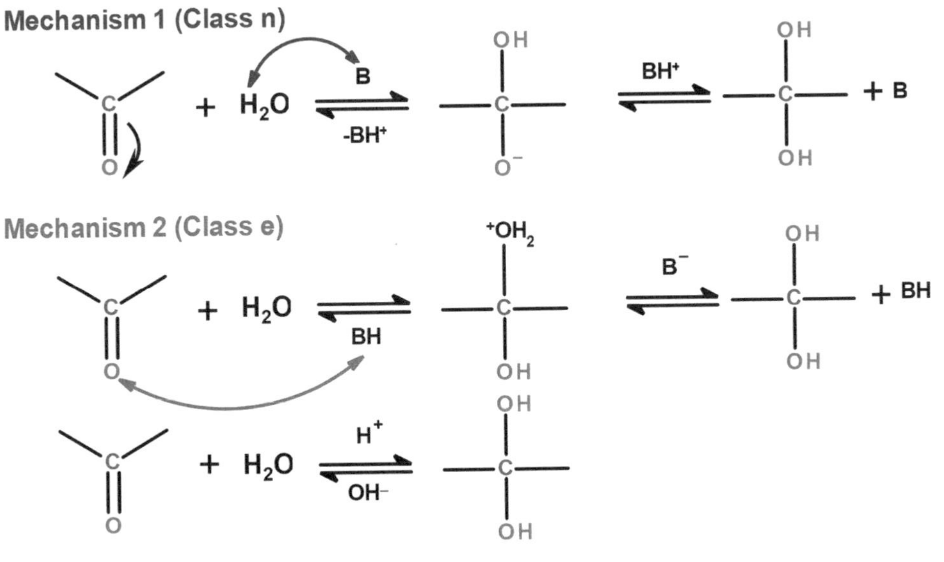

SCHEME 15.4 Class n and class e reaction.

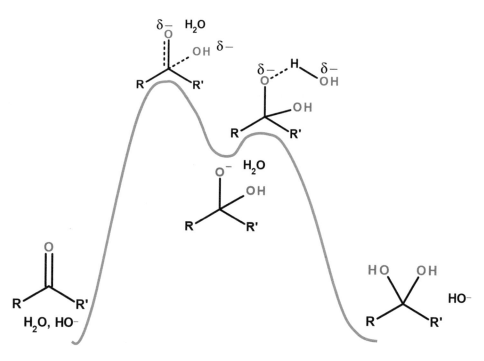

FIGURE 15.2   Mechanism of base-catalyzed hydration.

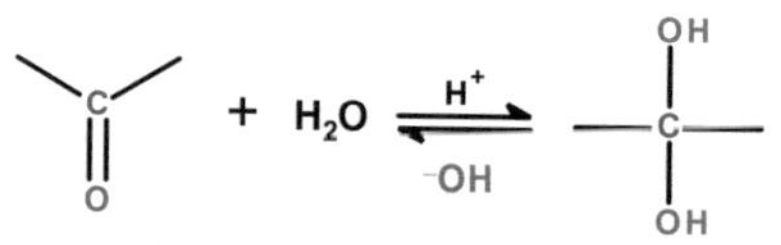

$$\text{C=O} + H_2O \underset{\overset{-OH}{\rightleftharpoons}}{\overset{H^+}{\rightleftharpoons}} \begin{matrix} OH \\ | \\ -C- \\ | \\ OH \end{matrix}$$

$$K_{hydr} = [\text{hydrate}]/[\text{C=O compound}]$$

Carbonyl compound	K_{hydr}	% conversion
H-(C=O)-H	2300	> 99.9
CH$_3$-(C=O)-H	1.0	50
CH$_3$-(C=O)-CH$_3$	0.0014	0.14
CF$_3$-(C=O)-CF$_3$	22,000	>99.999

FIGURE 15.3   K for hydration.

- Carbonyl addition is more favorable when electron-withdrawing groups are present on the carbonyl carbon (larger K_{eq}).
- Carbonyl addition is less favorable when electron-donating groups are present on the carbonyl carbon (smaller K_{eq}).
- Carbonyl addition is less favorable when there is conjugation along with the carbonyl group (smaller K_{eq}).
- Large size of groups attached to the carbonyl carbon makes addition less favorable (smaller K_{eq}).

15.3.3   INDUCTIVE EFFECT AND HYDRATION

When we move from formaldehyde to trimethylacetaldehyde, we observe a decrease in the hydration constant (Figure 15.4).

Why increasing size reduce hydrate formation? The hybridization at the carbonyl carbon changes from sp^2 in the reactant (aldehyde or ketone) to sp^3 in the product (hydrate). In other words, the bond angles at the carbonyl carbon change from 120° to 109.5°. Due to bond angle reduction, the groups are now placed in close proximity which leads to crowding in the hydrate. This results in small K_{hydr} when the groups are bigger.

15.3.4   ORBITAL APPROACH TO CARBONYL ADDITION

In carbonyl compounds, π^* antibonding orbital is the LUMO. This π^* antibonding orbital is polarized toward the carbon atom (Figure 15.5).

The carbonyl carbon can have various substituents. If X: substituents are present, they will raise the energy of the LUMO and if Z and "C" substituents are present, they will lower the LUMO energy. The nucleophile will approach the carbonyl carbon from the direction that is perpendicular to the plane of the carbonyl group. The product of the reaction, the "tetrahedral intermediate," has an σ-bond between the carbon and the Nu. As we have seen earlier with respect to addition vs substitution, the nature of the substituents on the carbonyl carbon decides what will be the course of the reaction. For aldehydes and ketones, where we do not have a good leaving group, addition takes place. The approach of the nucleophile to the carbon via a path perpendicular to the planar face of the carbonyl is relatively unhindered. If bulky groups are present on the carbonyl compounds, due to steric hindrance, the reaction may not proceed efficiently. This is one of the reasons, why aldehydes undergo nucleophilic additions much faster than ketones.

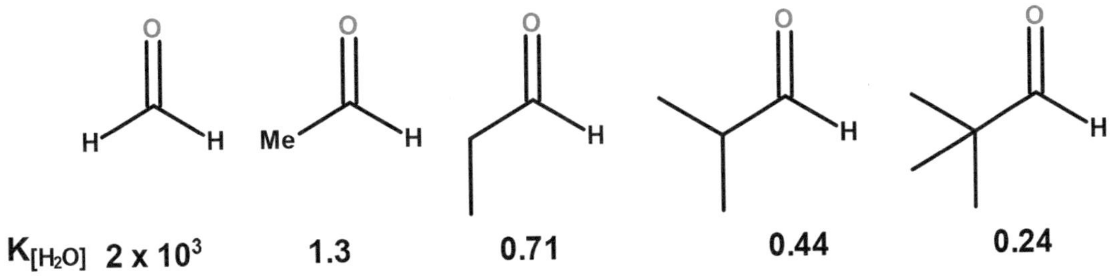

FIGURE 15.4   Inductive effect and hydration.

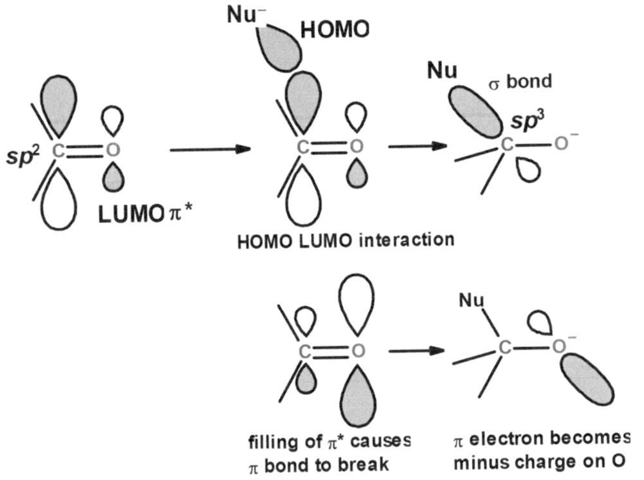

FIGURE 15.5   Orbital approach to carbonyl addition.

15.4   EXAMPLES OF CARBONYL ADDITION REACTIONS

15.4.1   ADDITION OF ALCOHOLS AND THIOLS TO ALDEHYDES AND KETONES

Aldehydes and ketones, when reacted with alcohols in the presence of acid catalysts (Lewis acids – $TiCl_4$) give acetals and ketals, respectively. Cyclic acetals like dioxolanes are formed when ethylene glycol is used as an alcohol. The reaction may be performed under microwave irradiation with *p*-toluenesulfonic acid as a catalyst. Due to the reversibility of this reaction, the acetals and ketals can be hydrolyzed by treatment with acid. Applying Le Chatelier principle, to prepare ketals, or acetals of larger molecules, the equilibrium is shifted to the right by the continuous removal of water. Azeotropic distillation, ordinary distillation, or the use of a drying agent such as Al_2O_3 or a molecular sieve are some of the methods by which water can be removed.

15.4.2   ADDITION OF AMINES TO ALDEHYDES AND KETONES

In contrast to ammonia, primary, secondary, and tertiary amines can add to aldehydes and ketones to give different kinds of products. Primary amines give imines. When there is an aryl group, the compounds are quite stable. They are

usually called Schiff bases, if the neighboring atom has hydrogen, then water or RNH_2 can be lost in that direction to give an enamine.

15.4.3   ADDITION OF HYDRIDE AND ALKOXIDE TO ALDEHYDES AND KETONES

Consider, two simple carbonyl addition reactions by (1) hydride and (2) alkoxide. The $LiAlH_4$ reduction of carbonyl compounds gives alcohols, where the hydride ion adds to the carbonyl carbon. In the case of alkoxide ion, the nucleophile adds to the carbonyl carbon giving the product.

15.4.4   CARBONYL ADDITION REACTIONS WITH HCN

Neutral HCN, being a poor nucleophile and less reactive, is seldom used. On the other hand, if we use the cyanide nucleophile it is, more reactive. The reaction is base catalyzed, and is also reversible. In this reaction addition of ^-CN is the rate-determining step. Both RCHO and RCOR reacts well. If we use ArCHO, the reaction is called Benzoin condensation (Chapter 12.1.1). If it is aryl ketones, they do not react like their aliphatic counterparts.

Converting aldehydes and ketones to cyanohydrins is of synthetic value for two reasons: (1) a new carbon–carbon bond is formed and (2) the cyano group in the product can be converted into a carboxylic acid (CO_2H) by hydrolysis or to an amine by reduction.

15.4.5   CARBONYL ADDITION WITH ACTIVE METHYLENE COMPOUNDS

In the aldol reaction, the α carbon of one aldehyde or ketone molecule adds to the carbonyl carbon of another. The entire reaction is an equilibrium (including the dehydration step), and α,β- unsaturated and β-hydroxy aldehydes and ketones can be cleaved by treatment with water.

The product is a β-hydroxy aldehyde (called an aldol) or ketone, which in some cases is dehydrated during the course of the reaction.

1. When two molecules of the same aldehyde react, we get self-condensed products.
2. When two molecules of the same ketone react, the self-condensed reaction hardly proceeds.

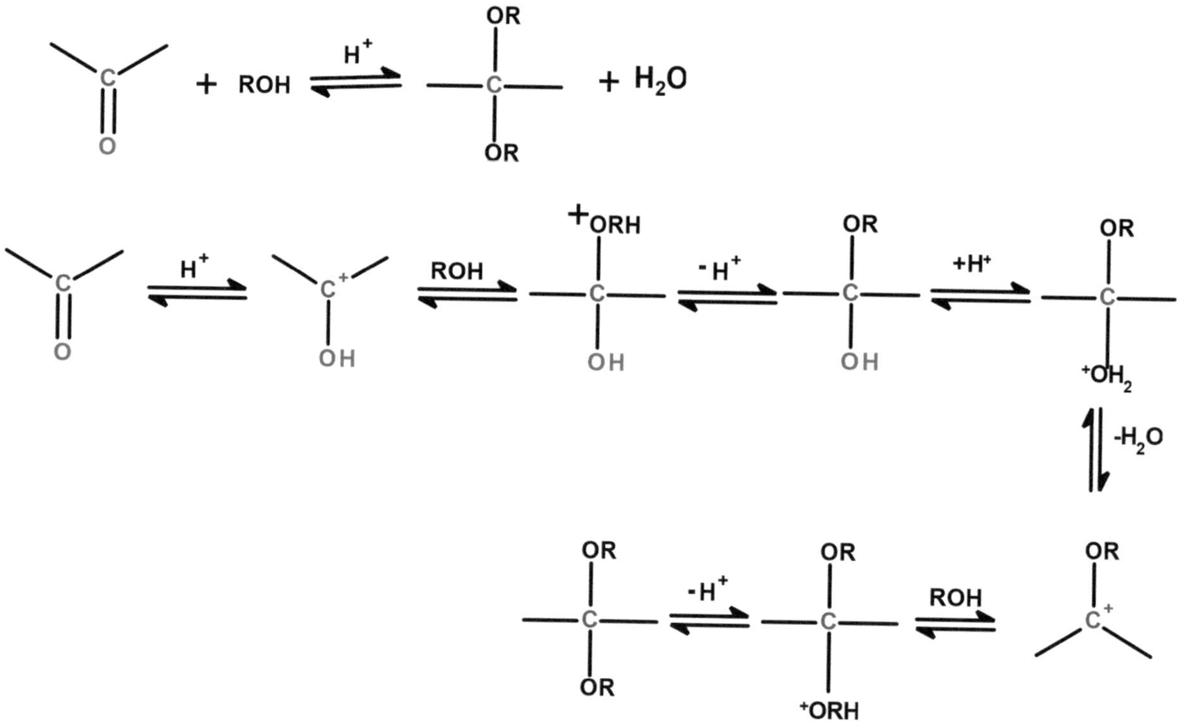

SCHEME 15.5   Addition of alcohols to aldehydes and ketones.

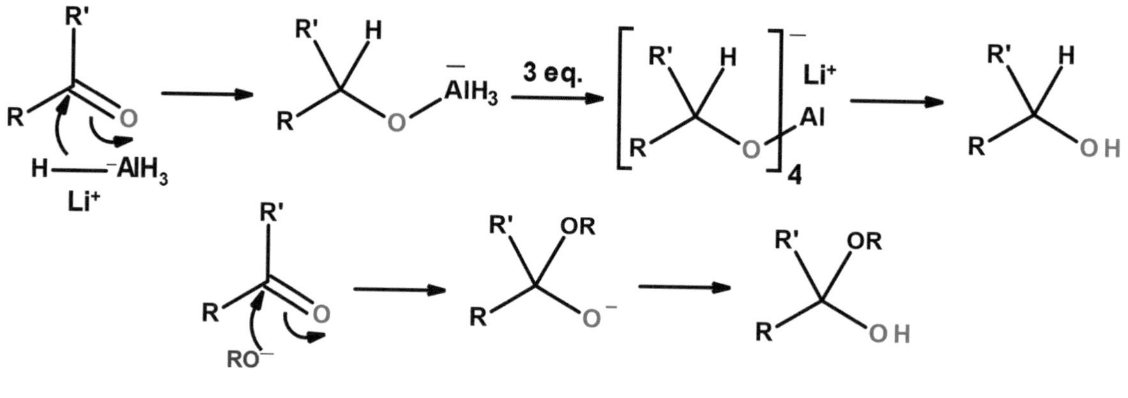

SCHEME 15.6   Addition of hydride and alkoxide to aldehydes and ketones.

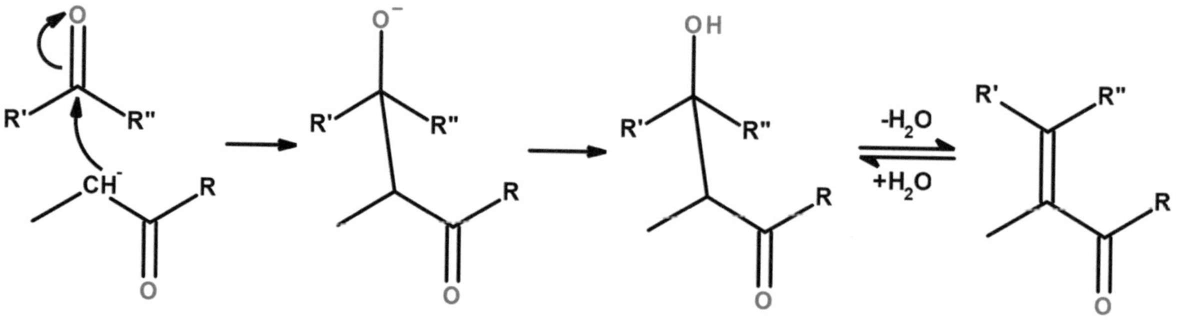

SCHEME 15.7   Carbonyl addition with active methylene compounds.

3. When two different aldehydes react, we get two self-condensed as well as two cross condensed products. The Claisen–Schmidt reaction is an example of crossed aldol reaction.
4. When two different ketones react, we hardly get any product.
5. Reaction between an aldehyde and a ketone, is of practical use only when the aldehyde has no α hydrogen. Otherwise, self-condensation of the aldehyde only takes place.
6. If the aldehyde has an α hydrogen, it is the α carbon of the ketone that adds to the carbonyl of the aldehyde. The reaction can be made regioselective by preparing an enol derivative of the ketone separately and then adding this to the aldehyde (or ketone), which assures that the coupling takes place on the desired side of an unsymmetrical ketone.

15.5 BASE-CATALYZED CONDENSATION REACTIONS

Here is the overview of many base-catalyzed condensation reaction of carbonyl compounds (Table 15.1).

15.5.1 Cannizzaro Reaction

Disproportionation of aldehydes lacking α hydrogen into an acid anion and an alcohol in the presence of a base is called Cannizzaro reaction. Hydride ion transfer takes place from one aldehyde to another aldehyde without an intermediate

formation (no D incorporation with D_2O); for example, HCHO, R_3CCHO, and ArCHO. We get an oxidized and reduced product from aldehyde.

15.5.2 Grignard Reaction

In Grignard reaction, the alkyl or aryl carbon of the halide acts as the nucleophile. The sp^2-hybridized carbonyl carbon is converted to sp^3-hybridized carbon. Here we see a simple one-step mechanism. We also have another cyclic six-membered TS mechanism (Figure 15.6).

If the Grignard reagent has β hydrogen or the carbonyl compound has a proton adjacent to the carbonyl carbon, one of the products is an alkene as shown here.

15.6 ENOLATE ANIONS

Why study about enols and enolates? Enolate anions are an important class of carbanions. Enolate is nucleophilic due to oxygen's small atomic radius and the presence of a formal negative charge. An important tool for an organic chemist will be a new carbon–carbon bond formation by the reactions of enolate ions with C-centered electrophiles.

15.6.1 Acid-Catalyzed Halogenation of Carbonyl Compounds

Aldehyde/ketone+halogen à α-halo aldehyde/ketone+HX

The reaction is acid catalyzed. In this reaction, only the α-hydrogen is replaced by halogen. The reaction can be

TABLE 15.1
Base-Catalyzed Condensation Reactions

Name reaction	Substrate 1	Substrate 2	Product
Aldol condensation	Enol or an enolate ion	Carbonyl compound	Aldol/conjugated enone
Knoevenagel condensation	Active methylene	Carbonyl compound	αβ-unsaturated ketone
Claisen condensation	Ester	Ester/Carbonyl compound	β-keto ester or a β-diketone
Dieckmann	Diesters	–	Cyclic β-keto esters
Claisen–Schmidt condensation	Aromatic carbonyl compound (no α-H)	Carbonyl compound with α-hydrogen	Dibenzylidene acetone
Perkin reaction	Acid anhydride	Aromatic aldehyde	Cinnamic acids
Henry reaction	Nitroalkane	Carbonyl compound	Nitro-aldol

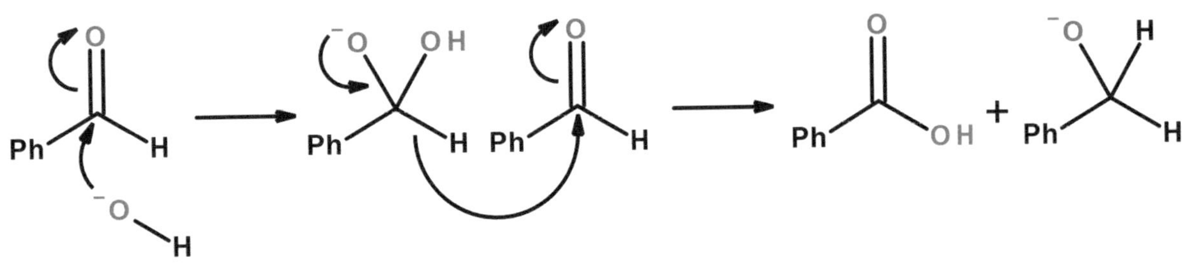

SCHEME 15.8 Cannizzaro reaction.

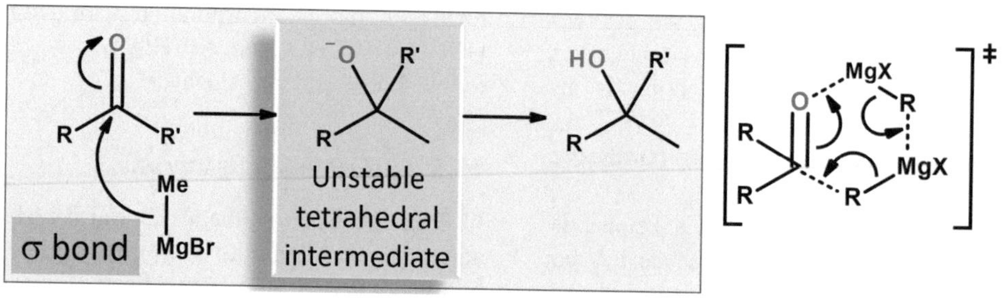

FIGURE 15.6   Grignard reaction.

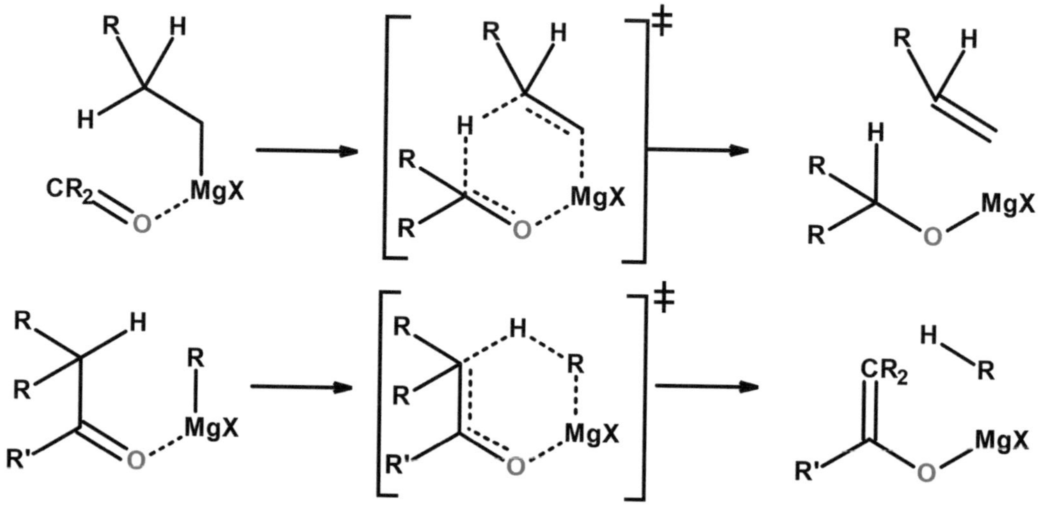

SCHEME 15.9   Alkene formation from Grignard reaction.

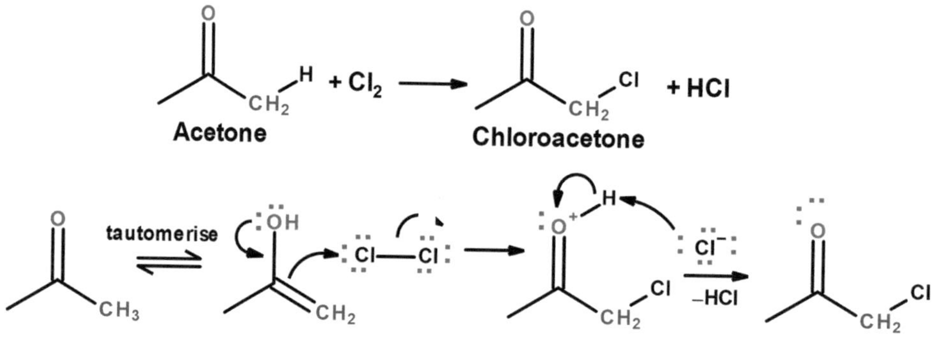

SCHEME 15.10   Acid-catalyzed halogenation of carbonyl compounds.

carried out in protic or aprotic solvents. It was found that the rate of the reaction is the same for all the halogens (first order). Moreover, the rate is independent of the concentration of halogen. In other words, the halogen is not involved in the RDS.

The reaction starts with tautomerization of the ketone. Once the enol is formed, this alkene double bond undergoes electrophilic substitution. The enolic proton is finally abstracted by the halide to give HX. The initial formation of enol is the RDS. This enol reacts rapidly with halogen. A lone pair of oxygen is crucial for this reaction.

15.6.2   BASE-CATALYZED HALOGENATION OF CARBONYL COMPOUNDS

Aldehyde/ketone + halogen à α-halo aldehyde/ketone + HX

Consider halogenation under basic catalysis. Similar to the acid-catalyzed reactions, here also α-hydrogen alone is replaced. Unlike acid-catalyzed reactions, when methyl ketones are used, the end product is *haloform* and not trihalomethyl ketone. The rate of reaction is the same for all

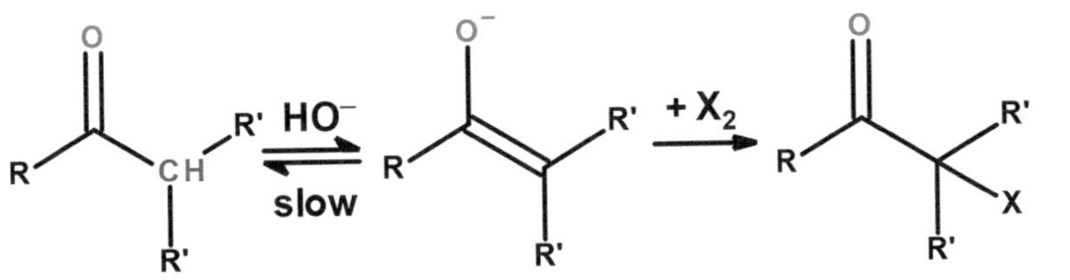

SCHEME 15.11   Base-catalyzed halogenation of carbonyl compounds.

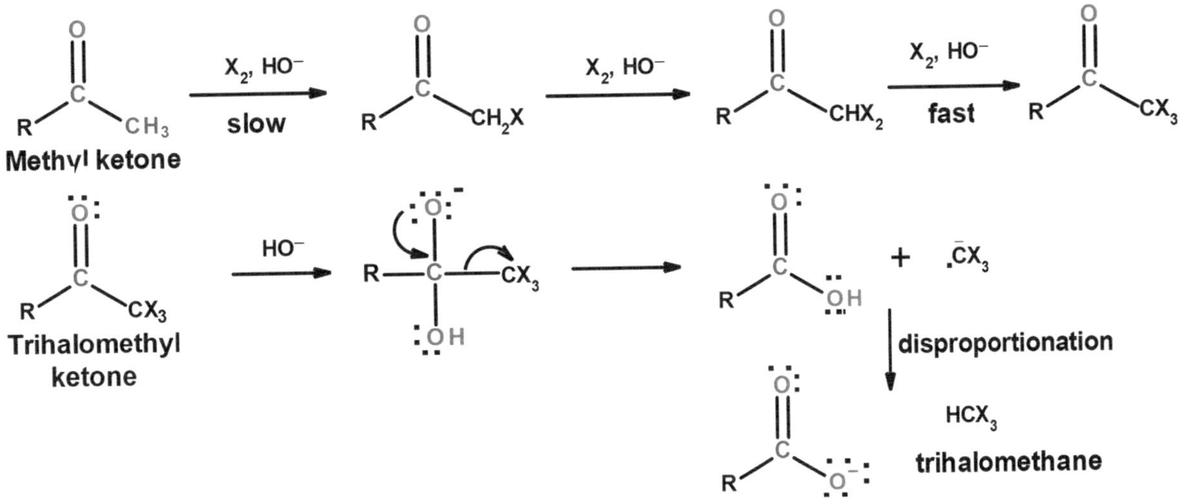

SCHEME 15.12   Haloform reaction.

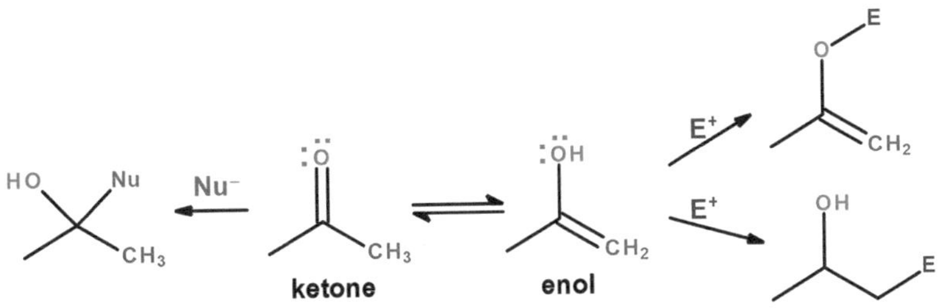

SCHEME 15.13   C vs O reactivity of enols.

halogens (first order). The rate is independent of the concentration of halogen (not involved in RDS).

When we compare these two reactions it was observed that the base-catalyzed reaction is much faster than the acid-catalyzed reaction.

15.6.3   HALOFORM REACTION

As it was mentioned earlier, when methylketones are used, we get trihalo-substituted ketones. This is because the addition of each halogen increases the acidity of the alkyl proton and so is easily removed.

The hydroxide ion present in the medium reacts with the trihalo derivative to give the carboxylate anion and

haloform. Since a free carboxylic acid cannot exist in alkaline medium, the initially formed carboxylic acid loses a proton to trihalo carbanion to give the carboxylate ion and haloform.

Just to give a brief overview, ketone and enol exist as tautomers, depending on the conditions. Ketone has electrophilic carbonyl carbon, which can undergo nucleophilic addition. Enol is nucleophilic, so it can undergo electrophilic addition. Due to the ambident nature of the enols, we have C vs O reactivity, in other words, we can have two different products. When the ketone reacts, we have a change in geometry from planar to tetrahedral. In the case of enols, the geometry changes only when the reaction takes place at the carbon.

208

A Foundation Course for College Organic Chemistry

15.6.4 Keto-Enol Tautomers

By virtue of its powerful electron-withdrawing nature, the carbonyl group increases the acidity of protons on the adjacent carbons. The presence of keto-enol tautomers can be confirmed by NMR. When we look at the NMR of dimedone, there is a broad signal around δ 8.15, representing the presence of an OH group, and a sharp singlet at δ 5.5 in the olefinic region. While the ^{13}C spectrum shows *two different sp^2 carbon atoms*. In the keto-enol tautomer, one of the methylene hydrogens is transferred to the oxygen. Although there was a shift of proton from carbon to oxygen, there is no net change in the pH. A good example of stabilization gained by 1,3-dicarbonyl compound is given by ethyl acetoactate. As shown here, there is an extension of conjugation between the olefinic bond and the carbonyl bond. There is also a cyclic six-membered hydrogen bonding between the enolic proton (O*H*) and the carbonyl oxygen. Of course, you are aware that keto and enol forms are isolable entities and they are not resonance structures (Figure 15.7).

Here we will focus on enols and enolates only. Later, we will be studying about enol ethers separately.

The base abstracts a proton from the α-carbon of the carbonyl resulting in the formation of an anion. Due to the presence of the carbonyl group, the electronegative oxygen pulls the electron toward itself, thereby enhancing the stability of the anion. The resonance forms contain enolate ions. If the protonation occurs before enolate formation, then we will get back the ketone (C–H bonds are not very acidic).

We already know, the presence of an electron-withdrawing group, increases the acidity of the proton. So, when we compare alkanes and aldehydes or ketones, the latter compounds have more acidic protons due to the presence of the electron-withdrawing carbonyl group. In fact, it was observed that pK$_a$'s for enolate formation from simple aldehydes and ketones are in the 16–20 range. We can explain this based on the presence of electronegative oxygen in the neighborhood of the anion. Due to the presence of oxygen, there is delocalization of the negative charge in aldehydes and ketones. Hence those α-protons are more acidic. In the case of alkanes, we need strong bases like n-BuLi to remove the acidic proton. But for aldehydes and ketones due to higher acidity, we can employ bases like hydroxide ions or alkoxide ions to generate enolate ions.

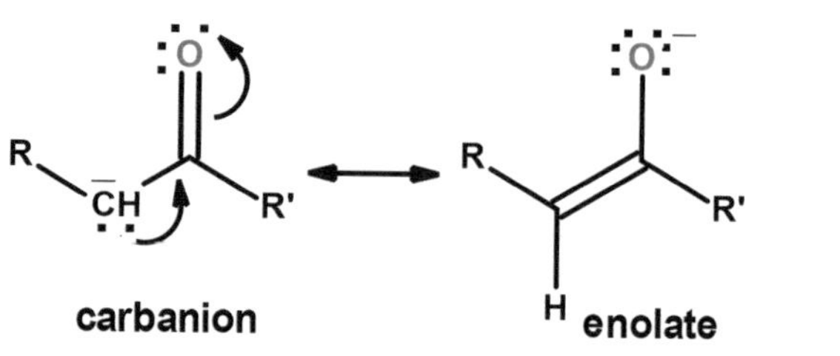

carbanion   **enolate**

SCHEME 15.14   Enolate formation.

When we look at 1,3-dicarbonyl compounds, they are even more acidic. The central methylene group is flanked by the two electron-withdrawing carbonyl groups. This increases the acidity of the proton. Here we observe enhanced stabilization of the negative charge due to the presence of two carbonyl both of which can stabilize the negative charge.

15.6.5 Stereochemical Outcome of Enolization

The only limitation of enolate formation is that the α-carbon loses the chirality. Due to the formation of enolate, the α-carbon geometry changes from tetrahedral to planar.

15.6.6 Choice of Base

We need to choose the base according to pK$_a$ of the proton to be abstracted. We need to allow the equilibrium to be established. To establish the equilibrium, the base used should not completely deprotonate the carbonyl compound. The base should be non-nucleophilic (n-BuLi vs LDA). Because, enolates are nucleophiles and ketones are electrophiles. Therefore, there is always the potential problem of self-condensation. The use of a non-nucleophilic base is important to avoid potential chemoselectivity problems. Given below are some non-nucleophilic bases (Figure 15.8).

Generally, bulky bases where the nucleophilic center is surrounded by bigger substituents are non-nucleophilic (LDA, LHMDS). We need to carry out the reaction at low temperatures. Sodium hydride (the pK$_a$ of its conjugate acid is 35) and K-*tert*-butoxide (pK$_a$ of the conjugate acid is 17) are some useful bases that are employed in organic chemistry.

When I say sodium hydride is a base, but LiAlH$_4$ or NaBH$_4$ is a reducing agent. You may be wondering, why NaH is not acting like a reducing agent.

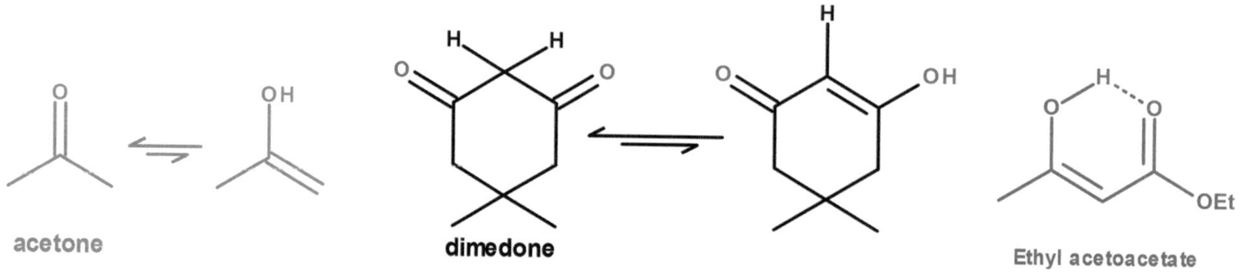

acetone   **dimedone**   **Ethyl acetoacetate**

FIGURE 15.7   Keto-enol tautomers.

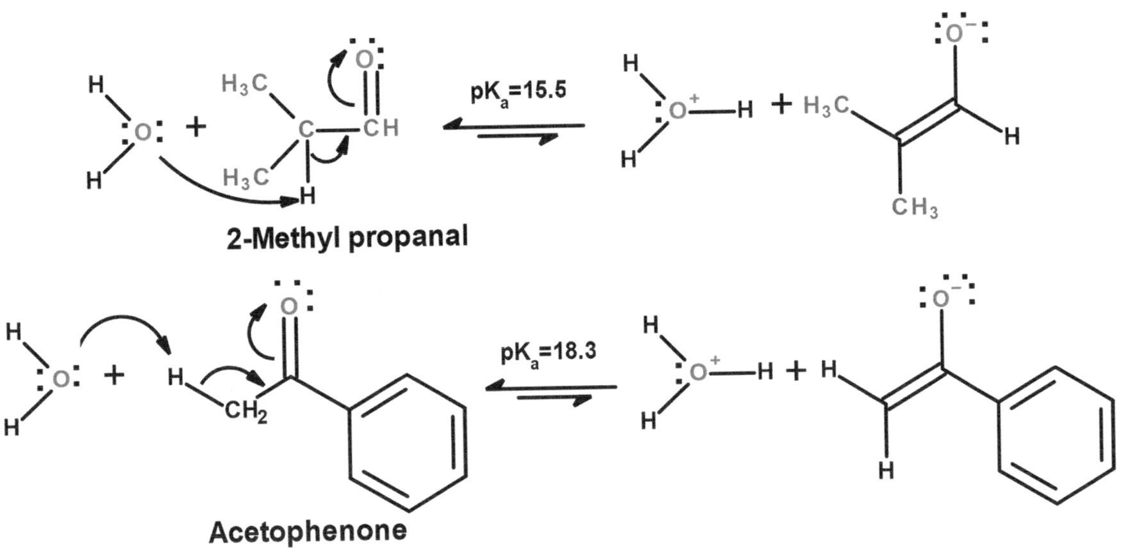

SCHEME 15.15   Acidity of α-proton and enolate formation.

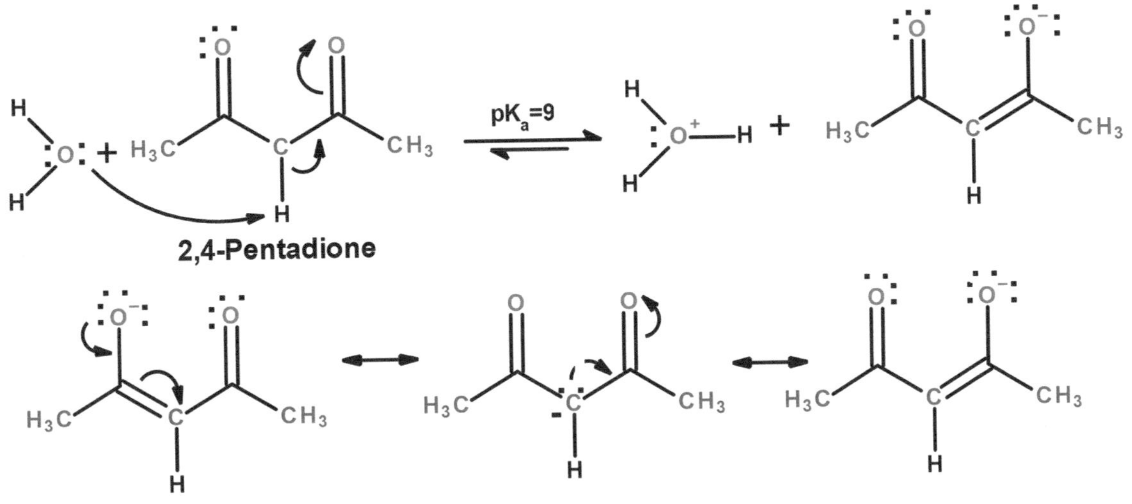

SCHEME 15.16   1,3-dicarbonyl compounds and enolate formation.

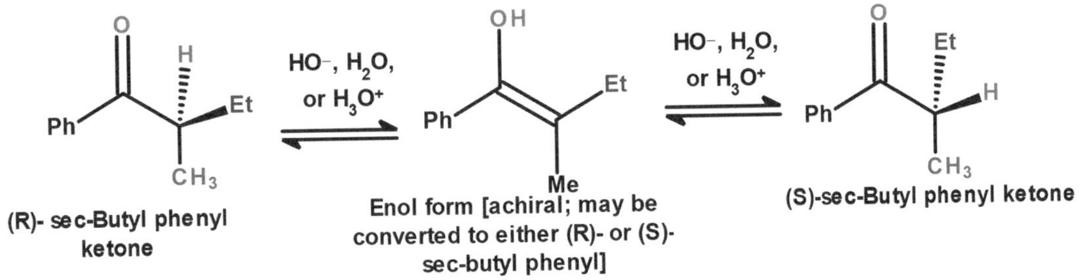

SCHEME 15.17   Enolate formation and loss of the chirality.

The reason is LiAlH$_4$ and NaBH$_4$ are nucleophilic, they transfer hydride ions, whereas NaH is non-nucleophilic so it acts only like a base and not as a nucleophile. Of course, there is a recent paper that uses NaH as a reducing agent in the presence of NaI/LiI (Shunsuke Chiba et al. *Angew. Chem. Int. Ed. Engl.* 2016; 55(11): 3719–3723.)

15.6.7   STERIC EFFECT OF R ON ENOLATE FORMATION

For enolate formation using LDA at −78 °C, the size of the R group plays a vital role in *cis* or *trans*-enolate formation. It is observed that when the size of the R group is big, they tend to form the *cis*-enolate. When R is *t*-Bu,

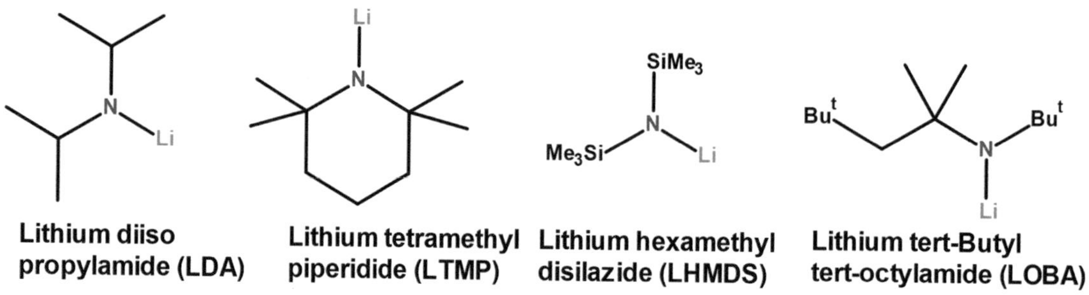

FIGURE 15.8   Examples of few bases.

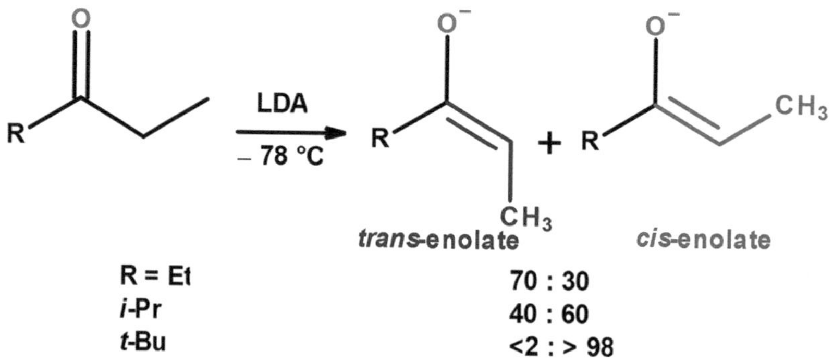

R = Et	70 : 30
i-Pr	40 : 60
t-Bu	<2 : > 98

SCHEME 15.18   Steric effect of R on enolate formation.

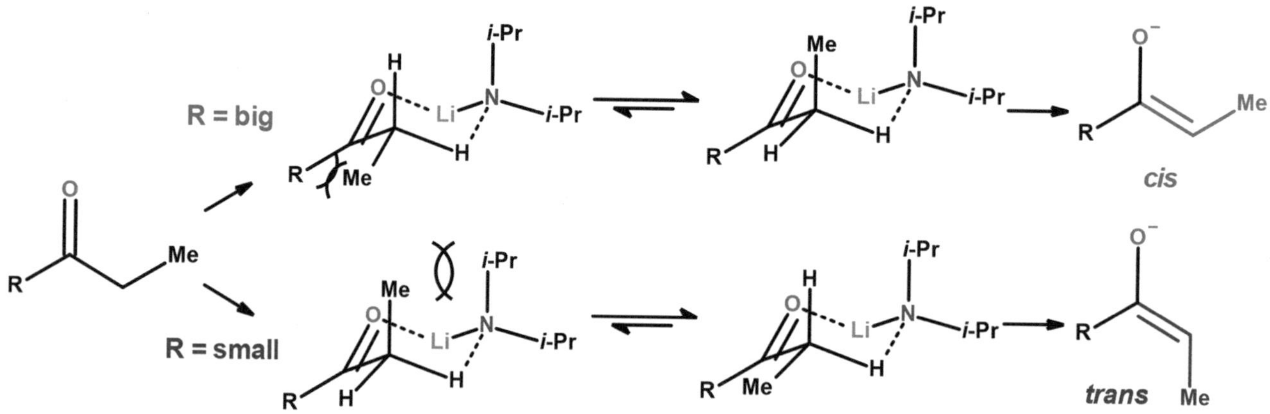

SCHEME 15.19   Transition state for stereoselective formation of one of the enolates.

the *cis*-enolate is exclusively formed in 98%. When R is ethyl, a small group, the *trans*-enolate is formed in large quantities.

How do we explain the stereoselective formation of one of the enolates? In the transition state, when the R group is large there is a steric repulsion between the R group and the methyl. So, the conformation of the molecule changes in such a way that the methyl group is far away from the R group. This leads to the *cis*-enolate. On the other hand, when the R group is small, there is steric repulsion between the base (LDA) *iso*-propyl group and the methyl group.

This puts the methyl group in the same side of R group, which leads to the *trans*-enolate.

15.6.8   Nature of the Base on Enolate Formation

1. LHMDS due to the presence of silicon generally provides the *cis*-enolate as the major product
2. LTMP (very bulky) affords the *trans*-enolate as the major product
3. LDA gives intermediate results.
4. Use of HMPA as a strongly Lewis basic donor-co-solvent can reverse selectivity.

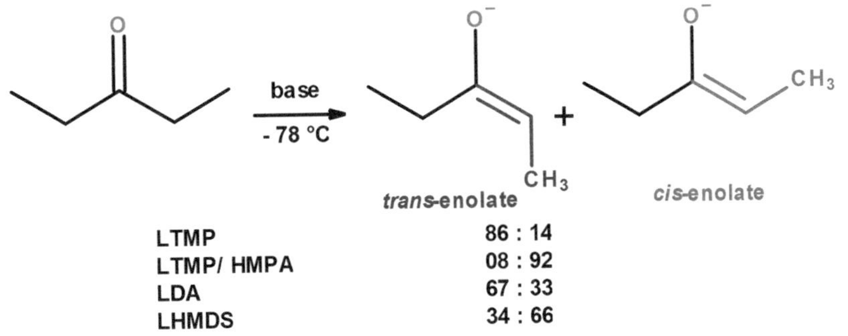

SCHEME 15.20   Nature of the base on enolate formation.

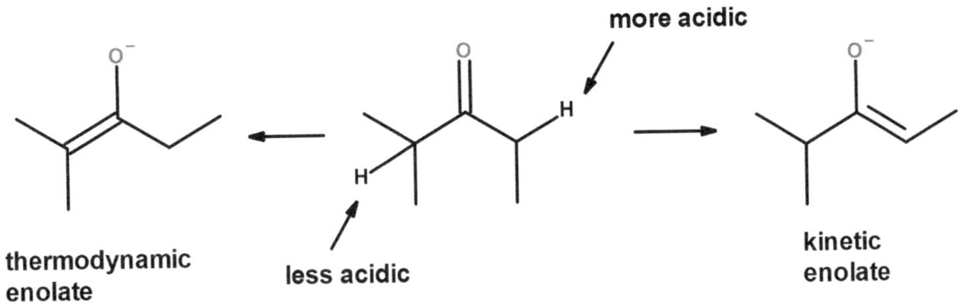

SCHEME 15.21   Regioselectivity in enolate formation.

15.6.9   REGIOSELECTIVITY IN ENOLATE FORMATION

Regioselectivity in enolate formation is also called thermodynamic vs kinetic control of enolate formation. When we have two different protons, one is more acidic and terminal or easily reachable proton; another one is less accessible but gives a more substituted double bond or the thermodynamically stable enolate.

15.6.10   ENOLATE FORMATION

What parameters favor kinetic over thermodynamic product formation? Aprotic solvent, low temperature, stronger and bulky bases, and shorter reaction time favor kinetic enolate. The use of 1–1.05 equivalence of the base favors kinetic product. This deprotonation is an irreversible process. When the opposite conditions prevail, we end up with thermodynamic enolate. Unlike kinetic enolate formation, the thermodynamic one is a reversible process (Figure 15.9).

We have various bases like alkyl lithiums, metal hydrides, amides based on ammonia and amines, silazides, and alkoxides. Alkyl lithiums and amides based on ammonia and alkoxides are nucleophilic. In other words, they are smaller in size, hence they are nucleophilic. Other bases are non-nucleophilic. The conjugate acids and their pK_a are mentioned below (Figure 15.10).

Kinetic products are formed at low temp, whereas thermodynamic products are formed at high temp. Another important thing is kinetic product is sensitive to structure. In the cyclohexanone derivative, whether the R is H or Ph,

Enolate formation

	Kinetic	Thermodynamic
Solvent	Aprotic	Protic
Temperature	Low	High
Base strength	Strong	Weak
Reaction time	Shorter	Longer
Size of the base	Bulky	Smaller
Equivalence	Stoichiometric	Sub Stoichiometric
Deprotonation	Irreversible	Reversible

FIGURE 15.9   Parameters that favor kinetic vs thermodynamic enolate formation.

kinetic conditions favor the kinetic product. On the other hand, when we take the open-chain example, the kinetic conditions favor the thermodynamic product due to the extension of conjugation or stability.

15.6.11   MICHAEL ADDITION

Michael's addition is a nucleophilic addition of a nucleophile to an α,β-unsaturated carbonyl compound. It is also called 1,4-conjugate addition. The nucleophile adds to the 4th place and hydrogen to the third place.

Bases

Types	Examples	Reactivity	Conjugate acid	pK_a
Alkyl lithium	BuLi, MeLi	Nucleophilic	Alkane	45
Metal hydride	NaH, KH	Non nucleophilic	Hydrogen	35
Amides (NH_3)	KNH_2, $NaNH_2$	Nucleophilic	Ammonia	35
Amides (amine)	LDA	Non nucleophilic	Amine	35
Silazide	LiHMDS	Non nucleophilic	Silyl Amine	26
Alkoxides	Kt-BuO	Nucleophilic	Alcohol	17

FIGURE 15.10   Comparison between various bases.

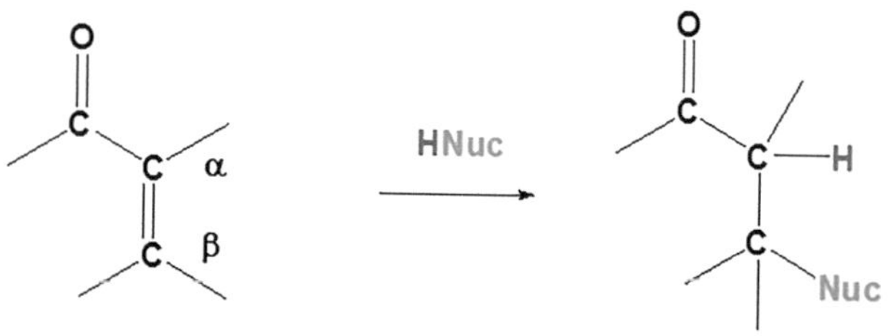

SCHEME 15.22   Michael addition.

The acidic proton from the Michael donor is removed by the base to produce the nucleophile. This attacks the α,β-unsaturated compound to give the product (Michael adduct). The complete mechanism with the enolate resonance is shown here. After the addition followed by charge reversal provides the product.

15.7   CONJUGATE ADDITION OF ORGANOCOPPER REAGENTS

Next, we look into the conjugate addition of organo-copper reagents. The lithium dialkylcuprates undergo conjugate addition to α, β-unsaturated aldehydes and ketones.

15.7.1   ALKYLATION OF ENOLATES

Enolate ions can undergo alkylation using alkyl halides. Alkylation occurs by an S_N2 mechanism. This reaction is difficult to carry out with simple aldehydes and ketones because aldol condensation competes with alkylation. However, 1,3-dicarbonyl compounds can be alkylated easily using this method.

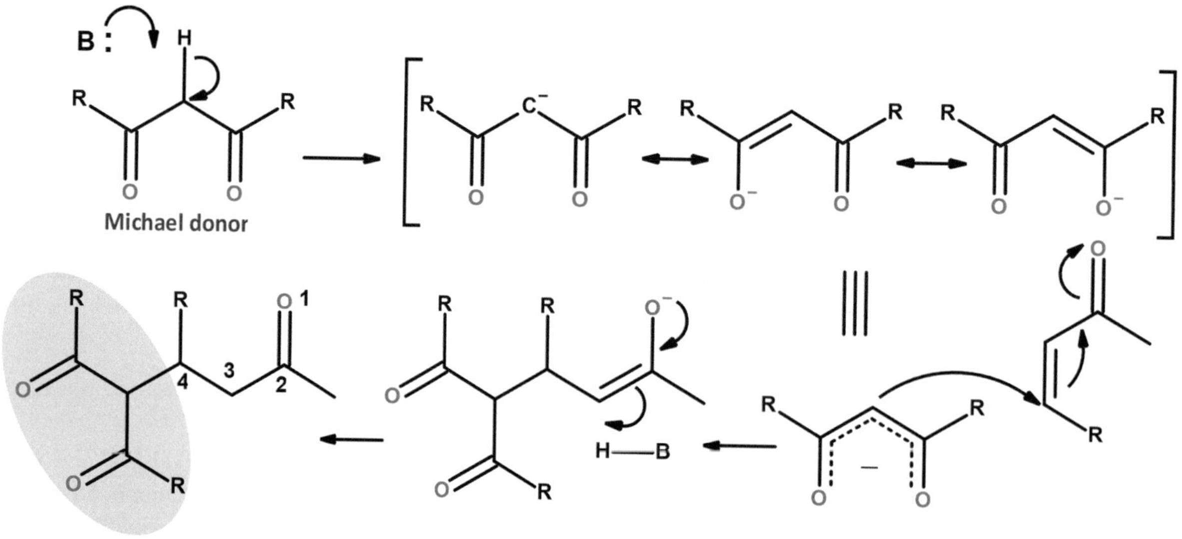

SCHEME 15.23   Mechanism of Michael addition.

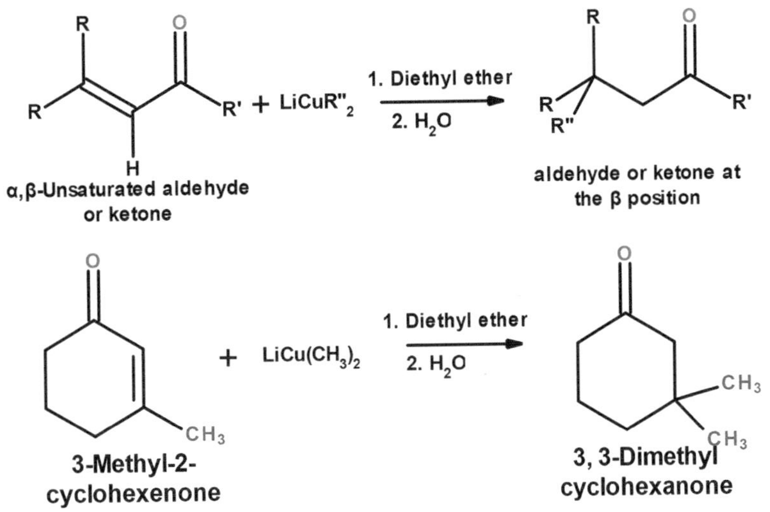

SCHEME 15.24 Conjugate addition of organocopper reagents.

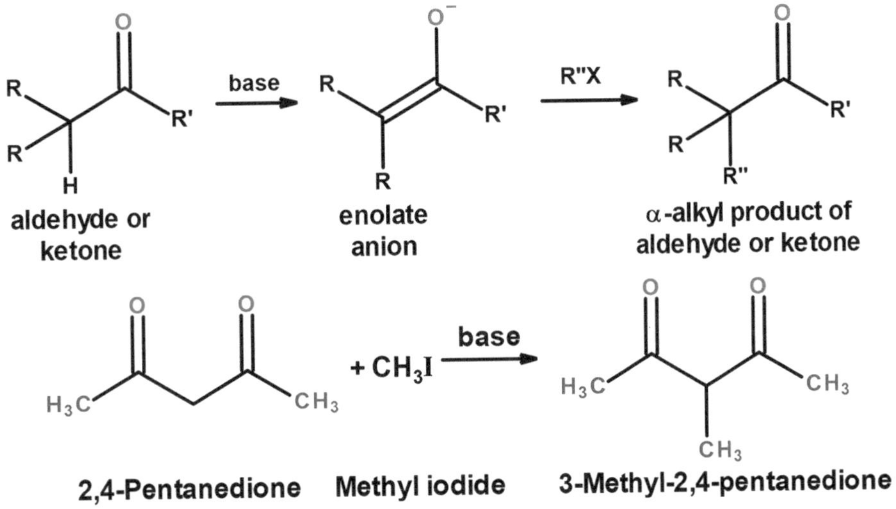

SCHEME 15.25 Alkylation of enolates and 1,3-dicarbonyl compounds.

15.8 STOBBE CONDENSATION

Stobbe condensation is similar to Claisen condensation. Instead of two esters, here we have a succinic ester and another carbonyl compound. The mechanism is given here.

15.9 ENAMINES

Next, we will study enamines. They are nitrogen counterpart to enols but are more nucleophilic than enols.

Why enamines? Carbonyl addition reactions using enolates have some limitations. We use strong bases like LDA to generate the enolate. These enolates are also strong Brønsted bases. They can also react with the carbonyl compounds from which they are formed, that is, they can undergo self-condensation. The enolates are unstable and highly reactive, so they cannot be isolated. In many instances controlling the regio- and stereo-chemistry is very difficult. On the other hand, enamines are less nucleophilic than enolates but have stronger nucleophile than enol. It is possible to employ milder conditions for the enamine formation. We can avoid the use of strong bases like LDA, due to which many functional groups can be present in the molecule which are not affected while enamines are prepared. Another aspect is enamines can be isolated and purified. Enamine reactions avoid polyalkylation and they also avoid O-alkylation (Figure 15.11).

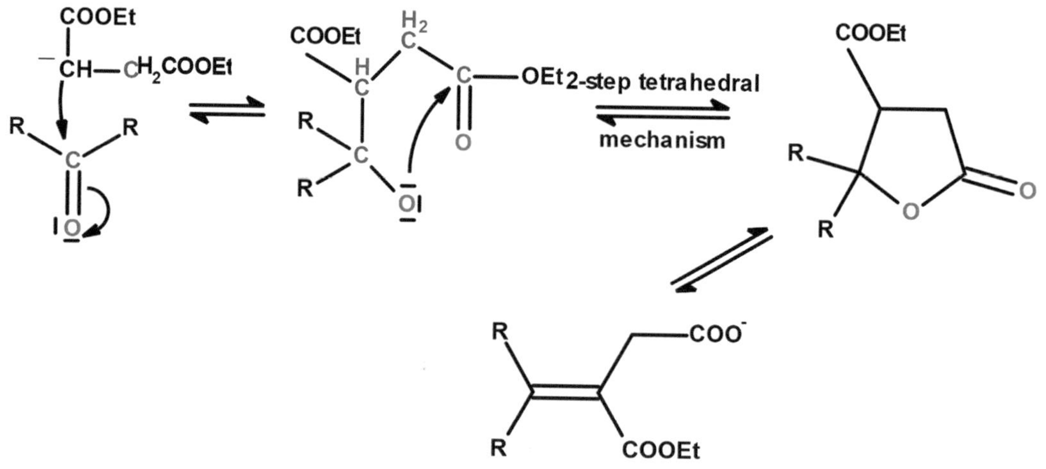

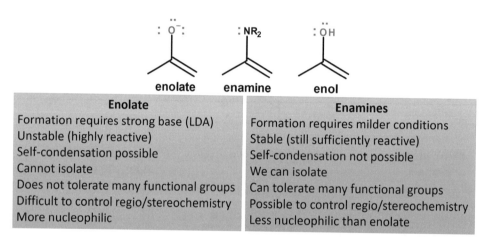

SCHEME 15.26 Stobbe condensation.

Enolate	Enamines
Formation requires strong base (LDA)	Formation requires milder conditions
Unstable (highly reactive)	Stable (still sufficiently reactive)
Self-condensation possible	Self-condensation not possible
Cannot isolate	We can isolate
Does not tolerate many functional groups	Can tolerate many functional groups
Difficult to control regio/stereochemistry	Possible to control regio/stereochemistry
More nucleophilic	Less nucleophilic than enolate

FIGURE 15.11   Comparison of enolates and enamines.

15.9.1   WHY ENAMINES ARE IMPORTANT?

The lone pair on nitrogen can readily participate in resonance with a double bond. Because of this, the β-carbon becomes susceptible to electrophilic attack. After addition of the electrophile to the β-carbon, the α-carbon becomes more electropositive. Now the nucleophile can attack the α-carbon. In many cases, the electrophile added to the β-carbon is a proton, because proton is a frequently used catalyst in many organic reactions (Figure 15.12).

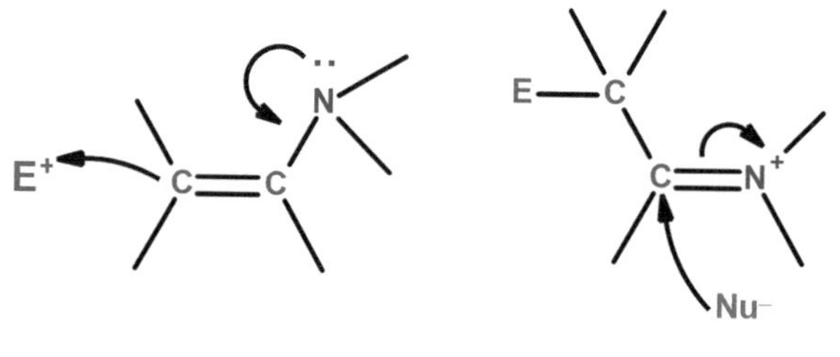

There are some unique features associated with this transformation: (1) the carbon can act like a low-energy carbanion by the lone-pair donation from nitrogen, (2) the formation of a true carbanion is a much higher-energy process, and (3) this helps the enamines to form a new bond to a carbon atom.

In the second reaction, we have some important things happening. We know nitrogen is not strongly electronegative compared with oxygen. Due to this, nitrogen can act as an effective electron sink. Nitrogen can also accept a proton very readily to form unsaturated cationic adducts. This is one of the reasons why charged compounds are formed readily even though nitrogen is not strongly electronegative

FIGURE 15.12   Electrophilic and nucleophilic reactions of enamines.

like oxygen. So nitrogen forms a cation at neutral pH much easier than oxygen which is the crucial factor for this transformation to occur.

15.9.2   IMINES AND ENAMINES

Imine and enamine can be formed from aldehydes. The difference lies in the amine. Based on whether we have primary or secondary amine, we get an imine of enamine. If the amine is primary, the initially formed aminol loses water. The proton that is attached to the nitrogen is lost and we get imine. Here there is a double bond between carbon and nitrogen. When we use secondary amine, there is no proton on the aminal nitrogen, so for the loss of water molecule, a proton is lost from the neighboring carbon leading to an enamine formation. Here we have the double bond between two carbons.

15.9.3 Resonance Structure of Enamines

The electronic structure of an enamine can be most simply represented by the Lewis structure with contributing resonance forms indicated by A and B. It also tends to imply a geometric coplanarity of the alkene and amine portions of the system which is usually absent.

Normally an amine has a pyramidal geometry with bond angles of 109.5° around the nitrogen. This implies sp^3-type hybridization involving three of the hybrid orbitals forming σ-bonds with the three substituents, and the fourth sp^3-hybrid orbital being occupied by the lone-pair electrons. Normally, the two unsaturated carbons and their four substituent atoms in an alkene describe a plane. Each of the two alkyl groups attached to the alkene carbons is sp^3-hybridized, with the π-bond consisting of a molecular orbital described by orthogonal, overlapping p-orbitals. This p-molecular orbital is occupied by a pair of electrons.

To achieve *full* delocalization of the lone pair of electrons in nitrogen into the alkene π-system for the enamine ground state, the plane formed by the nitrogen and its two nonalkene substituents must be coplanar with the two carbons and four substituents of the alkene. Such coplanarity requires sp^2-hybridization for the nitrogen atom.

The three sp^2-hybrid orbitals form σ-bonds with the three substituents, and the lone-pair electrons occupy the remaining p-orbital.

This p-orbital should be parallel to the p-orbital of the π bond and perpendicular to the plane where the C-N atoms are present. This allows for maximum overlap of the parallel p-orbitals and delocalization of the nitrogen lone-pair electrons into the alkene π-system.

The name enamine usually refers to α,β-unsaturated tertiary amines, >C=C-N<. Although this is the customary usage, the name enamine really implies any C=C-N linkage.

15.9.4 Spectral Evidences for Enamines

Photoelectron (PE) spectra of simple enamines show two broad distinct bands. This shows there are localized electrons. The lower-energy band is attributed to the lone-pair electrons on the nitrogen, and the higher-energy band is due to the alkene electrons.

The first ionization potential (IP) of amines and enamines can be correlated with the amount of s-character possessed by the nitrogen lone-pair orbital. If the IP decreases, the amount of s-character in the lone-pair electron orbital also decreases.

Rotational barriers (C-N bond 4–6 kcal/mol), are another good method for determining conjugative interaction in enamines.

The introduction of an amine auxochrome onto the C=C chromophore of an alkene produces a bathochromic or red-shift (shift to longer wavelength) for the π to π* electronic transition. Isolated alkenes show absorption maxima for π to π * transition around 190± 10 nm and normal saturated amines show l_{max} at 210± 5 nm. Whereas enamines show absorption maxima at 230± 10 nm. This shows that the amine group has greater conjugative interaction with the C=C π-system.

Bond distance is related to n-π interaction. As the s-character of a bond increases, the bond becomes shorter. In enamines as the amount of n- π interaction increases, the s-character of the σ-bond on nitrogen increases; that is, they approach sp^2-hybridization. Hence the C—N bond distance is shorter. Calculations on the simple primary enamine, vinylamine, have shown it to be nonplanar.

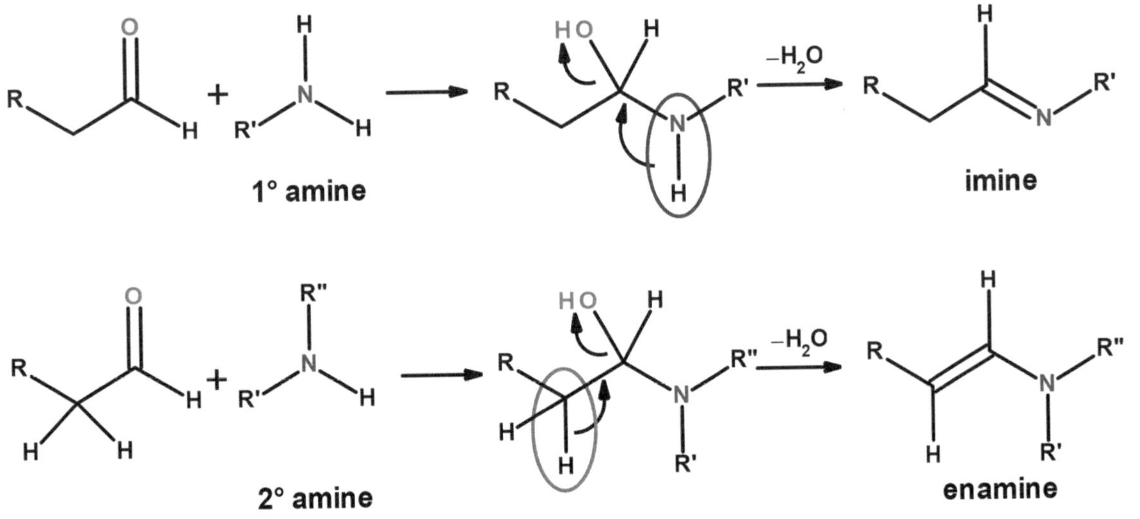

SCHEME 15.27   Imine and enamine formation.

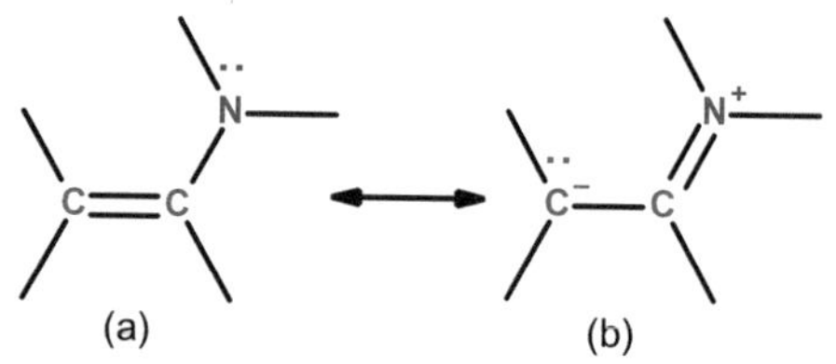

SCHEME 15.28   Resonance structure of enamines.

15.9.5   FORMATION OF ENAMINES

The most versatile method for preparing enamines involves the condensation of aldehydes and ketones with secondary amines. Earlier scientists used potassium carbonate at 0 °C or calcium oxide at a higher temperature for dehydration to give the enamines.

But the most widely used method now is the removal of water by the Dean–Stark azeotropic distillation method. Solvents like benzene toluene or xylene can be used. Many acid catalysts like p- toluenesulfonic acid (p-TSA), Dowex-50, and acetic acid facilitate this reaction. For the secondary amine part, heterocyclic secondary amines like pyrrolidine, piperidine, hexamethylenimine, and morpholine are used.

The overall reaction pathway usually presented for the preparation of an enamine from an aldehyde bearing an α-hydrogen and a secondary amine is given. Intermediates, which can be isolated in some cases, are called aminal. Although animals are produced when aldehydes and secondary amines react, the aminals are not necessarily the direct precursor of the enamine.

15.9.6   FORMATION OF ENAMINES FROM KETONE

15.9.6.1   Uncatalyzed Reaction

Although aminal formation is readily possible in the case of aldehydes, such intermediates are not common for reactions

of ketones with secondary amines. The only direct evidence for this is the infrared spectra of the reaction mixtures produced when dimethyl- or diethylamine was allowed to react with cyclohexanone or cyclopentanone in ether. The spectra revealed the presence of the enamine double bond ($1,640\,cm^{-1}$) before distillative workup. General mechanisms for the noncatalyzed and acid-catalyzed reactions have been offered.

15.9.6.2   Formation of Enamines from Ketone Under Acidic Conditions

The various steps involved in the enamine formation are depicted here. The hydroxyl of initially formed aminal can be protonated and can be lost as a water molecule. Concomitant loss of a proton gives the enamine. What can happen also is, the aminal can be lost as a water molecule generating an amino cation. This can be resonance stabilized as shown here. This cationic intermediate can lose a proton to give the final enamine.

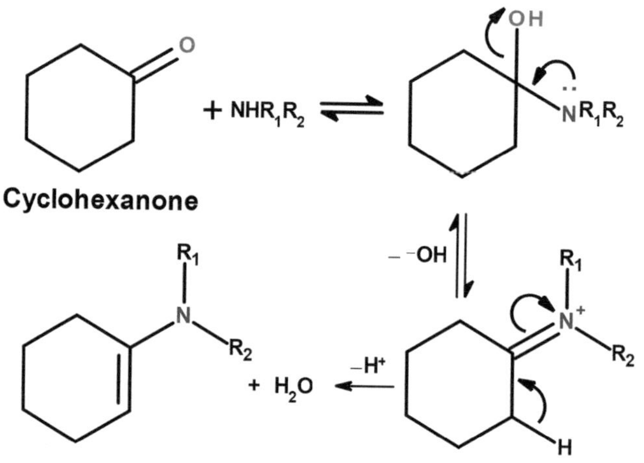

Cyclohexanone

SCHEME 15.30   Formation of enamines from ketones, uncatalyzed reaction.

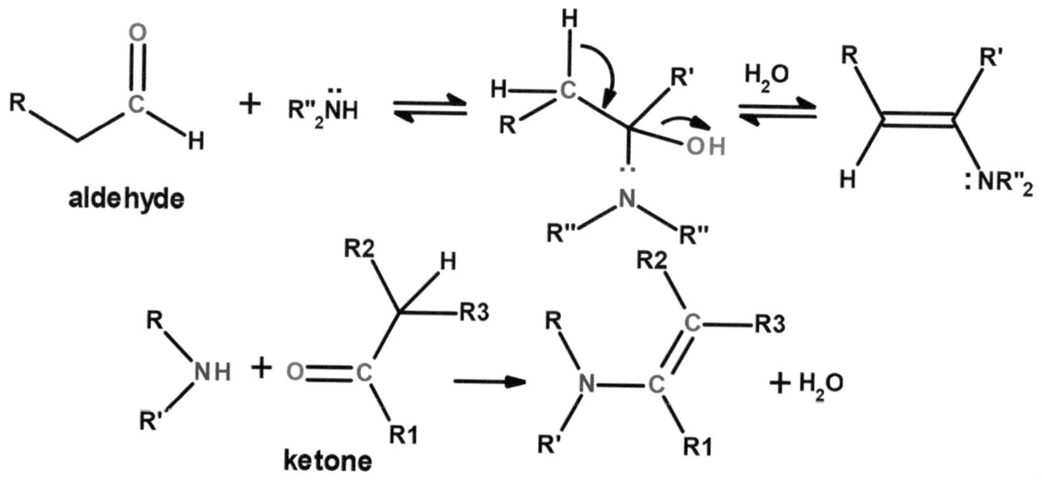

aldehyde

ketone

SCHEME 15.29   Formation of enamines from aldehyde or ketones.

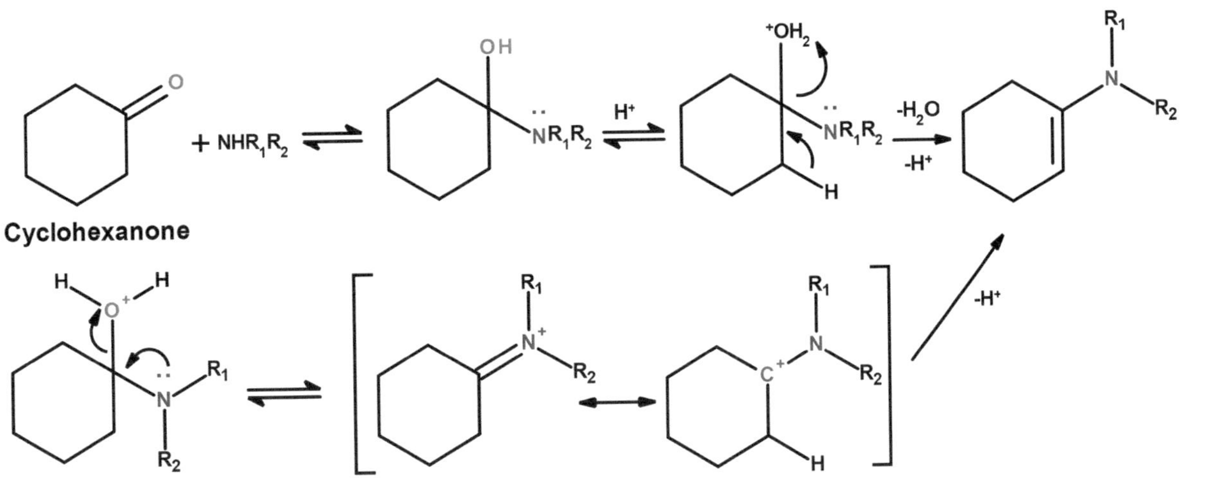

Cyclohexanone

SCHEME 15.31 Formation of enamines from ketones, acid-catalyzed reaction.

15.9.7 Factors That Affect Enamine Formation

Rate of Formation of Enamines – the nature and environment of the carbonyl group play a vital role in this. The effect of the ring size in the case of cyclic ketones is important. Cyclopentanone reacts most rapidly, followed by cyclohexanone which is faster than the seven- and higher-membered ring ketones.

Two factors on amines affect the rate: (1) the basicity and (2) steric environment of the secondary amino group. Among the secondary amines used, pyrrolidine gives a higher reaction rate than the weakly basic morpholine. While cyclic amines generally produce enamines faster than open-chain analogs. When the basicity and steric environment are very close like pyrrolidine (five-membered) and piperidine (six-membered), the dehydration step determines the rate of formation of the enamine.

We have two steps. One is carbinolamine formation and the other is the dehydration. If carbinolamine formation is important, then cyclohexanone will form its enamine faster than cyclopentanone. If dehydration is important, then C7 ketones will react faster than C6 ketones. But the observed rate is different from this order, we can say overall rate is not solely dependent on any one of the reversible steps.

Pyrrolidine is used in reactions of ketone enamines with alkyl halides and electrophilic olefins. Morpholine is used in acylation reactions, and electrophilic olefins with ketone and aldehyde enamines. Piperidine is used in electrophilic olefins with aldehyde enamines.

15.9.8 Structure of the Enamines from Unsymmetrical Ketones

The less substituted enamine is formed from unsymmetrical ketones. The kinetic product is formed in most cases. This is in contrast to the formation of enols or enolates where we get the thermodynamic product or the more substituted product. In the case of enamines, the double bond and the

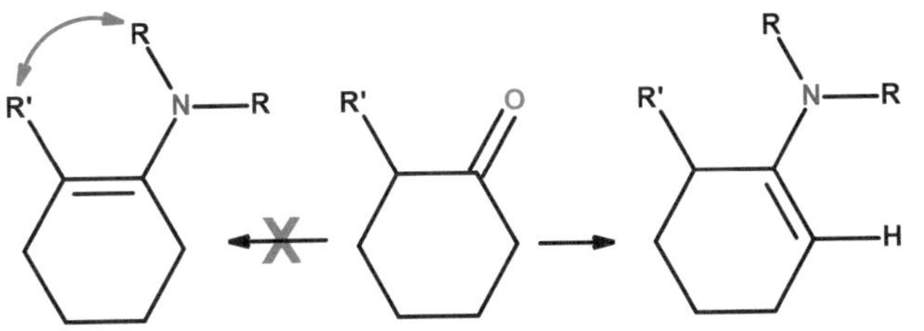

SCHEME 15.32 Structure of the enamines from unsymmetrical ketones.

lone pair on nitrogen have to overlap for better stability. This arrangement puts the α-substituents in close proximity to substituents on nitrogen and increases the steric strain. In the other structure, such effect is absent. Hence the less substituted enamine is formed.

15.9.9 Separation of Ketones (Using Enamines)

We have three ketones, 4-*tert*-butylcyclohexanone, and its mono and dimethyl analogs. The rate of enamine formation is greatly reduced by increasing the amount of substitution at the position α to the carbonyl group of the precursor ketone.

This fact has been used as a basis for the separation of a mixture of unsubstituted, mono- and dimethylated ketones which were obtained by alkylation of enamines followed by hydrolysis. The separation is accomplished by first allowing the mixture of ketones in refluxing benzene to react with gradually increasing amounts of morpholine until gas chromatographic analysis indicated that all of them had disappeared owing to formation of enamine. It is separated. Then the remaining mixture of ketones is treated with increasing amounts of the more reactive amine until gas chromatography showed all of the pyrrolidine had disappeared as it formed enamine. The remaining ketone could then be distilled.

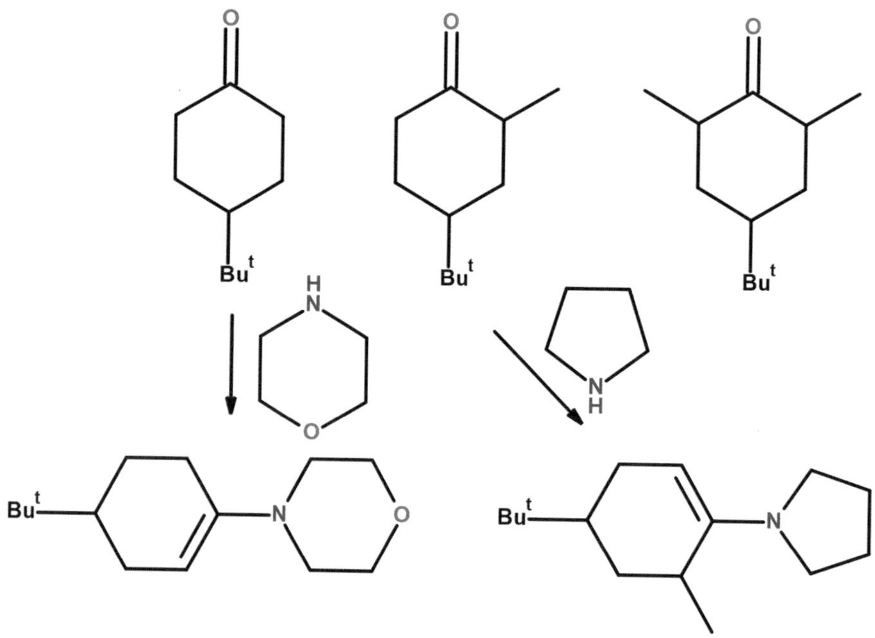

SCHEME 15.33   Separation of ketones using enamines.

15.10   REACTIONS OF ENAMINES

Electrophilic reactions include protonation, alkylation, acylation, and halogenation. The proton can add to the nitrogen to give an enammonium ion, which is a kinetically favored product. Or the proton can add to the β-carbon atom, resulting in the formation of an iminium ion, which is a thermodynamically most stable product.

15.10.1   ALKYLATIONS OF ENAMINES

An older method is very effective in controlling the kinetic-thermodynamic product. The method developed by Gilbert Stork in 1954 is the most used method. Here, cyclohexanone is reacted with pyrrolidine to give the corresponding enamine. As we mentioned earlier, enamine can be considered like a low-energy carbanion or a nitrogen enolate. Now this nitrogen enolate can be attacked by alkyl halides such as iodomethane to give an equilibrium mixture of the iminium salt and the alkylated enamine (Scheme 15.34). Hydrolysis of the reaction mixture using an

aqueous acid produces the alkylated ketone along with the amine, in this case, 2-methylcyclohexanone and pyrrolidine respectively. A more general sequence shows the behavior of a generic enamine reacting with a halide via the β-carbon to give α-alkyl iminium salt, via an S_N2 process. The intermediate iminium salt can isomerize under the reaction conditions to the less substituted enamine in an equilibrium process, and hydrolysis of the intermediate gives the corresponding α-alkyl carbonyl compound. Alkylation gives the best yields with reactive primary halides, since it is essentially an S_N2 reaction. This sequence is referred to as the *Stork enamine synthesis*. Enamines are usually formed by the reaction of a secondary amine with a ketone, in the presence of an acid catalyst.

Although the majority of the enamine formed from 2-methylcyclohexanone has the double bond oriented away from the methyl group (A) as evidenced by predominant formation of the 2-cyano-6-methylcyclohexanone (C), there is some orientation toward the methyl group (B), yielding a small amount of 2-cyano-2-methylcyclohexanone (D) (Figure 15.13).

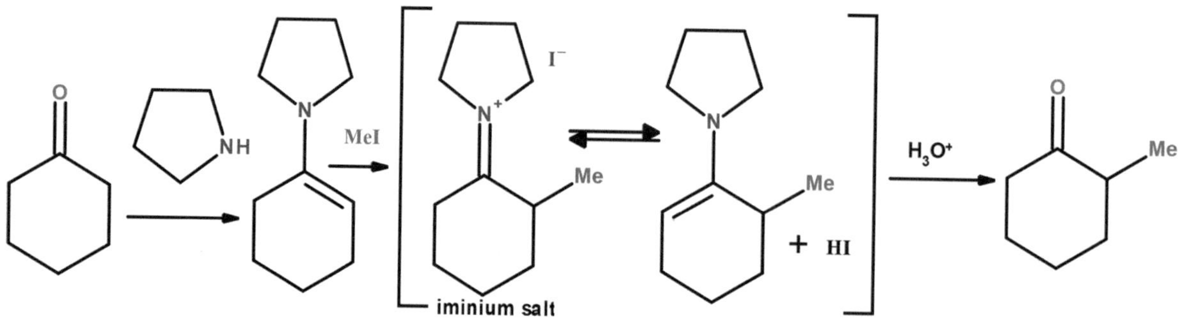

SCHEME 15.34   Alkylations of enamines.

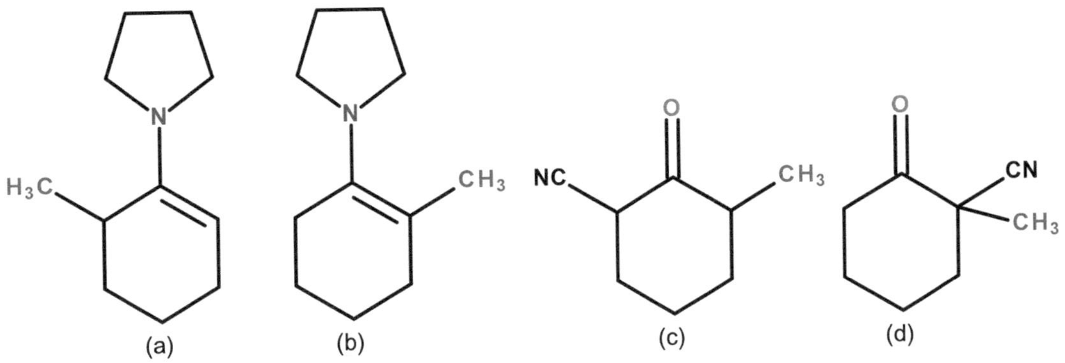

FIGURE 15.13 Different orientation of methyl group in alkylations.

15.10.2 WHY ENAMINES FOR ALKYLATIONS?

Alkylation of enamines can take place on carbon or on nitrogen. This depends on the ease of formation of a trigonal atom in the transition state, and on the nature of the enamine, the alkylating agent, and the solvent.

Issues with simple alkyl halide alkylation of carbonyl compounds are (1) N-alkylation predominates over C-alkylation; (2) mixtures of unalkylated, monoalkylated, and dialkylated products are obtained along with N-alkylated products; and (3) aldol condensation products can be obtained with aldehyde enamines.

Because of self-condensation under the conditions of the alkylation reaction, enamines derived from acetaldehyde or monosubstituted acetaldehydes cannot usually be alkylated with simple unactivated alkyl halides. Enamines derived from aldehydes disubstituted on the β-carbon, such as those derived from *iso*-butyraldehyde, are alkylated on nitrogen by alkyl halides. One method that has been used to achieve C-alkylation with aldehyde enamines using simple unactivated alkyl halides is the use of a bulky secondary amine to make the enamine.

When we talk about the reactivity of enamines, we should understand that enamines are bidentate nucleophiles similar to enolates. In enamines, both nitrogen and carbon can function as nucleophiles. When a highly reactive alkyl halide (methyl iodide or allyl bromide), which is small, is used for enamine alkylation, we end up with a mixture of products. Nucleophilicity usually parallels that of the unsubstituted amine. In general, if the alkyl halide is small, there is more chance of kinetic nitrogen enolate (the less substituted enamine) formation.

In the orbital approach to enamine alkylation, we have two planar resonating structures. The lone-pair electrons on nitrogen are parallel with the π bond. We have two steric interactions. One is between R group on nitrogen and R′ group on alkene carbon and the other is between the methyl group on alkene carbon and R on nitrogen. If the steric interaction (R↔R′) is greater than the R↔Me interaction, there is a preference for that kinetic nitrogen enolate. When R on nitrogen is Me or Et moieties, these interactions are small, and a mixture of both the intermediates is obtained. As R and R′ increase in size on nitrogen and on alkene carbon respectively, the steric repulsion increases, favoring

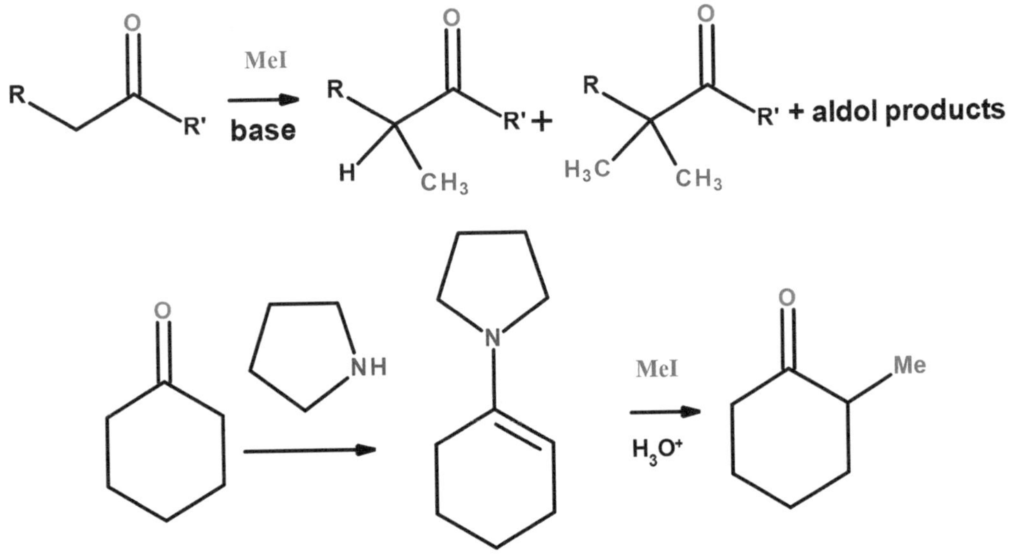

SCHEME 15.35 Different alkylation of carbonyl compounds.

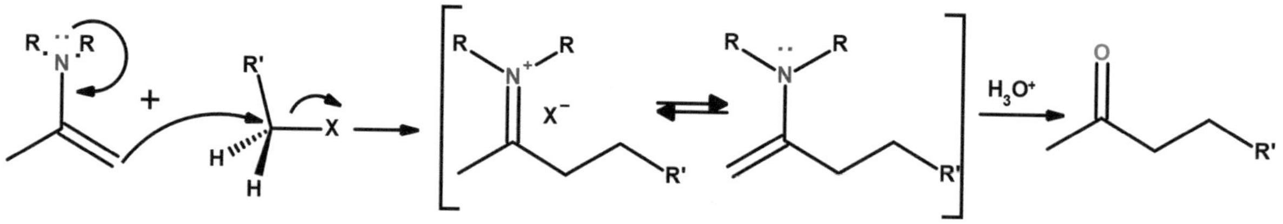

SCHEME 15.36 Alkylation at less substituted sites of carbonyl compounds.

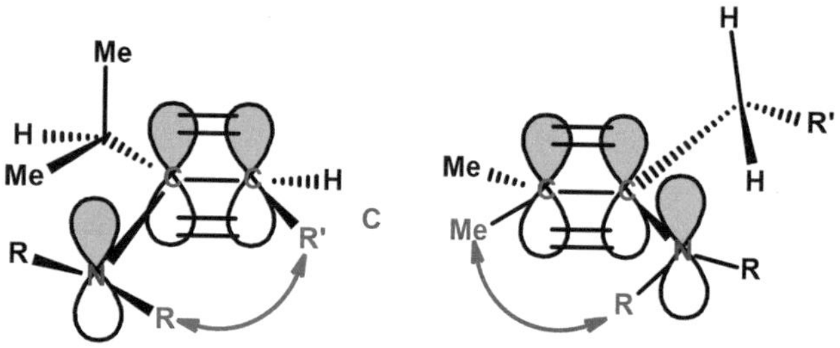

FIGURE 15.14 Orbital approach to enamine alkylation.

the latter. Pyrrolidine, piperidine, morpholine, or diethyl-amine are the most common amine precursors to enamines (Figure 15.14).

When an aldehyde such as the one given below reacts with pyrrolidine (secondary amines), the initial reaction gives an iminium salt. Reaction with excess of pyrrolidine leads to aminal rather than to the enamine. Treatment with a base such as sodium carbonate is required to get the enamine. Here elimination of the amine (pyrrolidine in this case), leads to the enamine.

15.10.3 ALKYLATIONS WITH UNSATURATED ELECTROPHILES

The series of electrophilic olefins that typically undergo Michael addition to enamines are α,β-unsaturated ketones, α,β-unsaturated nitriles, α,β-unsaturated esters, diesters, or amides, α,β-unsaturated sulfones, and nitroolefins. The typical reaction for the pyrrolidine enamine of cyclohexanone with these electrophilic olefins is shown here.

Similar to Robinson's annulation sequence, enamines can also be employed to make annulated products. Methyl vinyl ketone has been condensed with substituted cyclohexanones to give 2-octatlones derivatives.

15.10.4 ACYLATIONS OF ENAMINES

Selectivity of C vs N acylation, stability of C vs N-acylated product and more reactive vs less reactive are some of the features we are going to study here.

Acylation of enamines can take place on carbon or on nitrogen. In contrast to the N-alkylated products, the N-acylated products either are not stable or are acylating

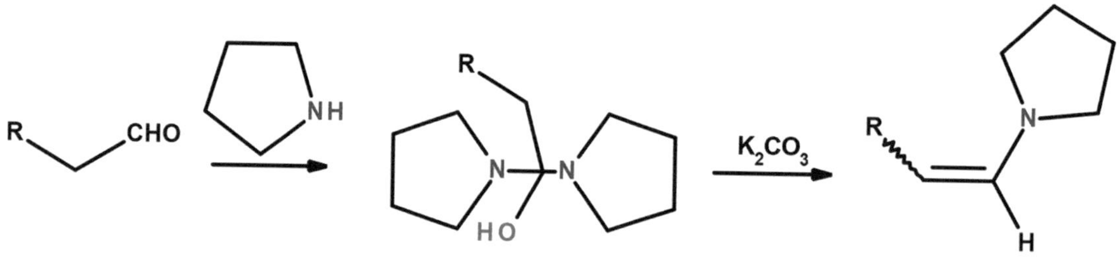

SCHEME 15.37 Side reactions in enamine formation.

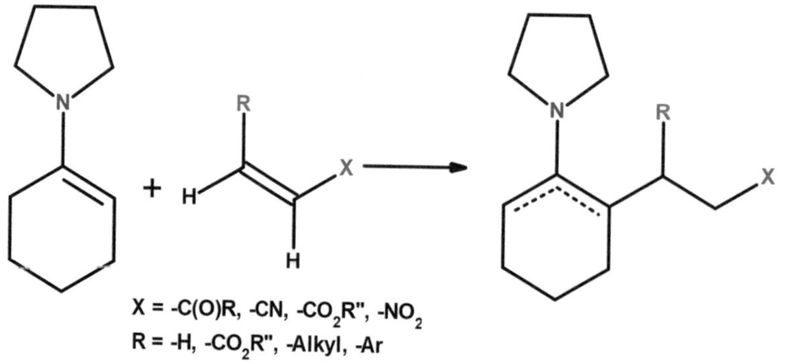

X = -C(O)R, -CN, -CO$_2$R", -NO$_2$
R = -H, -CO$_2$R", -Alkyl, -Ar

SCHEME 15.38 Alkylations with unsaturated electrophiles.

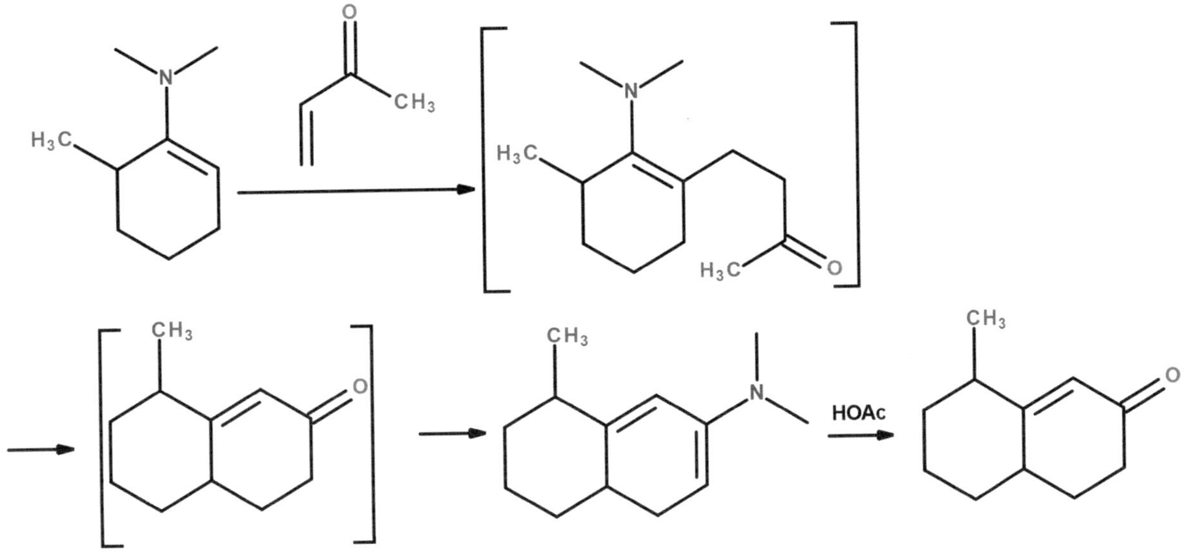

SCHEME 15.39   Annulation using enamines.

agents, so C-acylation is the normal mode of reaction. The enamines are stronger bases than the acylated enamines, so that in acylation with acid chlorides half an equivalent of the enamine is lost to the acylation reaction by salt formation. Loss of enamine in this manner can be avoided by the addition of an organic base such as triethylamine. It has also been shown that the less reactive morpholine enamines give better yields of acylated products. The acylation of enamines with acid anhydrides or acid chlorides having no α-hydrogen atom appears to be a straightforward reaction at the enamine carbon.

15.11   ENOL ETHERS

In the enolates and their reactions, we used strong bases like LDA to generate the carbanion and subsequent carbonyl addition under basic conditions. We will see the complimentary reactions that is the reaction of carbonyl compounds under neutral or acidic conditions. This necessitates the use of new functionality. Enol ethers are one such functional group. In comparison with simple alkenes, enol ethers exhibit enhanced susceptibility to attack by electrophiles such as Brønsted acids.

Previously, we compared enolates with enamines. What are the differences between enolate and enols ethers? In the

case of enolates, the major problem is the regioselective formation of ketone enolate.

15.11.1   Why Enol Ethers?

When 2-methylcyclohexanone was subjected to enolate formation, due to the presence of 2-methyl substituent, the less hindered C-6 proton was removed and the less stable kinetic enolate was formed. This enolate undergoes a reaction with electrophile to produce the 2,6-disubstituted product. Although we saw the thermodynamic product can be formed as a major product by change of reaction conditions, it was not the exclusive product but a mixture with one of the isomers formed in large quantities.

To get the kinetically inaccessible (but not necessarily thermodynamically more stable) enolates as the exclusive product, we need other approaches. This problem can be solved when we use enol silyl ether, a method pioneered by Prof. Stork, as shown here. As we mentioned earlier, the enolate is a highly reactive intermediate and was never isolated. In the case of enol ethers, they are stable products and can be isolated. As shown here the reaction of 2-methyl cyclohexanone with trimethylsilyl chloride (TMSCl) and triethyl amine (TEA) in dimethyl formamide (DMF) heating gives a mixture of two regio-isomeric silyl enol ethers.

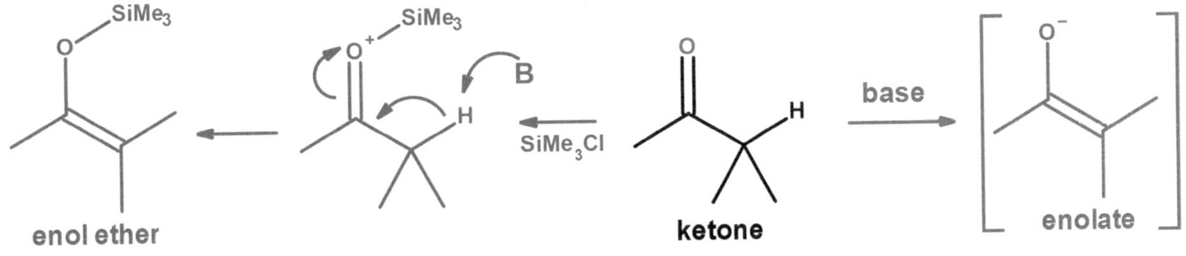

SCHEME 15.40   Formation of enol ethers.

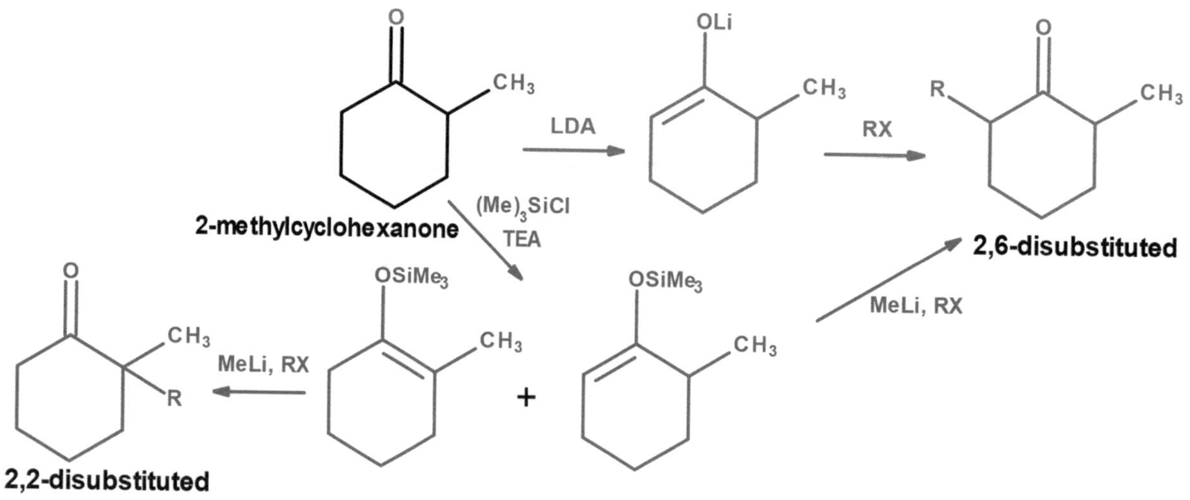

SCHEME 15.41 Formation of enol ethers under different conditions.

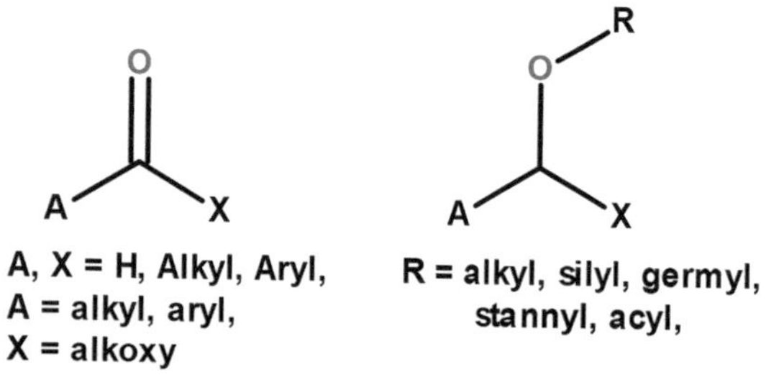

A, X = H, Alkyl, Aryl,   R = alkyl, silyl, germyl,
A = alkyl, aryl,   stannyl, acyl,
X = alkoxy

FIGURE 15.15 Different enol ethers.

1. Formation of alkyl enol ethers
2. Formation of carbocyclic enol ethers
3. Formation of enol silyl ethers
4. Formation of enol esters

Methods 1–3 are the most commonly used methods. Enol esters are rarely used.

15.12.1 FORMATION OF ALKYL ENOL ETHERS

Although many methods are available, the most widely used method is based on open-chain acetal or ketals. Conversion of aldehydes and ketones to the corresponding open-chain acetals followed by elimination of an alcohol moiety is a general method to prepare enol ethers. The elimination is usually conducted under mildly acid-catalyzed conditions. Although this is a good method for symmetric compounds, when the ketone is unsymmetrically substituted, good regiocontrol of the enol ether formed is difficult to achieve.

Another method is also based on the conversion of acetal to enol ether. The only difference is that this is not acid-catalyzed. We can form the dimethyl acetal by the reaction of methanol and the corresponding ketone, in this case, cyclopentanone. This is treated with trimethylsilyl trifluoromethanesulfonate (TMSOTf) and 1.2–1.7 equiv of DIPEA. This leads to the formation of the enol ether. We can assume that mechanistically, this process occurs through a stepwise reaction. In the first equilibrium step, the intermediate

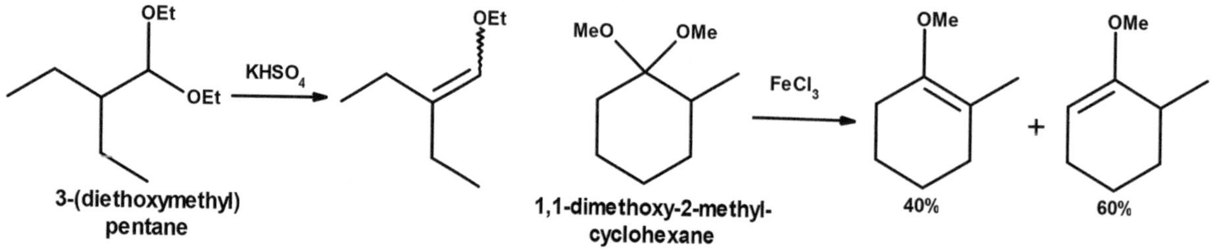

These two can be separated by distillation and each isomer can be converted to either 2,6-disubstituted product or 2,2-disubstituted product as shown here. This is the major advantage of using enol ethers.

The enol ethers considered include alkyl, silyl, germyl, and stannyl ethers, and to a small extent enol esters. The carbonyl compounds encompass aldehydes, ketones, esters, and their functional equivalents (Figure 15.15).

The various types of enol ethers that can be formed are alkyl enol ethers, carbocyclic enol ethers, enol silyl ethers, and enol esters. Here, we will focus mainly on carbon and silicon derivatives. We will not study germyl or stannyl ethers.

15.12 FORMATION OF ENOL ETHERS-ESTERS

For the alkyl enol ether preparation, we have four methods:

SCHEME 15.42 Formation of alkyl enol ethers.

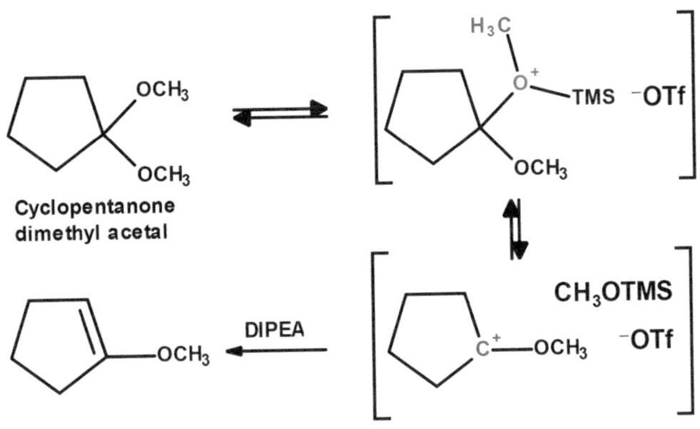

SCHEME 15.43 Conversion of acetal to enol ether.

is formed. It is an oxonium ion. The driving force for the formation is the strong oxygen-silicon bond. This may lose the trimethylsilyl ether of methanol and will give the second intermediate which is a carbocation. Deprotonation of the carbocation by a non-nucleophilic base would then produce the final enol ether.

This method works even in the case of cyclic acetals. Various ethylene glycol and propylene glycols gave cyclic acetals (1,4-dioxolanes and 1,5-dioxalanes) of ketones. These cyclic acetals were subjected to the same condition that is treatment of acetals with TMS triflate followed by treatment with a base to give the enol ethers. Unlike the previous open-chain acetals case where the TMS group was lost, in the cyclic acetals the final enol ethers have retained the TMS group.

Alkyl enol ethers can be conveniently prepared by the alkylation of α-methoxyvinyllithium and related metallated enol ethers. Carbometallation of alkynic ethers with organocopper reagents can also give enol ethers as shown in Scheme 15.44.

15.12.2 FORMATION OF CARBOCYCLIC ENOL ETHERS

An important method to prepare carbocyclic enol ethers is the Birch reduction of aryl ethers. The mechanism for this reaction has already been given under radical reactions (Chapter 13.15).

15.12.3 FORMATION OF ENOL SILYL ETHERS

Enol silyl ethers can be prepared readily from the parent carbonyl compounds by silylation of the corresponding enolate anions. Particularly useful is the observation that with unsymmetrical ketones it is often possible to generate the <u>less substituted</u> enol silyl ether under conditions of <u>kinetic control</u>, and the **more substituted** enol silyl ether by equilibration, that is, **thermodynamic** control.

Generation of silyl enol ethers from acyclic ketone precursors can be accomplished using the same kind of reagents. Depending on the reaction conditions, stereoselective formation of either the *(E)*- or the *(Z)*-isomer of the enol silyl ethers can be prepared. An *in-situ* method of generating the enolate anion with lithium dialkylamides in the presence of trimethylchlorosilane leads to enhanced selection for the kinetically preferred enol silyl ether. Lithium t-octyl-t-butylamide (LOBA) is shown to be superior to LDA in the regioselective generation of enolates and the stereoselective formation of (E)-enolates in this *in-situ* method.

15.12.4 FORMATION OF ENOL ESTERS

So far, we have seen *O*-alkylated products. On the other hand, *O*-acylation of enols or enolate anions is also possible. Enol esters can therefore be prepared readily from the parent carbonyl compounds. Cyclooctanone is converted to

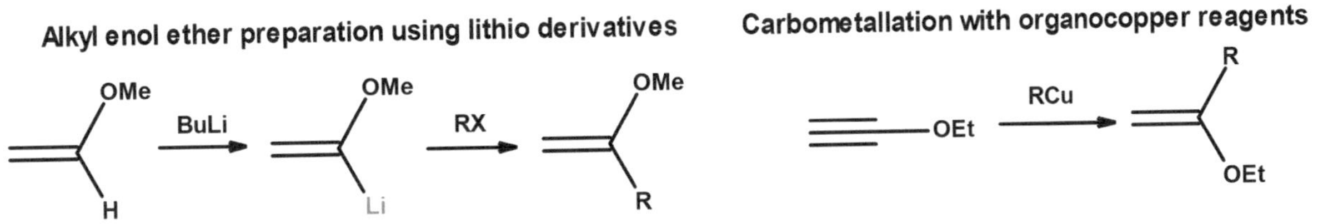

Alkyl enol ether preparation using lithio derivatives

Carbometallation with organocopper reagents

SCHEME 15.44 Preparation of alkyl enol ethers using lithium or copper reagents.

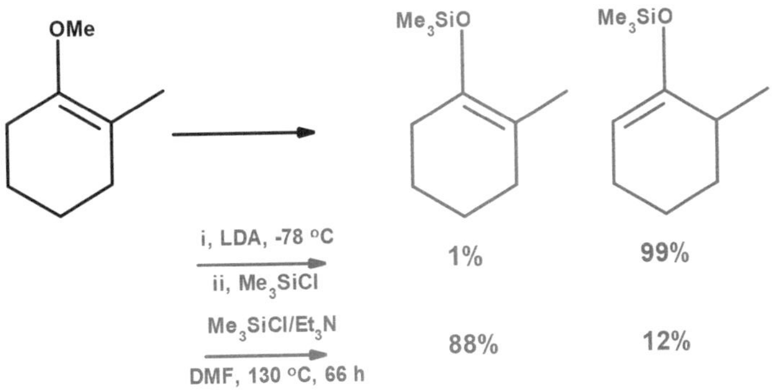

SCHEME 15.45 Formation of enol silyl ethers.

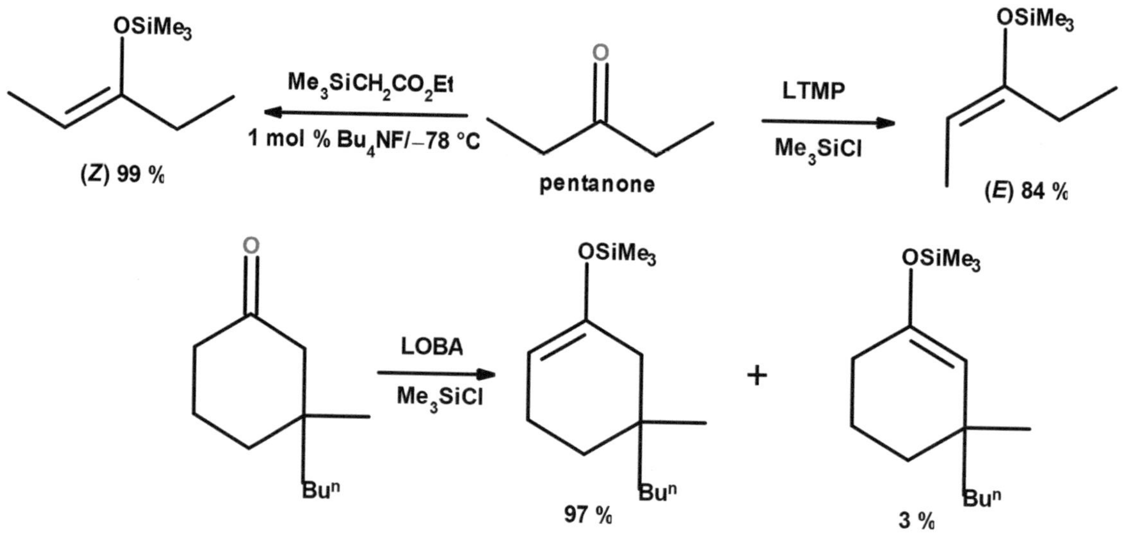

SCHEME 15.46   Stereoselective formation of enol silyl ethers.

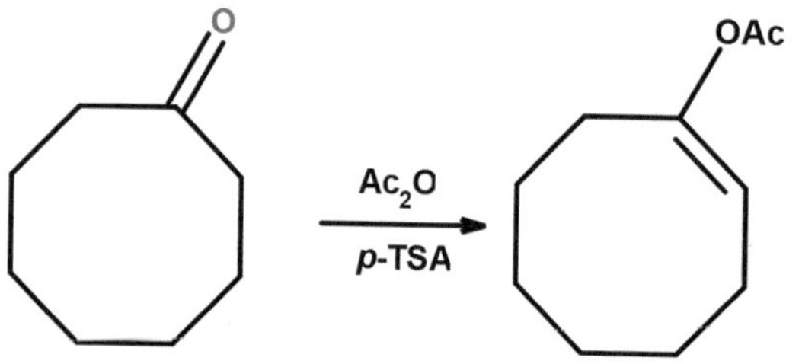

SCHEME 15.47   Formation of enol esters.

enol acetate by heating with acetic anhydride in the presence of p-toluenesulfonic acid.

15.13   REACTIONS OF ENOL ETHERS

The following are a few reactions of enol ethers:

1. Reactions with C-X π- bonds and equivalents
2. Catalyzed Addition to C-X π bonds and equivalents
3. Reactions of enol silyl ethers (Mukaiyama aldol reaction)
4. The cyclopropanation of silyl enol ethers

15.13.1   REACTIONS WITH C-X π- BONDS AND EQUIVALENTS

Various enol ethers (insufficiently nucleophilic) do not directly react with carbonyl compounds (C-X-bonds) in the absence of a catalyst to give the aldol condensation.

In comparison, stannyl enol ethers are sufficiently nucleophilic, similar to metal enolates, and can react readily with aldehydes. These stannyl enol ethers undergo aldol condensation.

15.13.2   CATALYZED ADDITION TO C-X π BONDS AND EQUIVALENTS

Enol ethers react with acetals or ketals, promoted by Lewis acids, to give aldol-type adducts. The relative reactivities of acetals and orthoesters in Lewis acid-catalyzed reactions with enol ethers have been investigated. For BF$_3$-Et$_2$O-catalyzed reactions with methyl vinyl ether, the following order has been found; saturated acetals are less reactive compared with methylorthoformate which is less reactive compared with benzaldehyde acetals. Finally α,β-unsaturated acetals are highly reactive.

Aminohemiacetals can also condense with enol ethers or enol esters to give the cross-aldol products.

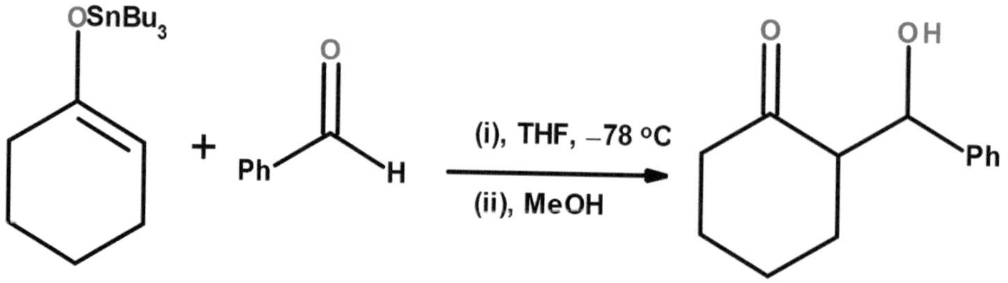

SCHEME 15.48   Reactions of stannyl enol ethers with aldehydes.

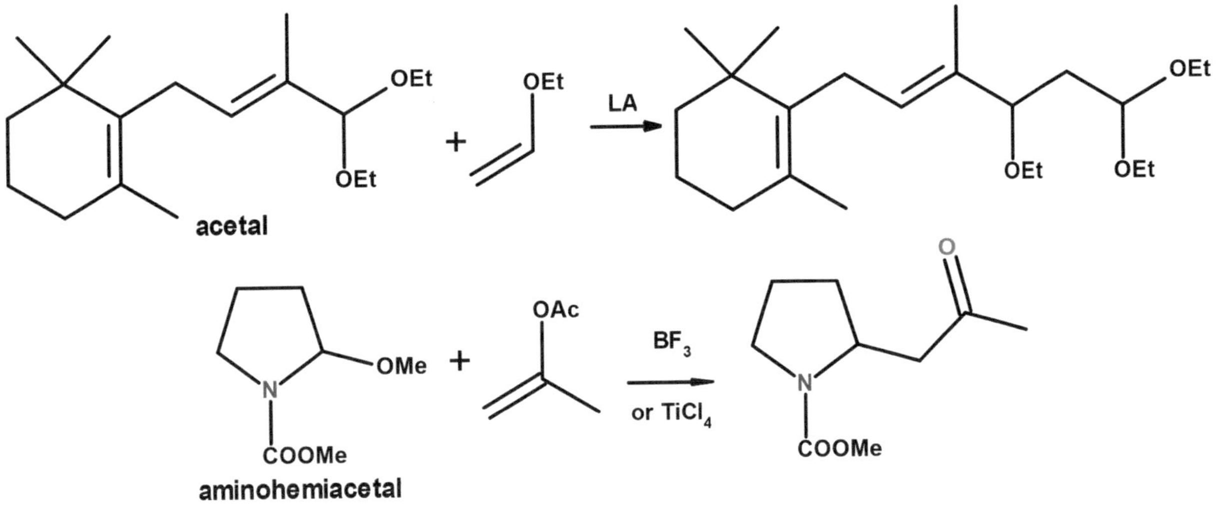

SCHEME 15.49   Reactions of acetals or aminohemiacetals with vinyl ether or enol ethers.

Enol esters, like enol ethers, can also react with various acetals and aldehydes in the presence of Lewis acids such as $TiCl_4$, $AlCl_3$, $SnCl_4$, $ZnCl_2$, and BF_3-Et_2O to afford the corresponding aldol-type addition products. A typical example is the reaction of isopropenyl acetate and γ-phenylpropionaldehyde acetal (Scheme 15.50).

15.13.3   MUKAIYAMA ALDOL REACTION

Mukaiyama and his coworkers reported that the Lewis acid promoted condensation of enol silyl ethers with carbonyl compounds to give the cross-aldol products.

A variety of Lewis acids have been used for the catalyzed reaction of enol silyl ethers with carbonyl compounds.

TMS enol ethers are less nucleophilic than boron or lithium enolates so they are less reactive. Lewis acid complexation increases the electrophilicity of aldehydes and this is

sufficient to allow reaction. The reaction proceeds through a transition state.

Prof. Mukaiyama had shown that activation of carbonyl compound by $TiCl_4$ or other Lewis acids increases the electrophilicity of the carbonyl carbon by coordination of carbonyl oxygen with $TiCl_4$. The intermediate can lose TMSCl forming a cyclic six-membered chelate intermediate. This is the driving force for the reaction. The final hydrolysis of the chelate produces the aldol product (Scheme 15.52).

Treatment of cyclohexanone and trimethylsilyl chloride gives the silylenol ether of cyclohexanone. This intermediate undergoes aldol condensation with benzaldehyde in the presence of various Lewis acids. It was observed by Mukaiyama that $TiCl_4$ worked excellently and gave the maximum product. In fact, at low temp. $ZnCl_2$ did not react and it reacted only at room temp, but again the yields were not very high (Figure 15.16).

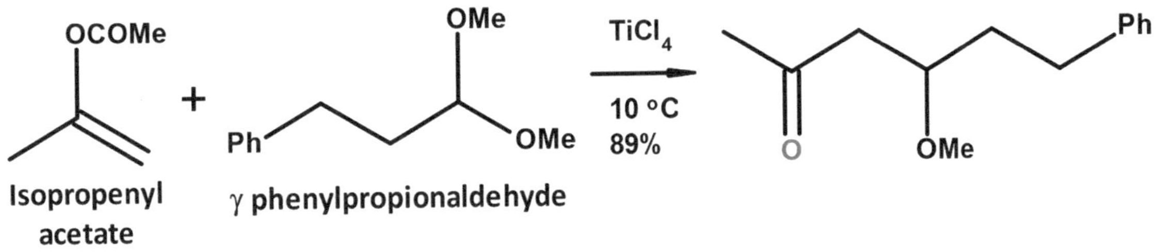

SCHEME 15.50   Reactions of enol esters.

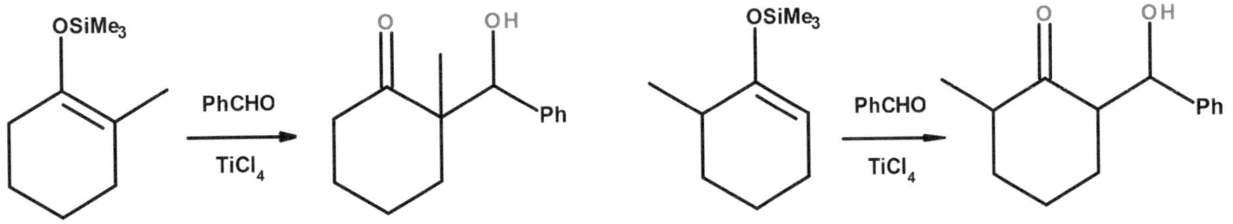

SCHEME 15.51   Mukaiyama aldol reaction.

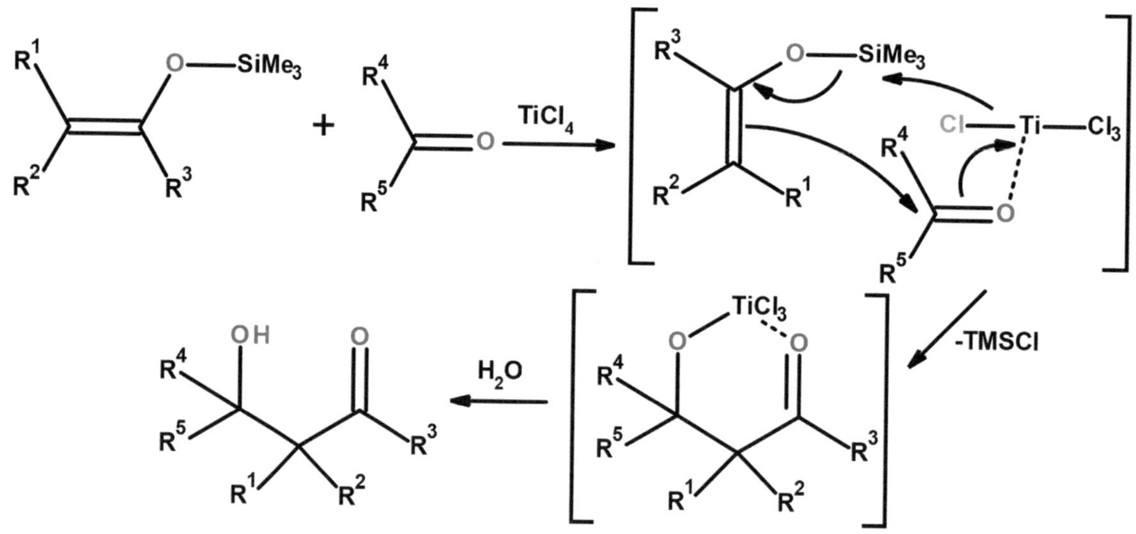

SCHEME 15.52 Mechanism of Mukaiyama aldol reaction.

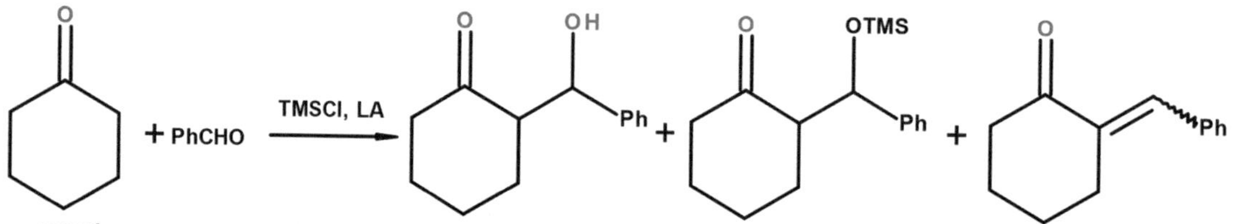

TiCl$_4$	−78 °C	92	0	0
SnCl$_4$	−78 °C	83	trace	trace
BF$_3$.Et$_2$O	−78 °C	80	12	0
ZnCl$_2$	RT	69	8	3

FIGURE 15.16 Reaction of cyclohexanone with benzaldehyde with different Lewis acids.

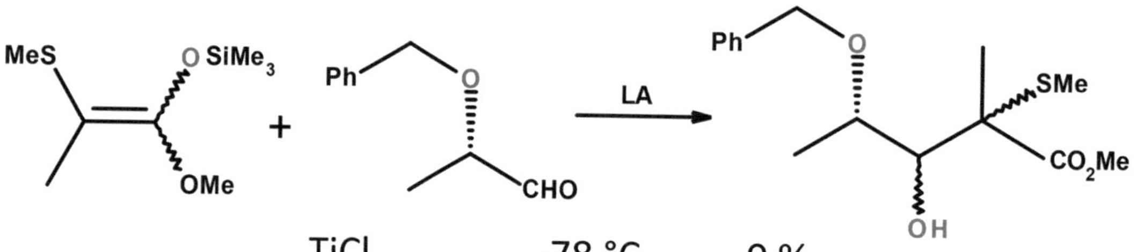

TiCl$_4$	−78 °C	0 %
SnCl$_4$	−78 °C	65 %
MgBr$_2$	−40 °C	60 %
ZnCl$_2$	−40 °C	45 %

FIGURE 15.17 Different product ratio with different Lewis acids.

Although TiCl$_4$ worked excellently in some reactions but is not always successful. In some cases, SnCl$_4$ worked better. So, it is always advisable to screen various catalysts with different solvents at different temperatures to identify the best reagent for the particular reaction (Figure 15.17).

15.13.4 Choice of Catalyst

Aldol-type condensation of enol silyl ethers and acetals or orthoesters can be accomplished by the use of trimethylsilyl trifluoromethanesulfonate (TMSOTf). In these reactions,

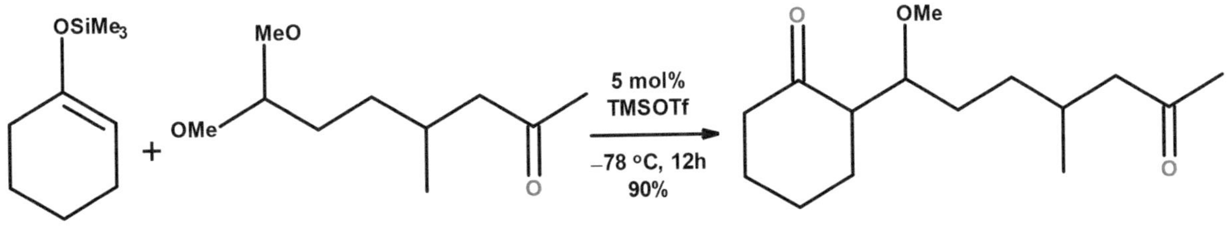

SCHEME 15.53 TMSOTf catalyzed aldol-type reaction of enol silyl ethers.

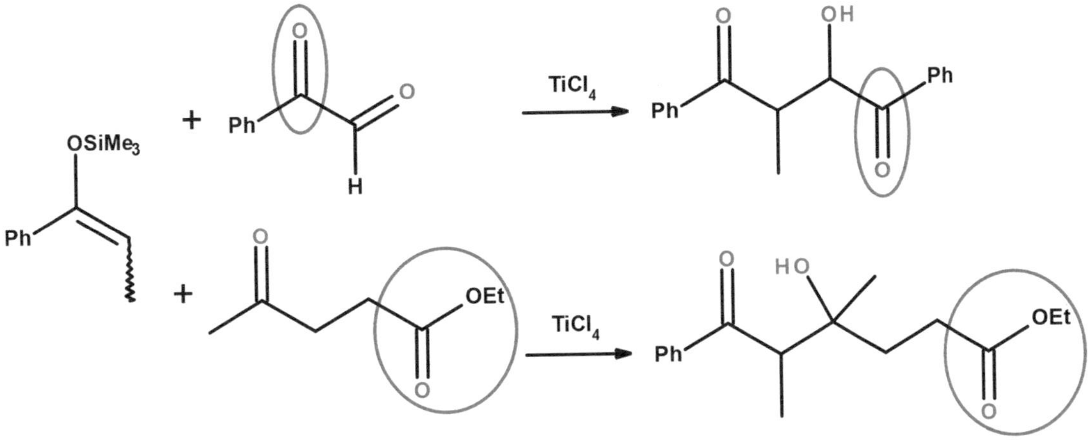

SCHEME 15.54 Chemoselectivity reaction of enol silyl ethers.

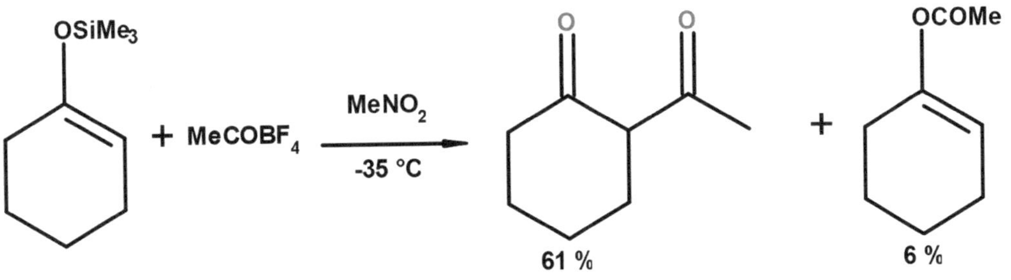

SCHEME 15.55 Regioselectivity in reactions of enol silyl ethers.

TMSOTf acts as a true catalyst and is required in 1–5 mol %. The reactions show interesting chemoselectivity in that acetals are highly reactive receptors of enol silyl ethers but the parent aldehydes and ketones do not react under these conditions.

15.13.5 CHEMOSELECTIVITY

So far what we have seen is that different kinds of carbonyl functions will react differently with enol silyl ethers depending on the catalyst we are using. Under the same reaction conditions, chemoselectivity is observed with acceptors having two or more different kinds of carbonyl electrophilic sites in the same molecule.

In general, with TiCl$_4$ as the Lewis acid, aldehydes are more reactive than ketones, which are in turn more reactive than esters. The same silyl enol ether reacts differently with a keto-aldehyde (phenylglyoxal) and keto esters

(ethyl levulinate). In the first case, aldehyde is more reactive so the ketone is unaffected. In the second case, ketone is more reactive than ester so it undergoes a reaction.

15.13.6 REGIOSELECTIVITY C- VS O-ACYLATION

Although enol silyl ethers are ambident species capable of reacting either at carbon or at oxygen, nearly all the aldol-type condensations occur at carbon. The exception is in the reaction with acylating agents, where either the *O*- or the C-acylated products can be obtained depending on the reaction conditions

15.13.7 CYCLOPROPANATION OF SILYL ENOL ETHERS

Trimethylsilyl enol ethers of both saturated and unsaturated aliphatic or alicyclic carbonyl compounds, when submitted to cyclopropanation by an improved Simmons-Smith

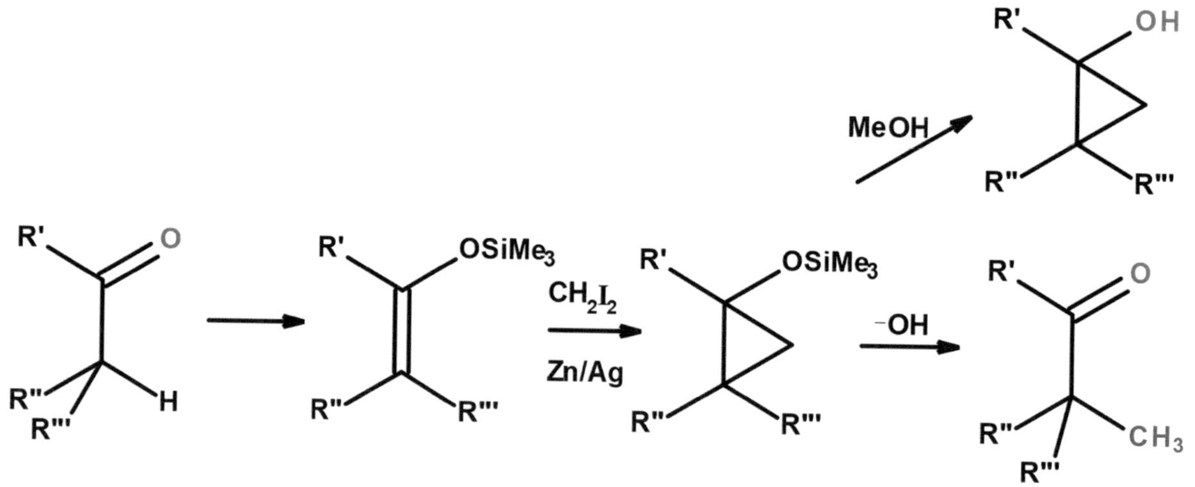

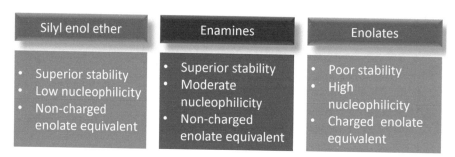

SCHEME 15.56   Cyclopropanation of silyl enol ethers.

reaction (Zn/Ag couple is used and hydrolysis of the product is replaced by pyridine workup), lead to siloxycyclopropane derivatives, which are useful synthetic intermediates.

15.13.8   COMPARISON OF REACTIVITY

Silyl enol ether: superior stability (non-charged enolate equivalent) (low nucleophilicity) (reacts with strong electrophiles, H⁺).

Enamines: neutral conditions mitigate self-condensation side reactions (non-charged enolate equivalent) (reacts with moderate electrophiles, RCOCl and RCHO).

Enolates: quick/easy generation, predictable regioselectivity (high nucleophilicity) (reacts with weak electrophiles, RI, RCO₂R) (Figures 15.18 and 15.19).

Silyl enol ether	Enamines	Enolates
• Superior stability • Low nucleophilicity • Non-charged enolate equivalent	• Superior stability • Moderate nucleophilicity • Non-charged enolate equivalent	• Poor stability • High nucleophilicity • Charged enolate equivalent

FIGURE 15.18   Comparison of reactivity between enolates, enamines, and silyl enol ethers.

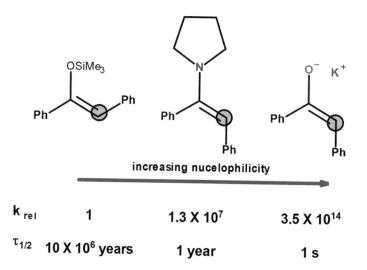

FIGURE 15.19   Comparison of nucleophilicity between enolates, enamines, and silyl enol ethers.

15.14 QUESTIONS

1. Write a short note on the comparison of C=C and C=X (X=O, N, S).
2. What are class e and class n reactions? Give some examples.
3. With suitable reaction energy profile explain the mechanism of base-catalyzed hydration.
4. Write a short note on the addition of alcohols and thiols to aldehydes and ketones.
5. Why carbonyl addition reactions with HCN are not favored compared to $^-$CN.
6. Write a short note on haloform reaction.
7. With a suitable example explain stereoselective enolate formation.
8. Write briefly about the steric effect of R on enolate formation.
9. Why it is important to know the nature of the base in enolate formation.
10. With a suitable example explain conjugate addition of organocopper reagents.
11. Why enamines are important?
12. What are the differences between imines and enamines? How they are synthesized?
13. Write the mechanism for the formation of enamines from ketone under acid catalysis.
14. Give a few examples of alkylations of enamines.
15. Why enamines are better than enolates for alkykations?
16. Why enol ethers are important in organic synthesis?
17. Write a brief note about the formation of alkyl enol ethers.
18. What are the different methods of formation of enol silyl ethers?
19. What is Mukaiyama aldol reaction?
20. Explain cyclopropanation of silyl enol ethers.

BIBLIOGRAPHY

Carey, F. A., & Sundberg, R. J. *Advanced Organic Chemistry, Part B-Reactions and Synthesis*, 4th Ed. Kluwer Academic/Plenum Publishers, 2000, 723.

Cook, G. *Enamines: Synthesis: Structure, and Reactions*. CRC Press, 1988.

Gassman, P. G., Burns, S. J., & Pfister, K. B. (1993). Synthesis of cyclic and acyclic enol ethers (vinyl ethers). *The Journal of Organic Chemistry*, 58(6), 1449–1457.

Hans Reich, Carbonyl Chemistry, (https://organicchemistrydata.org/hansreich/resources/carbonyl/?page=carbonyl00%2F) (accessed, 15-Apr-2024)

Hickmott, P. W. (1982). Enamines: Recent advances in synthetic, spectroscopic, mechanistic, and stereochemical aspects—I. *Tetrahedron*, 38(14), 1975–2050.

Ian Hunt, University of Calgary, Chapter 17: Aldehydes and Ketones. Nucleophilic Addition to C=O (https://www.chem.ucalgary.ca/courses/350/Carey5th/Ch17/ch17-3-0.html) (accessed, 15-Apr-2024)

Ian Hunt, University of Calgary, Chapter Chapter 18: Enols and Enolates (https://www.chem.ucalgary.ca/courses/351/Carey5th/Ch18/ch18-1.html) (accessed, 15-Apr-2024)

Jencks, W. P. (1976). Enforced general acid-base catalysis of complex reactions and its limitations. *Accounts of Chemical Research*, 9(12), 425–432.

John D. Roberts and Marjorie C. Caserio, California Institute of Technology, 16.5: Typical Carbonyl-Addition Reactions (https://chem.libretexts.org/Bookshelves/Organic_Chemistry/Book%3A_Basic_Principles_of_Organic_Chemistry_(Roberts_and_Caserio)/16%3A_Carbonyl_Compounds_I%3A_Aldehydes_and_Ketones._Addition_Reactions_of_the_Carbonyl_Group/16.4%3A_Typical_Carbonyl-Addition_Reactions) (accessed, 15-Apr-2024)

Kuwajima, I., & Nakamura, E. (1985). Reactive enolates from enol silyl ethers. *Accounts of Chemical Research*, 18, 181–187.

Leonov, A. I., Timofeeva, D. S., Ofial, A. R., & Mayr, H. (2019). Metal enolates – enamines – enol ethers: How do enolate equivalents differ in nucleophilic reactivity?. *Synthesis*, 51, 1157–1170.

Mukaiyama, T. (1984). Metal enolate mediated aldol reactions. *Israel Journal of Chemistry*, 24, 162–166.

Mukaiyama, T., Banno, K., & Narasaka, K. (1974). New cross-aldol reactions. Reactions of silyl enol ethers with carbonyl compounds activated by titanium tetrachloride. *Journal of the American Chemical Society*, 96, 7503.

Rauk, A. *Orbital Interaction Theory of Organic Chemistry*, 2nd Ed. John Wiley & Sons, Inc., 2001, 121.

Roger W. Binkley (Cleveland State University) and Edith R. Binkley (Cleveland Heights-University Heights school system), II. Intramolecular Addition of Carbon-Centered Radicals to Aldehydo and Keto Groups (https://chem.libretexts.org/Bookshelves/Organic_Chemistry/Book%3A_Radical_Reactions_of_Carbohydrates_(Binkley)/II%3A_Radical_Reactions_of_Carbohydrates/10%3A_Aldehydes_and_Ketones/02._Intramolecular_Addition_of_Carbon-Centered_Radicals_to_Aldehydo_and_Keto_Groups) (accessed, 15-Apr-2024)

Smith, M. B., & March, J. *March's Advanced Organic Chemistry Reactions, Mechanisms and Structure Chapter 5, Carbocations, Carbanions, Free Radicals, Carbenes, and Nitrenes*, 5th Ed. John Wiley & Sons, Inc., 2001, 1190.

Stork, G., Brizzolara, A., Landesman, H., Szmuszkovicz, J., & Terrell, R. (1963). The enamine alkylation and acylation of carbonyl compounds. *Journal of the American Chemical Society*, 85, 207.

Stork, G., Terrell, R., & Szmuszkovicz, J. (1954). A new synthesis of 2-alkyl and 2-acyl ketones. *Journal of the American Chemical Society*, 76, 2029.

Trost, B. M., & Tak-Hang, C. *2.3 Formation and Addition Reactions of Enol Ethers, Editors-in-Chief, Ian Fleming, Comprehensive Organic Synthesis*. Pergamon Press, 1991, 595–628.

William Reusch, Michigan State University, The Mechanism of Enamine Formation. (https://www2.chemistry.msu.edu/faculty/reusch/virttxtjmL/Questions/MechPrb/enamine.htm) (accessed, 15-Apr-2024)

West, J. A. (1963). The chemistry of enamines. *Journal of Chemical Education*, 40(4), 194.

16 Name Reactions

16.1 SHARPLESS EPOXIDATION

In this chapter we will be looking at what is Sharpless epoxidation, click reactions, copper-catalyzed azide-alkyne cycloaddition, its importance, and Diels–Alder reaction.

We need to know why Sharpless epoxidation is an important reaction. To make chiral molecules that are used as drug materials, we need to have a single enantiomer as the starting material (Figure 16.1). How do we make one of the enantiomers exclusively and not make its enantiomer? These kinds of problems will come when we go for kinetic resolution of racemic mixtures or any achiral reaction that is used. During resolution, 50% of the material is lost which is not economical. To avoid this chiral separation technique, Sharpless has devised a new method in which a simple epoxidation can be carried out using a chiral catalyst to produce highly pure enantiomeric epoxides. Although there are many epoxidation methods like Jacobsen–Katsuki epoxidation, Payne epoxidation, porphyrin-based, Salen-based, dioxirane-based methods, and nucleophilic epoxidations are available, we will be studying about Sharpless method only.

16.1.1 REACTIONS THAT GENERATE CHIRAL CENTERS

Although this is not an exhaustive list of reactions that are available for an organic chemist, only a handful of them are listed which include asymmetric hydrogenation of C=C, cyanation of C=O, asymmetric epoxidation of C=C (Figure 16.2).

16.2 EPOXIDATION

Epoxidation is an addition of oxygen across a double bond to give a cyclic ether. When we have an olefinic system that is treated with the peroxy compound that leads to the formation of epoxides. In this reaction, two new C–O bonds are formed. Various epoxidizing agents like peroxy acetic acid, *tert*-butyl hydrogen peroxide, *meta*-chloroperoxybenzoic acid (m-CPBA) are known.

Both a cyclic system and an open-chain system can form epoxide. In the open-chain system like ethylene, we don't have any specificity, whereas in the case of cyclic systems, we are getting *cis*-epoxide, so this is the importance of cyclic systems stereochemistry.

Prof. K. B. Sharpless got Nobel Prize for his pioneering work on chiral catalyzed epoxidation reaction.

16.2.1 SHARPLESS EPOXIDATION

This is an enantioselective chemical reaction to prepare epoxides from primary and secondary allylic alcohol. The Sharpless reagent is a combination of *tertiary*-butyl hydroperoxide, titanium isopropoxide plus or minus isomer of diethyl tartrate. All these three reagents combined to form the active catalyst, which facilitates the reaction with allylic alcohol to give only one enantiomer the epoxide, based on which tartrate is used.

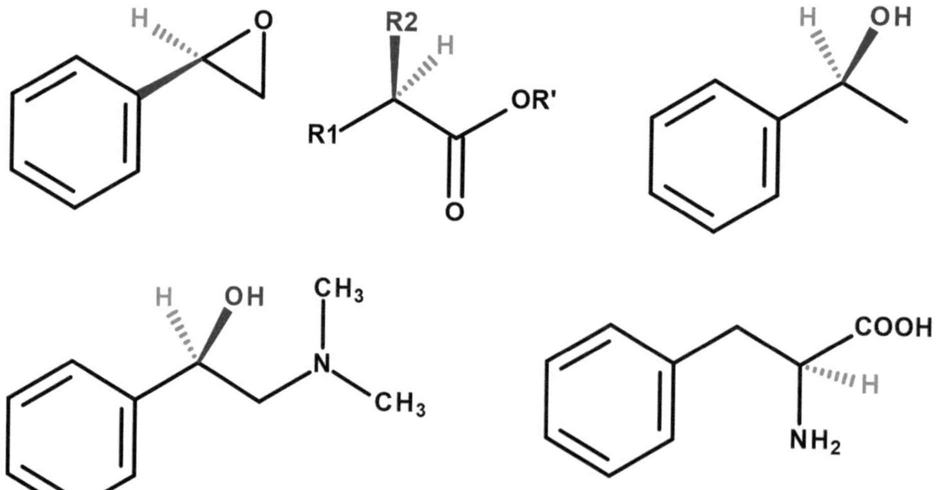

FIGURE 16.1   Examples of a few chiral compounds.

DOI: 10.1201/9781032631165-16

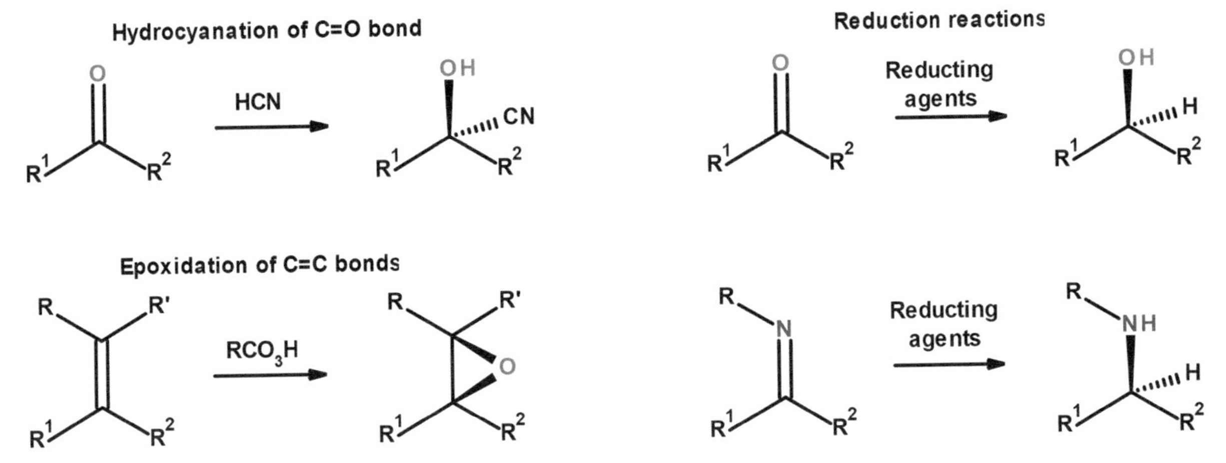
Hydrocyanation of C=O bond
HCN
Reduction reactions
Reducting agents
Epoxidation of C=C bonds
RCO3H
Reducting agents

FIGURE 16.2 Examples of a few reactions producing chiral compounds.

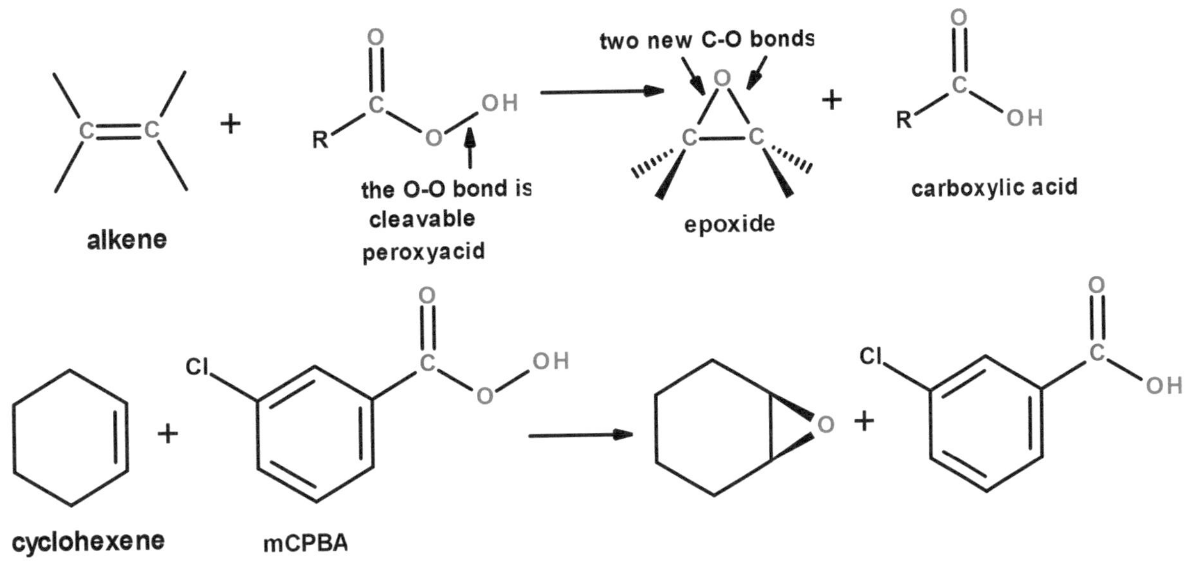
two new C-O bonds
alkene
the O-O bond is cleavable
peroxyacid
epoxide
carboxylic acid
cyclohexene
mCPBA

SCHEME 16.1 Examples of an epoxidation reaction.

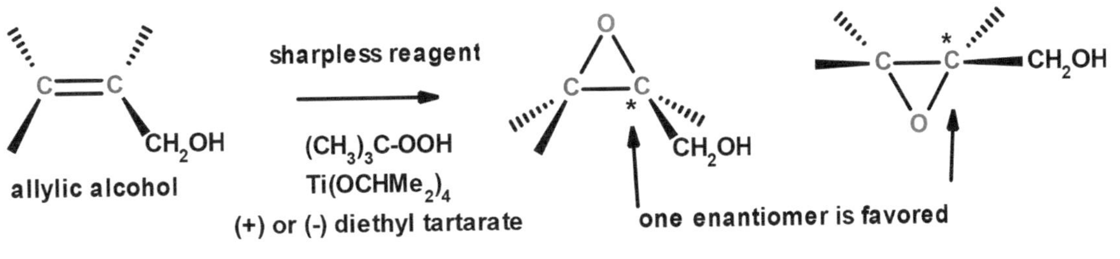
sharpless reagent
allylic alcohol
(CH3)3C-OOH
Ti(OCHMe2)4
(+) or (-) diethyl tartarate
one enantiomer is favored

SCHEME 16.2 Examples of Sharpless epoxidation reaction.

16.2.2 Advantages of Sharpless Epoxidation

The epoxides can be easily converted to diols, amino alcohols, ethers, and many other functional groups. We can easily predict the stereochemistry of the product. Another important feature is industrial application. The starting material and the reagents are very cheap. For example, tartrate salt is very cheap and the other reagents are not expensive. This reaction tolerates various functional groups and the product formation is not affected by groups on the hydroxy carbon. The reaction is very selective for allyl alcohol.

When we look at the types of allyl alcohols, many substituents can be on both the alkene carbons, and the hydroxy carbon may be substituted, or unsubstituted (1°). Various compatible groups include aldehydes, ketones, esters, amides, acetals, ketals, ethers, cyanides, azides, nitro, alkenes, and alkynes. Phenols, thiols, amines, and carboxylic acids are some of the incompatible groups.

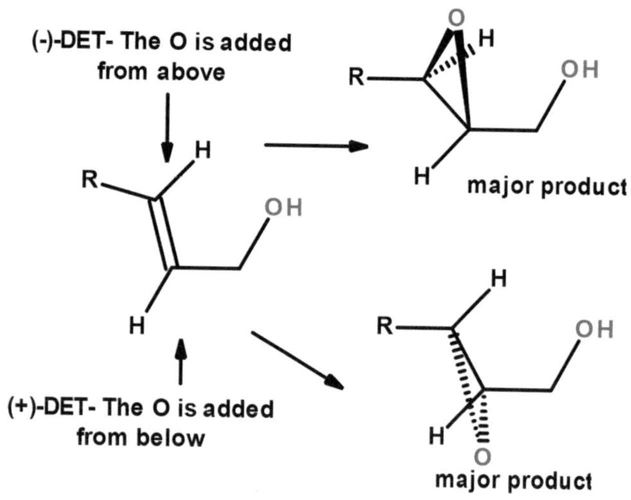

(-)-DET- The O is added from above

(+)-DET- The O is added from below

major product

major product

SCHEME 16.3 Use of DET isomer and stereospecificity of epoxidation.

16.2.3 STEREOSPECIFIC EPOXIDATION

This reaction is a stereospecific epoxidation. The addition of oxygen takes place either from the top or bottom depending on the type of chiral tartrate. The addition of oxygen takes place from one side, which is why this is a *syn/cis* addition. This is a stereospecific reaction. If the starting material is a *cis*-alkene, we get epoxide with the *cis*-substituents, and *trans*-alkene gives *trans* products. The nature of the tartrate salt, + or – isomer determines which enantiomer will be formed.

If we assume that the allyl alcohol is in the plane of the paper. Although we can draw it in many ways. For our easy understanding of identifying the stereospecificity, we keep the hydroxyl group in the bottom right corner of the plane. If (–)-diethyl tartrate is used, then the oxygen is added from the top, and if (+)-diethyl tartrate is used, the addition of oxygen takes place from the bottom.

How we can easily remember is by using the simple mnemonics MA. M for Minus A for Above. If it is a minus isomer, it gives the product where the epoxide oxygen is at the top. If it is the plus isomer, it gives the product with oxygen at the bottom.

There are some examples given here. If you look at both the reactions, (+)-DET isomers give 97.5% product with oxygen below the plane, and in the case of (–)-DET isomer gives the product with oxygen at the top in 97.5% so this is how we can easily predict the type of product that will be formed during Sharpless epoxidation reaction.

16.3 CLICK REACTIONS

We will move on to the next topic, click reactions which is used in both drug discovery and chemical biology.

Click reaction (similar to the click of the computer mouse) is a modular reaction and it has a wide scope. The chemical yields are more than 90%. In most of the cases, the side products are simple molecules like water or ammonia. This reaction has to be very stereospecific. These are all the requirements for a click reaction. Any reaction, if it has to be classified as a click reaction, has to follow at least a few of these parameters. For example, it has to have a high atom economy in other words every atom in the starting material has to be consumed or transformed into products. Some of the examples include nucleophilic ring-opening reactions of epoxides and aziridines, carbon-carbon multiple bonds additions, Michael additions, cycloaddition reactions, and so on.

16.3.1 AZIDE-ALKYNE CYCLOADDITION REACTION

We will now look at an important azide-alkyne cycloaddition reaction. Although the non-metal-based reaction or the thermal Huisgen 1,3-dipolar cycloaddition of alkynes and azides requires elevated temperatures and often produces

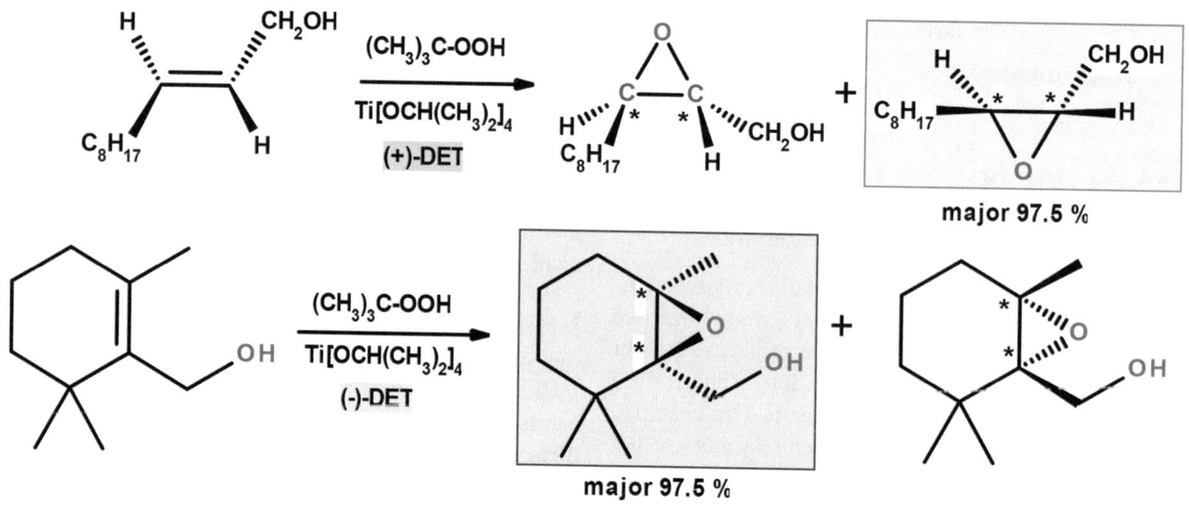

major 97.5 %

major 97.5 %

SCHEME 16.4 Some examples of Sharpless epoxidation.

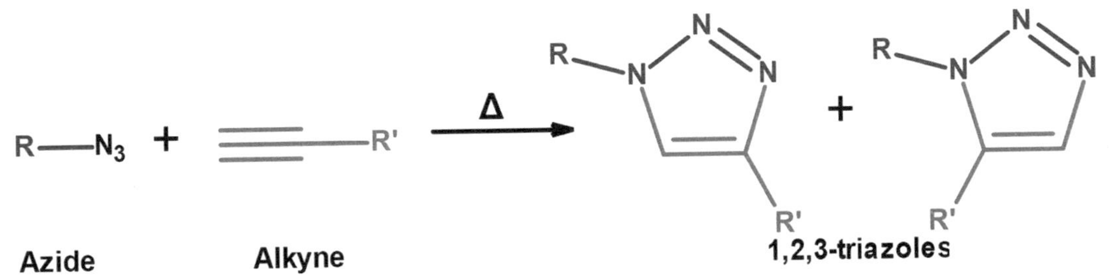

SCHEME 16.5 Example of click reaction.

mixtures of two regioisomers when using unsymmetric alkynes. Although this reaction does conform many click reaction requirements, the formation of two regioisomers does not make it a true click reaction.

16.3.2 Cu Catalyzed Azide-Alkyne Cycloaddition

A variant of azide-alkyne cycloaddition reaction in the presence of copper catalyst also gives a cyclic product 1,2,3-triazole. Unlike the thermal reaction here we are getting only one product. This is called the click reaction.

What is a click reaction? It is a reaction of dipolarophile with 1,3-dipolar compounds leading to the five-membered heterocycle formation. Some dipolarophiles which are used are alkenes and alkynes, and the carbonyls or the nitriles are the functional groups that are introduced in the final product (Figure 16.3).

Azide undergoes cyclization with an alkene to give a cyclic product. The reagent that is generally used for the cyclization is sodium ascorbate and copper sulfate is used as the catalyst. The cyclic product is generally 1,2,3-triazoles having 1,4-disubstituted product. The mechanism given here is a simplified version of the actual mechanism.

We will see here what the characteristics of click reactions are. This involves the formation of carbon hetero

atom bonds and it is strongly exothermic. So that means the reaction is very fast. Click reactions are usually fusion processes (leaving no by-products) or condensation processes (producing water as a by-product). Many click reactions are highly tolerant of, and often accelerated by, the presence of water.

16.4 DIELS–ALDER REACTION

We will be looking at Diels–Alder reaction its significance, rules governing DAR, regioselectivity, and so on.

Diels–Alder reaction is one of the important reactions in organic chemistry. Diels and Alder both found 1,3-diene and the dienophile reacts to form a cyclic system. They were awarded the Nobel Prize for their discovery in 1950 for this reaction.

DAR is the cycloaddition reaction between a 1,3-diene and an alkene which is called a dienophile to form a six-membered ring. The crucial part of a Diels–Alder reaction is the formation of a cyclic product from open-chain reactants. In this reaction, two new σ bonds are formed and one new π bond is also formed. If you look at the starting material you have three π bonds, which are getting converted into two new σ bonds and one π bond.

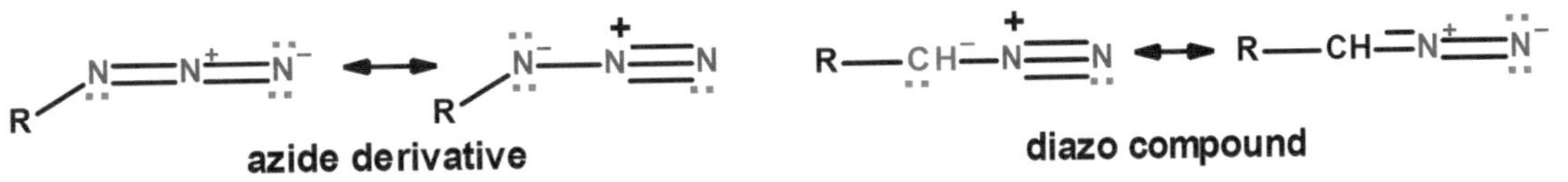

FIGURE 16.3 Examples of resonance structures of azido and diazo derivatives.

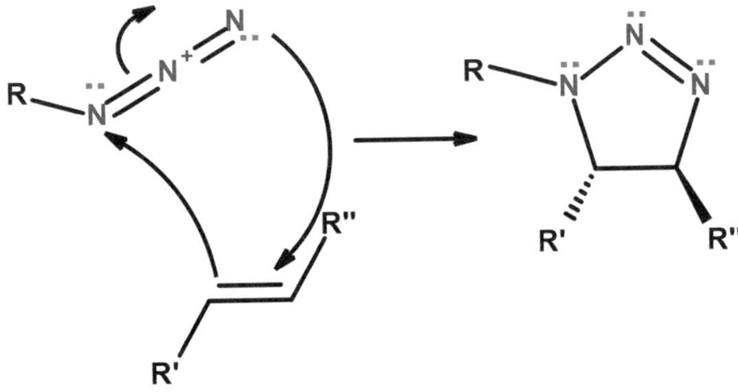

SCHEME 16.6 Example of a click reaction giving 1,2,3-triazoles.

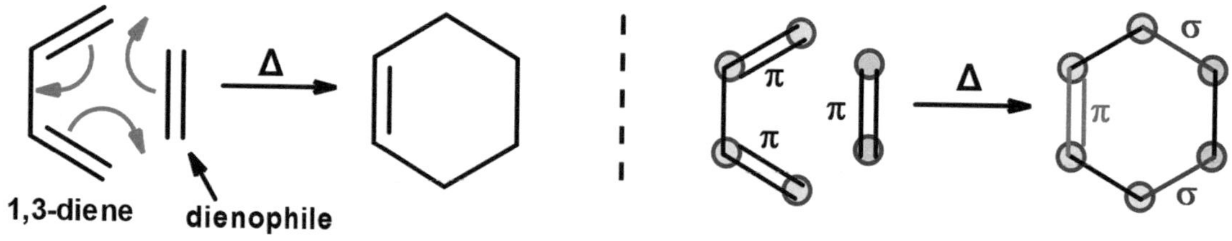

SCHEME 16.7   Example of a Diels–Alder reaction.

In conjugated diene, the double bonds are in conjugation with each other and the dienophile is electron-seeking (diene seeking).

In Diels–Alder reaction, even alkyne can also function as a dienophile. Diene reacts with the dienophile, which is an alkyne to give a cyclic product. When alkynes are used, we get two double bonds in this product.

Some examples of diene are butadiene, substituted butadiene, 1,3-cyclohexadiene, and cyclopentadiene. So, these are various dienes which are used as the starting materials.

Figure 16.4 depicts a few examples of dienophiles. When you look at all these compounds there is one commonality you can observe. All these double bonds have some electron-withdrawing group. We will see the important role played by these electron-withdrawing groups in Diels–Alder reaction. If we look at these dienophiles, they all have a double or a triple bond and carbonyl groups. Some of the electron-withdrawing groups present are ketone, an ester or even anhydride, cyanides.

16.4.1   Importance of the DAR

It is an important reaction for the formation of many six-membered rings. In this reaction, new C-C or C-X σ bonds are formed simultaneously and one π bond is also formed. The most striking feature of this reaction is stereospecificity and regioselectivity. This reaction can be carried out under thermal conditions, that is initiated by heat

and it follows a concreted mechanism, that is, a single-step reaction.

16.4.2   Stereospecificity and Regioselectivity

If we take a *cis* dienophile, the products have substituents in the *cis* orientation and the *trans* dienophile gives the products where the substituents are in *trans* orientation. We can say there is stereospecificity in Diels–Alder reactions. When we take alkynes, it always gives a *cis* product. Similarly, the diene stereochemistry is also retained in the product (Figure 16.5).

Although both electron-donating and electron-withdrawing groups can be present in the starting materials, the presence of electron-withdrawing group in the dienophile (aldehydes, carboxyls, esters, nitro, and cyano) and the presence of electron-donating groups in the diene (alkyl groups, ethers, amino groups) are highly favored.

16.4.3   Rules Governing Diels–Alder Reaction

Following are the rules governing DAR

1. The diene should have **s-cis** conformation
2. EWG on **dienophile** increases the reaction rate (conjugated diene acts as a nucleophile and the dienophile acts as an electrophile)

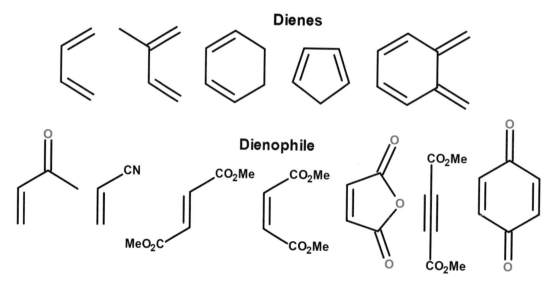

FIGURE 16.4   Example of a diene and dienophile.

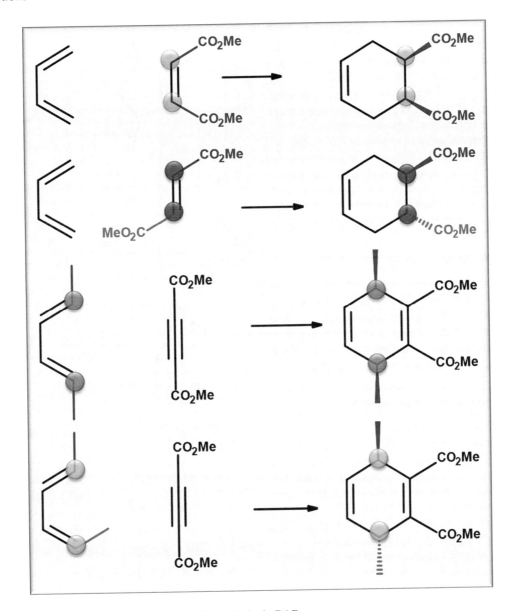

FIGURE 16.5 Examples of stereospecificity and regioselectivity in DAR.

3. The **stereochemistry** of the dienophile is conserved
4. When **endo** and **exo** products are possible **endo** is preferred

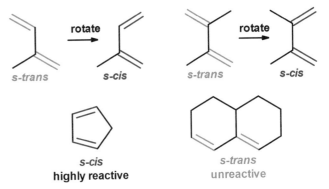

FIGURE 16.6 Examples of s-cis and s-trans isomers.

The first rule is that double bonds should have a s-*cis* conformation. 'S' can be given a name either for a single bond or a σ bond both start with s, so you can remember with either the single bond or with the σ bond. In either way, it refers to the single bond that separates the dienes of both the double bonds. This single bond is free to rotate so when it is free to rotate even though we write this in the *trans* orientation in the second diene, the single bond can easily rotate to give the s-*cis* conformation. This is the most important thing for Diels–Alder reaction to occur, both the double bonds should be in this *cis* orientation (Figure 16.6).

If you look at the cyclic system as shown for cyclopentadiene, these two double bonds are on the same side so this is also a s-*cis* conformation. This can undergo Diels–Alder reaction very efficiently. But if you look at the bicyclic diene, here we have two double bonds they are separated by again one single bond. But if you look at the conformation this is now a *trans* orientation so when it is the *trans* orientation, the particular Diels–Alder reaction will never occur. That is one of the criteria for Diels–Alder reaction.

We talked about *exo* and *endo* products, *exo* means outside. When the double bond is formed, if the substituent lies away from the double bond, then it is called the *exo* product. If it is close to the double bond then it is called the *endo* product. For the Diels–Alder reaction under kinetic control mainly the *endo* orientation of the dienophile is favored.

16.4.4 Exo and Endo Naming

How do we name *exo* and endo so when the diene reacts with a dienophile z is the group that is used as reference? If z is the functional group that we are looking at, if it is close to the double bond, it is called *endo* when the z group is far away from the double bond it is called *exo*. This is how the nomenclature is given for the *exo* and *endo* products.

16.4.5 Stereochemistry of Dienophile

We have also seen the conservation of stereochemistry of dienophile. Starting with the *cis* isomer, maleic acid we end up with the *cis* product as the major one either both the carbonyl groups are pointing upwards or downwards. When we start with the *trans* isomer that is fumaric acid we end up with the *trans* product, that is, one of the carboxylic acids is above the plane another one is below the plane. This is how the dienophile stereochemistry is conserved.

There is also another feature, here in the ring junction the two hydrogens have *cis* orientation because in the starting material, they are in the *cis* orientation so that is maintained in the final product also.

If you look at *exo* and *endo* during the transition state we can actually visualize the two groups say the z group is below the two new sigma bonds so this is the *endo* product. If the z group is lying above the product, then it is called the *exo* product.

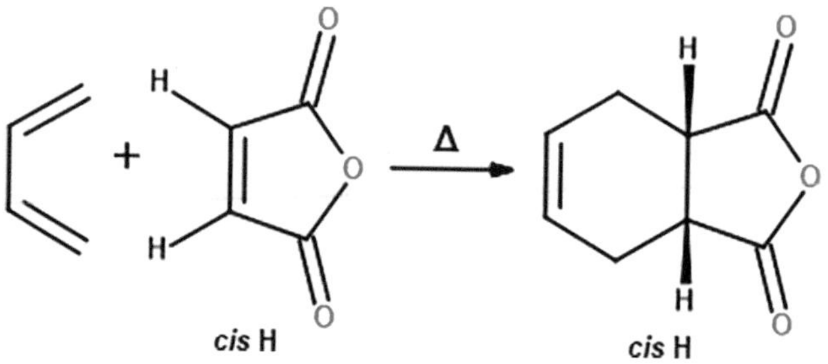

SCHEME 16.9   Conservation of stereochemistry of dienophile.

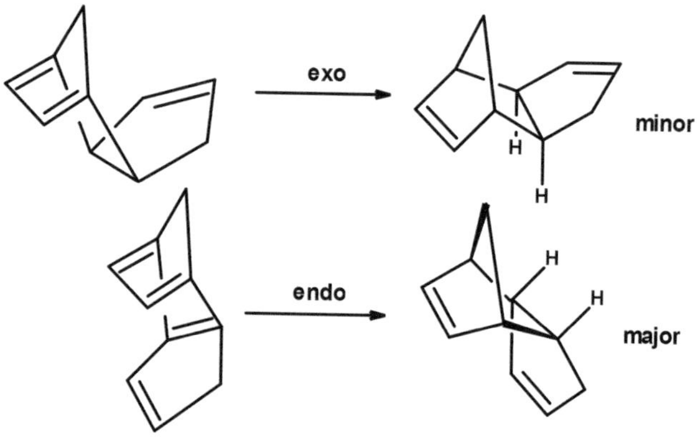

SCHEME 16.10   Example of two cyclic systems undergoing DAR.

Here is an example of two cyclic systems undergoing DAR. Here again, *endo* is the major product. Here the Z group is the double bond.

16.4.6 Regioselectivity (Ortho/Para)

If you look at regioselectivity that is *ortho* and *para*. Of course, this *ortho para* has nothing to do with aromatic systems but this helps you remember what type of a product can be formed.

We see here that methoxy is the electron-donating group. If you look at the methoxy, this is electron donating due to the lone pair of electrons flows toward the diene double bond. As shown below, the electron flow shifts the double bond.

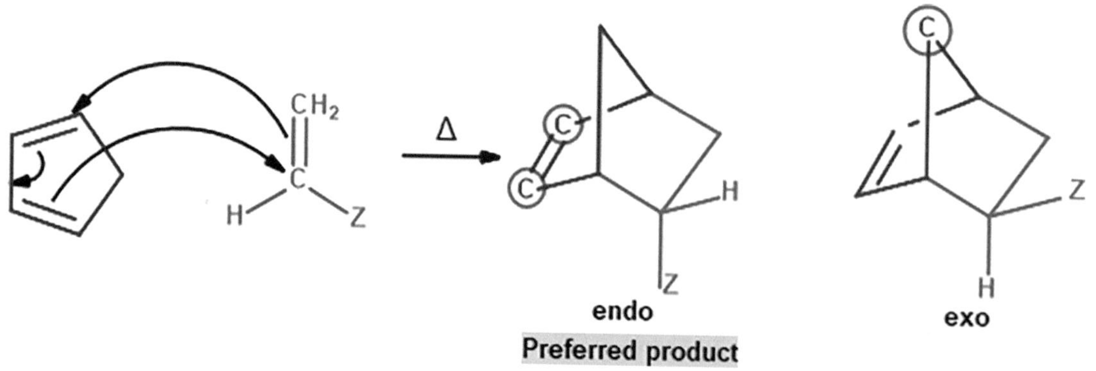

SCHEME 16.8   Exo and endo isomers.

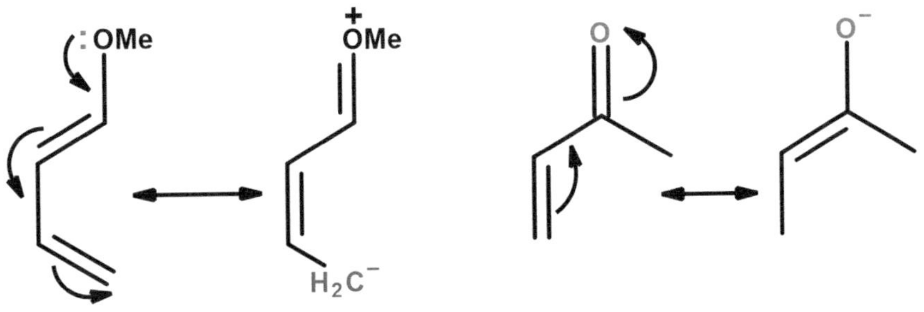

SCHEME 16.11 Polarization of bonds in DAR substrates.

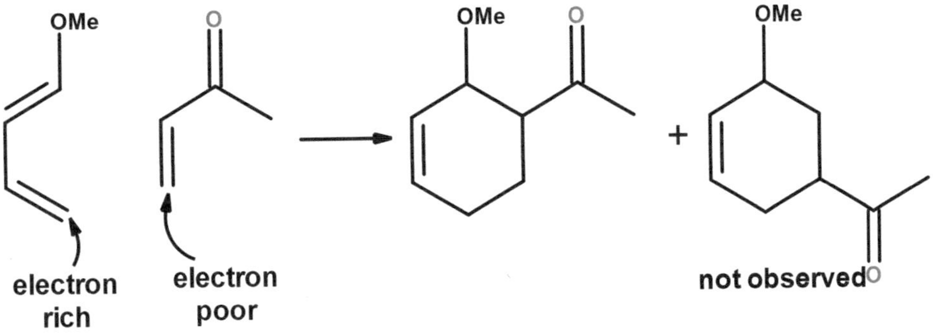

SCHEME 16.12 Example of *ortho* type product formation in DAR.

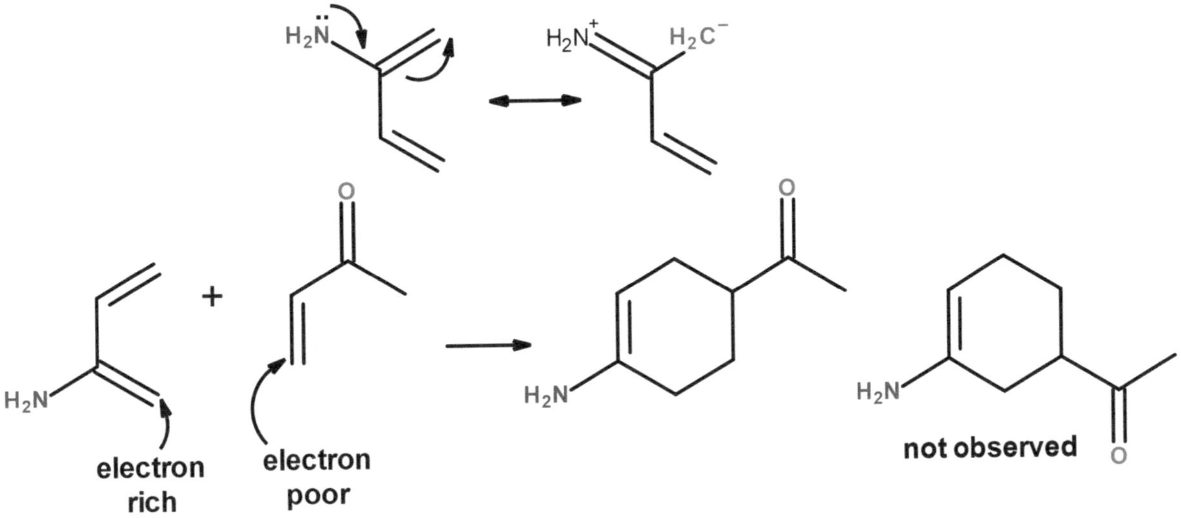

SCHEME 16.13 Example of *para* type product formation in DAR.

This double bond shift leads to another shift to the next double bond. In other words, the terminal diene carbon becomes electron-rich. This carbon is becoming more nucleophilic. When an electron-donating group is present at one of the terminal dienes, the other end of the diene becomes more nucleophilic.

Consider an electron-withdrawing group present on the dienophile (the simplest α,β-enone, methyl vinyl ketone). The electronegative oxygen pulls the electrons toward itself. That means this bond is shifted here and if you look at the oxygen, it becomes negative but the terminal carbon where the double bond was earlier present becomes electrophilic.

However, there is no change in the electron density for the carbonyl carbon. Before the shift and after the shift of bonds, there is no change in its electron density. We have one nucleophile on the diene and we have another electrophile on the dienophile. Now these two can easily interact with each other. We have an electron-rich center and an electron-deficient center. They both react to give this particular orientation product, the *ortho* product. If we actually rotate this upside down, we are supposed to get the *meta* product. However, this is never formed because this carbon never becomes electropositive. So that is the reason this reaction does not occur to give this particular product but

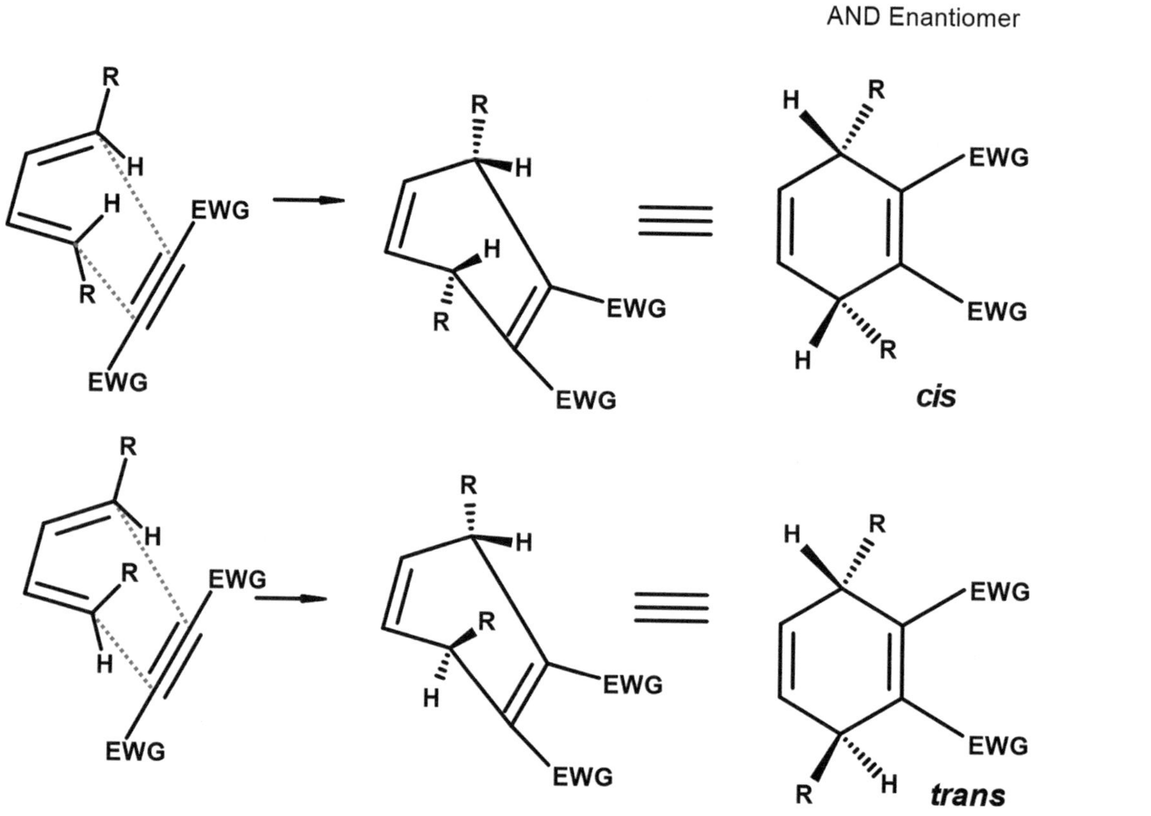

AND Enantiomer

SCHEME 16.14 Example of stereochemical outcome in DAR.

we always observe only the *ortho* product. Here the comparison is with respect to the aromatic system and is a *meta* product.

If we look at amines, the nitrogen is having lone pairs, and can donate these electrons. When it donates electrons, we have two systems one is if the electron is donated to this double bond this double bond becomes electron-rich. This electron power center reacts to give a *para*-substituted product. If you look at this one this is *para* and we don't see the reverse product that is the *meta* substituted product is not formed so this is what is called the *ortho para* product formation in Diels–Alder reaction.

Here is another example of the same thing and in most of the cases the *meta* product is not formed. Either the *ortho* or the *para*-substituted product is formed so with this we conclude about Diels–Alder reactions.

In the stereochemical outcome, we see the E,E-isomer reacts with alkyne to give the *cis* product. Both the R groups are in the *cis* orientation. If we start with the E,Z-isomer, we end up with the *trans* product. Here both the R groups are in the *trans* orientation. Here again, the stereochemistry is maintained.

16.4.7 FRONTIER MOLECULAR ORBITAL INTERACTIONS

If the difference between the HOMO and LUMO is very high the resulting overlap is poor and the stabilization

gained is very minimum. On the other hand, if the energy difference between HOMO and LUMO is less, which leads to better overlap which results in better stabilization (Figure 16.7).

Energy levels can help to identify the best combination. We have two dienes and dienophiles whose energies are given here. We need to identify the best pair for DAR. To do this we need to look at the high energy HOMO and low energy LUMO. The substituted cyclopentadiene has the high energy HOMO and the α,β- unsaturated ketone has the low energy LUMO (Figure 16.8).

From the above values, we can calculate the energy difference between LUMO and HOMO and the energy gained when the product is formed.

DE = 0.10006 − (−0.29698)
= 0.39704 Hartee = 246.76 kcal/mol

For the *exo* vs *endo* product formation, we have two types of control happening here. One is a thermodynamic product and the other is a kinetic product. The *exo* product is more stable by 1.9 kcal/mol. In other words, it is the thermodynamic product. The energy of activation is low for the *endo* product. In other words, *endo* product is the kinetic product. By controlling the reaction conditions, it is possible to obtain both *endo* and *exo* product using DAR (Figure 16.9).

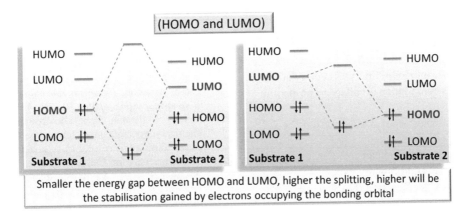

FIGURE 16.7 Frontier molecular orbital interactions in DAR.

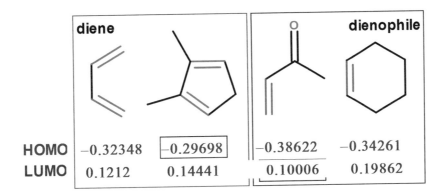

FIGURE 16.8 Identifying best pair energy level of dienes and dienophiles for DAR.

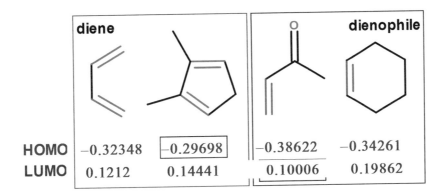

Exo product more stable by 1.9 kcal/mol
E_a lower for endo product by 3.8 kcal/mol

FIGURE 16.9 Identifying exo vs endo product formation in DAR.

16.5 QUESTIONS

1. Write few reactions that generate chiral centers.
2. What is Sharpless epoxidation?
3. What are the advantages of Sharpless epoxidation?
4. Why Sharpless epoxidation is called stereospecific reaction?
5. What is the Sharpless epoxidation catalytic cycle?
6. What is Click reaction?
7. Write about the mechanism of Cu catalyzed azide-alkyne cyclo addition reaction.
8. What is Diels–Alder Reaction (DAR)
9. Why DAR is important?
10. Comment about the stereospecificity and regioselectivity of DAR
11. What are the rules governing Diels–Alder reaction
12. Write a short note on regioselectivity (ortho/para)

BIBLIOGRAPHY

Amar Patel (UCD), 11.3.7 The Diels-Alder Reaction (https://chem.libretexts.org/Courses/Purdue/Purdue%3A_Chem_26605%3A_Organic_Chemistry_II_(Lipton)/Chapter_11.__Addition_to_pi_Systems/11.3%3A_Concerted_Additions/11.3.7_The_Diels-Alder_Reaction) (accessed, 15-Apr-2024)

Brieger, G., & Bennett, J. N. (1980). The intramolecular Diels–Alder reaction. *Chemical Reviews*, 80(1), 63–97.

Carey, F. A. *Organic Chemistry*, 5th Ed., McGraw-Hill, 2004, 409.

Carey, F. A., & Sundberg, R. J. *Advanced Organic Chemistry, Part B: Structure and Mechanisms*, Kluwer Academic/Plenum Publishers, 2000, 4th ed. 636.

Finn, M. G., & Sharpless, K. B. (1991). Mechanism of asymmetric epoxidation. 2. Catalyst structure. *Journal of the American Chemical Society*, 113(1), 113–126.

Gregoritza, M., & Brandl, F. P. (2015). The Diels–Alder reaction: A powerful tool for the design of drug delivery systems and biomaterials. *European Journal of Pharmaceutics and Biopharmaceutics*, 97(Part B), 438–453.

Hans Reich's Collection. Pericyclic Reactions (https://organic-chemistrydata.org/hansreich/resources/pericyclic/?page=pericyclic00%2F) (accessed, 15-Apr-2024)

Ian Hunt, University of Calgary, Chapter 10: Conjugation in Alkadienes and Allylic Systems (https://www.chem.ucalgary.ca/courses/350/Carey5th/Ch10/ch10-5.html) (accessed, 15-Apr-2024)

Jorgensen, K. A., Wheeler, R. A., & Hoffmann, R. (1987). Electronic and steric factors determining the asymmetric epoxidation of allylic alcohols by titanium-tartrate complexes (the Sharpless epoxidation). *Journal of the American Chemical Society*, 109, 3240–3246.

Myers, Sharpless Asymmetric Epoxidation Reaction (https://hwpi.harvard.edu/files/myers/files/22-sharpless_asymmetric_epoxidation_reaction.pdf) (accessed, 15-Apr-2024)

Rick L. Danheiser, Massachusetts Institute of Technology, Introduction to Strategies for the Synthesis of Complex Molecules, The Diels-Alder Reaction, 2007. (https://web.mit.edu/5.511/www/10-05-07.pdf) (accessed, 15-Apr-2024)

Smith, M. B., & March, J. *March's Advanced Organic Chemistry Reactions, Mechanisms and Structure*, 5th Ed. John Wiley & Sons, Inc., 2001, 1062 (normal), 1075 (hetero atom).

White, D. Practical Catalytic Asymmetric Epoxidations, 2006. (https://stoltz2.caltech.edu/seminars/2006_White.pdf) (accessed, 15-Apr-2024).

Xia, Q. H., Ge, H. Q., Ye, C. P., Liu, Z. M., & Su, K. X. (2005). Advances in homogeneous and heterogeneous catalytic asymmetric epoxidation. *Chemical Reviews*, 105(5), 1603–1662.

Index

For Product Safety Concerns and Information please contact our
EU representative GPSR@taylorandfrancis.com Taylor & Francis
Verlag GmbH, Kaufingerstraße 24, 80331 München, Germany